AF412178

ROLE OF THE SARCOPLASMIC RETICULUM IN SMOOTH MUSCLE

The Novartis Foundation is an international scientific and educational
charity (UK Registered Charity No. 313574). Known until September 1997
as the Ciba Foundation, it was established in 1947 by the CIBA company
of Basle, which merged with Sandoz in 1996, to form Novartis. The
Foundation operates independently in London under English trust
law. It was formally opened on 22 June 1949.

The Foundation promotes the study and general knowledge of
science and in particular encourages international co-operation in
scientific research. To this end, it organizes internationally
acclaimed meetings (typically eight symposia and allied open
meetings and 15–20 discussion meetings each year) and publishes
eight books per year featuring the presented papers and discussions
from the symposia. Although primarily an operational rather than
a grant-making foundation, it awards bursaries to young scientists
to attend the symposia and afterwards work with one of the other
participants.

The Foundation's headquarters at 41 Portland Place, London W1B 1BN,
provide library facilities, open to graduates in science and allied disciplines.
Media relations are fostered by regular press conferences and by articles
prepared by the Foundation's Science Writer in Residence. The Foundation
offers accommodation and meeting facilities to visiting scientists and their
societies.

Information on all Foundation activities can be found at
http://www.novartisfound.org.uk

Novartis Foundation Symposium 246

ROLE OF THE SARCOPLASMIC RETICULUM IN SMOOTH MUSCLE

2002

JOHN WILEY & SONS, LTD

Copyright © Novartis Foundation 2002
Published in 2002 by John Wiley & Sons Ltd,
Baffins Lane, Chichester,
West Sussex PO19 1UD, England

National   01243 779777
International (+44) 1243 779777
e-mail (for orders and customer service enquiries): cs-books@wiley.co.uk
Visit our Home Page on http://www.wiley.co.uk
or http://www.wiley.com

Other Wiley Editorial Offices

John Wiley & Sons, Inc., 605 Third Avenue,
New York, NY 10158-0012, USA

WILEY-VCH Verlag GmbH, Pappelallee 3,
D-69469 Weinheim, Germany

Jacaranda Wiley Ltd, 33 Park Road, Milton,
Queensland 4064, Australia

John Wiley & Sons (Asia) Pte Ltd, 2 Clementi Loop #02-01,
Jin Xing Distripark, Singapore 129809

John Wiley & Sons (Canada) Ltd, 22 Worcester Road,
Rexdale, Ontario M9W 1L1, Canada

Novartis Foundation Symposium 246
ix+285 pages, 57 figures, 2 tables

British Library Cataloguing in Publication Data

A catalogue record for this book is available from the British Library

ISBN 0 470 84479 5

Typeset in 10½ on 12½ pt Garamond by Dobbie Typesetting Limited, Tavistock, Devon.
Printed and bound in Great Britain by Biddles Ltd, Guildford and King's Lynn.
This book is printed on acid-free paper responsibly manufactured from sustainable forestry,
in which at least two trees are planted for each one used for paper production.

Contents

Participants

Mordecai P. Blaustein   Department of Physiology, University of Maryland School of Medicine, 655 W Baltimore St, Baltimore, MD 21201-1559, USA

Thomas B. Bolton   Department of Pharmacology & Clinical Pharmacology, St George's Hospital Medical School, Cranmer Terrace, London SW17 0RE, UK

Ludmila Borisova (*Novartis Foundation Bursar*)   Department of Muscle Biochemistry, A V Palladin Institute of Biochemistry, Leontovicha 9, 252030 Kiev, Ukraine

Alison F. Brading   University Department of Pharmacology, Mansfield Road, Oxford OX1 3QT, UK

Karen Bradley (*Novartis Foundation Bursar*)   Division of Neuroscience and Biomedical Systems, Institute of Biomedical and Life Sciences, West Medical Building, University of Glasgow, Glasgow G12 8QQ, UK

Theodore Burdyga   Physiological Laboratory, The University of Liverpool, Crown Street, Liverpool L69 3BX, UK

David A. Eisner (*Chair*)   Department of Medicine, University of Manchester, 1.524 Stopford Building, Oxford Road, Manchester M13 9PT, UK

Christopher Fry   Institute for Urology, Division of Applied Physiology, University College London, 48 Riding House Street, London W1W 7EY, UK

Per Hellstrand   Molecular and Cellular Physiology, Department of Physiological Sciences, BMC F12, SE-22184 Lund, Sweden

G. David Hirst   Department of Zoology, University of Melbourne, Parkville, Victoria 3010, Australia

Masamitsu Iino   Department of Pharmacology, Graduate School of Medicine, The University of Tokyo, Bunkyo-ku, Tokyo 113-0033, Japan

Ryuji Inoue[1]   Department of Pharmacology, Faculty of Medicine, Kyushu University, Fukuoka 812, Japan

Gerrit Isenberg   Department of Physiology, Martin Luther University, Magdeburger Strasse 6, D-06114 Halle, Germany

Michael I. Kotlikoff   Department of Biomedical Sciences, College of Veterinary Medicine, Cornell University, Ithaca, NY 14853-6401, USA

Anne-Marie Lompré   Tour D4, 2ème étage, Faculté de Pharmacie, 5 rue J-B Clément, 92296-Chatenay-Malabry, France

John G. McCarron   Neuroscience and Biomedical Systems, Institute of Biomedical & Life Science, West Medical Building, University of Glasgow, Glasgow G12 8QQ, UK

Noel McHale   Department of Physiology, Queen's University, Medical Biology Centre, 97 Lisburn Road, Belfast BT9 7BL, UK

Mark T. Nelson   Department of Pharmacology, University of Vermont, Given Medical Building, 89 Beaumont Avenue, Burlington, VT 05405-0068, USA

Graeme F. Nixon   Institute of Medical Sciences, Department of Biomedical Sciences, Foresterhill, Aberdeen AB25 2ZD, UK

Richard J. Paul   Department of Molecular & Cell Physiology, University of Cincinnati College of Medicine, 231 Albert Sabin Way, Cincinnati, OH 45267-0576, USA

Luc Raeymaekers   Laboratorium voor Fysiologie, Katholieke Universiteit Leuven, Campus Gasthuisberg O/N, Herestraat 49, B-3000 Leuven, Belgium

Kenton M. Sanders   Department of Physiology and Cell Biology, University of Nevada School of Medicine, Reno, NV 89557-0001, USA

Andrew P. Somlyo   University of Virginia, Molecular Physiology and Biological Physics, PO Box 800736, Jordan Hall, Charlottesville, VA 22908-0736, USA

---

[1]Unfortunately, Professor Inoue was unable to attend the symposium.

Colin W. Taylor   University of Cambridge, Department of Pharmacology, Tennis Court Road, Cambridge CB2 1TP, England

Cornelis van Breemen   The iCAPTUR^{4}E Center, University of British Columbia, St Paul's Hospital, 1081 Bernard Street, Vancouver, BC, Canada V6Z 1Y6

Michael P. Walsh   Department of Biochemistry & Molecular Biology, University of Calgary, 3330 Hospital Dr NW, Calgary, Alberta, Canada T2N 4N1

W. Gil Wier   University of Maryland School of Medicine, Department of Physiology, 655 W Baltimore St, Baltimore, MD 21201-1559, USA

Susan Wray   Physiological Laboratory, The University of Liverpool, Crown Street, Liverpool L69 3BX, UK

R. Young   Department of Obstetrics and Gynaecology, Medical University of South Carolina, 96 Jonathan Lucas Street, Suite 634, Charleston, SC 29425, USA

Chair's introduction

David Eisner

Unit of Cardiac Physiology, University of Manchester, 1.524 Stopford Building, Oxford Road, Manchester M13 9PT, UK

Smooth muscle is distributed throughout the body, largely around hollow structures such as blood vessels, the gastrointestinal tract and the genitourinary system. Normal function requires that the smooth muscles contract and relax at appropriate times, and abnormalities of contraction underlie such important pathologies as hypertension, incontinence and abnormal childbirth. Since contraction is initiated by an increase of cytoplasmic Ca^{2+} concentration then normal function requires appropriate Ca^{2+} handling.

In smooth muscle, as in other cell types, intracellular Ca^{2+} concentration ($[Ca^{2+}]_i$) can be controlled both by fluxes across the surface membrane and by fluxes into and out of intracellular organelles — of which the subject of this symposium, the sarcoplasmic reticulum (SR) is perhaps the most important. The SR has been extensively investigated in striated muscle (cardiac and skeletal). In these tissues it is responsible for releasing most if not all of the Ca^{2+} required for contraction (see Bers 2001 for a review of striated muscle excitation–contraction coupling). In striated muscle, Ca^{2+} is released from the SR via specialized release channels known as ryanodine receptors (RyRs). Cardiac and skeletal muscle differ in the mechanisms by which these RyRs are activated. In skeletal muscle, the surface membrane action potential changes the conformation of a protein (the dihydropyridine receptor) and this is directly coupled to the RyR, resulting in Ca^{2+} release from the SR. The SR contributes essentially all the Ca^{2+} required for contraction with the consequence that skeletal muscle can contract for long periods in the complete absence of external Ca^{2+}. In cardiac muscle, Ca^{2+} still leaves the SR through the RyR. However, the mechanism of control of this release is different. Ca^{2+} enters the cell from the extracellular fluid through the L-type Ca^{2+} channel. This results in an increase of $[Ca^{2+}]_i$ in the small space between the surface membrane and the RyR. This increase of $[Ca^{2+}]_i$ then opens the RyR leading to release of Ca^{2+} from the SR. This mechanism is known as Ca^{2+}-induced Ca^{2+} release, or CICR. A fundamental consequence of this is that Ca^{2+} release from the SR requires Ca^{2+} entry into the cell. In addition the state of loading of the SR with Ca^{2+} is determined, in part, by Ca^{2+} fluxes across the surface membrane.

Despite some differences, Ca^{2+} handling by skeletal and cardiac SR has two factors in common. (1) The release of Ca^{2+} occurs through the RyR. (2) The net action of Ca^{2+} release from the SR is to elevate $[Ca^{2+}]_i$ and thereby promote contraction. However, as discussed below, the situation is very different in smooth muscle.

In this meeting, we should begin by reminding ourselves of what is known and what is uncertain in smooth muscle. In considering smooth muscle, it must be remembered that this tissue is heterogeneous and, indeed, it can be argued that there is no single representative 'smooth muscle'. At one extreme are electrically excitable tissues such as myometrium and bladder. In these tissues there will be significant sarcolemmal Ca^{2+} fluxes due to Ca^{2+} entry through voltage-sensitive channels during the action potential (see Wray et al 2001, Fry & Wu 1997 for recent reviews). In contrast, arterial smooth muscle is non-excitable (e.g. Rembold 1992). I have listed below some of the questions which should be considered in this meeting, but it is important to note that the answers may vary between different smooth muscle types.

(1)   To what extent does the SR contribute to the rise of $[Ca^{2+}]_i$ that activates contraction? In other words, what are the relative contributions of the SR and the surface membrane? In contrast to the situation in striated muscle where inhibition of SR function abolishes most of contraction, there are several examples in smooth muscle of large amounts of force remaining under these conditions. The SR is an intracellular store of finite capacity. Release of Ca^{2+} from such a store is well suited to producing transient contractions. However, maintained contraction can be produced by steady state changes in Ca^{2+} fluxes across the surface membrane. Does the SR make different contributions during different phases of contraction?

(2)   What is the Ca^{2+} content of the SR and how does it vary during contraction? It is now possible to put Ca^{2+}-sensitive indicators into the SR and thereby obtain a measure of the free Ca^{2+} concentration, although uncertainty concerning the intra-SR Ca^{2+} buffering means that the *total* amount of Ca^{2+} in the SR available for release is not known. Direct measurement of SR Ca^{2+} during contraction provides the most direct way of seeing whether the SR does indeed release Ca^{2+} to produce contraction or not.

(3)   What is the role of intracellular Ca^{2+} waves due to Ca^{2+} release from the SR? In many cell types stimulation results in Ca^{2+} waves rather than a maintained increase of $[Ca^{2+}]_i$. Summation of such waves in many cells can result in a maintained contraction. Indeed, recent work suggests that these waves are implicated in the genesis of vascular tone (Peng et al 2001). It is important for us to consider how widespread in smooth muscle excitation–contraction coupling are such waves.

(4)   In smooth muscle, in addition to the RyRs, the SR also contains Ca^{2+} release channels activated by inositol-1,4,5-trisphosphate ($InsP_3$) (Nixon et al 1994). Agonist binding to the surface membrane results in the production of $InsP_3$ leading to $InsP_3$-induced Ca^{2+} release (IICR). One important question concerns whether the $InsP_3$ receptors and RyRs are on the same parts of the SR or whether there are regions of the SR which are sensitive to CICR and others to IICR (Yamazawa et al 1992, Flynn et al 2001). In addition to any functional heterogeneity of the SR due to the presence of different release channels, is there anatomical compartmentalization of separate regions of the SR (Blaustein & Golovina 2001)?

(5)   Recent work using confocal microscopy has found localized increases of $[Ca^{2+}]_i$ named Ca^{2+} 'sparks' which are due to the release of Ca^{2+} from one or a small number of RyRs (Jaggar et al 2000). These localized releases of Ca^{2+} activate Ca^{2+}-dependent channels in the surface membrane (Perez et al 2001). Activation of the Ca^{2+}-activated K^+ current will hyperpolarize the membrane potential (Herrera et al 2001) and thereby decrease Ca^{2+} entry into the cell on voltage-dependent Ca^{2+} channels. This provides a mechanism whereby Ca^{2+} release from the SR can *decrease* contraction. It is therefore important, in different smooth muscles, to consider to what extent SR Ca^{2+} release activates rather than decreases contraction. It is, of course, possible that, in the same smooth muscle, SR release may sometimes directly activate contraction and, at other times, decrease it by activating K^+ channels.

(6)   The Ca^{2+} sparks mentioned above are just one example of the fact that ionic concentrations are not the same throughout the cytoplasm. In particular, there is much evidence suggesting that the ionic concentrations in the space between the surface membrane and the SR may be very different from those in the bulk cytoplasm. This will have several consequences. First, it has been suggested that, during relaxation the SR may accumulate Ca^{2+} from the bulk cytoplasm and release it through release channels into this space from where it will be pumped out of the cell. This will tend to increase the rate of Ca^{2+} removal from the cell. Second, when Ca^{2+} enters the cell from the outside, much may enter first into this space and then be accumulated by the SR before it has a chance to enter the bulk cytoplasm. These interactions between SR and surface membrane have been referred to as the 'superficial buffer barrier' hypothesis (see van Breeman et al 1995 for review). In addition to this, there is also much evidence for specialized regions of the surface membrane named 'rafts', which are rich in cholesterol and may form a platform for cell signalling molecules (Babiychuk & Draeger 2000). Furthermore signalling molecules are particularly expressed in invaginations of the surface membrane termed caveolae. These caveolae are often located close to SR

and the interactions between caveolae and SR merit consideration (see Taggart 2001 for review).

(7) Discussion such as that above runs the risk of giving the impression that the structure and function of a given smooth muscle is invariant. This is not the case. For example, there are physiological changes in the size and excitability of the myometrium during pregnancy (cf. Wray 1993). Pathological changes include those of vascular smooth muscle in vascular disease and of bladder smooth muscle in incontinence (Turner & Brading 1997). Any attempt at a complete analysis of smooth muscle function must address these changes.

(8) Our knowledge of smooth muscle contraction depends on the use of model systems including cell lines, freshly cultured cells, multicellular tissues and whole organs. It is important to consider possible artefacts and, generally, the extent to which these models accurately reproduce the behaviour of smooth muscle *in vivo*.

I am sure that there are many other areas that should be covered. More importantly, I am equally certain that the group of eminent scientists gathered here is uniquely suited to defining and resolving these issues.

References

Babiychuk EB, Draeger A 2000 Annexins in cell membrane dynamics. Ca^{2+}-regulated association of lipid microdomains. J Cell Biol 150:1113–1124

Ber DM 2001 Excitation–contraction coupling and cardiac contractile force, 2nd edn. Kluwer Academic Publishers, Dordrecht/Boston/London

Blaustein MP, Golovina VA 2001 Structural complexity and functional diversity of endoplasmic reticulum Ca^{2+} stores. Trends Neurosci 24:602–608

Flynn ER, Bradley KN, Muir TC, McCarron JG 2001 Functionally separate intracellular Ca^{2+} stores in smooth muscle. J Biol Chem 276:36411–36418

Fry CH, Wu C 1997 Initiation of contraction in detrusor smooth muscle. Scand J Urol Nephrol Suppl 184:7–14

Herrera GM, Heppner TJ, Nelson MT 2001 Voltage dependence of the coupling of Ca^{2+} sparks to BK_{Ca} channels in urinary bladder smooth muscle. Am J Physiol Cell Physiol 280: C481–C490

Jaggar JH, Porter VA, Lederer WJ, Nelson MT 2000 Calcium sparks in smooth muscle. A J Physiol Cell Physiol 278:C235–C256

Nixon GF, Mignery GA, Somlyo AV 1994 Immunogold localization of inositol 1,4,5-trisphosphate receptors and characterization of ultrastructural features of the sarcoplasmic reticulum in phasic and tonic smooth muscle. J Muscle Res Cell Motil 15:682–700

Peng H, Matchkov V, Ivarsen A, Aalkjaer C, Nilsson H 2001 Hypothesis for the initiation of vasomotion. Circ Res 88:810–815

Perez GJ, Bonev AD, Nelson MT 2001 Micromolar Ca^{2+} from sparks activates Ca^{2+}-sensitive K^+ channels in rat cerebral artery smooth muscle. Am J Physiol Cell Physiol 281:C1769–C1775

Rembold CM 1992 Regulation of contraction and relaxation in arterial smooth muscle. Hypertension 20:129–137

Taggart MJ 2001 Smooth muscle excitation–contraction coupling: a role for caveolae and caveolins? News Physiol Sci 16:61–65

Turner WH, Brading AF 1997 Smooth muscle of the bladder in the normal and the diseased state: pathophysiology, diagnosis and treatment. Pharmacol Ther 75:77–110

van Breemen C, Chen Q, Laher I 1995 Superficial buffer barrier function of smooth muscle sarcoplasmic reticulum. Trends Pharmacol Sci 16:98–105

Wray S 1993 Uterine contraction and physiological mechanisms of modulation. Am J Physiol 264:C1–C18

Wray S, Kupittayanant S, Shmygol A, Smith RD, Burdyga T 2001 The physiological basis of uterine contractility: a short review. Exp Physiol 86:239–246

Yamazawa T, Iino M, Endo M 1992 Presence of functionally different compartments of the Ca^{2+} store in single intestinal smooth muscle cells. FEBS Lett 301:181–184

Role of the sarcoplasmic reticulum in uterine smooth muscle

Susan Wray, Sajeera Kupittayanant and Tony Shmigol

Department of Physiology, The University of Liverpool, Liverpool L69 3BX, UK

Abstract. The sarcoplasmic reticulum (SR) is present as an extensive network in uterine cells. In this chapter we examine its functional importance, relating in particular, to the control of contractility in pregnancy. The uterine SR has both ryanodine receptors (RyR) and inositol-1,4,5-trisphosphate InsP$_3$ receptors (InsP$_3$R). The RyR and subsequent Ca^{2+}-induced Ca^{2+} release play little role in either human or rat contractions or Ca^{2+} transients. There may be subtle, spatiotemporal effects at the single cell level. Caffeine, an agonist for RyR fails to release Ca^{2+} and indeed produces relaxation not contraction. InsP$_3$ clearly causes release of Ca^{2+} from the uterine SR and an increase in force, although these changes are only small and transient compared to those occurring due to external Ca^{2+} entry. Inhibition of the SR Ca-ATPase by cyclopiazonic acid, empties Ca^{2+} from the SR. This is associated with an augmentation of force and Ca^{2+} transient. Thus the SR normally functions in the uterus to limit, not increase contractions. The mechanism may involve vectoral release of Ca^{2+} from the SR and activation of surface membrane K$^+$ channels. This activation would tend to decrease L-type Ca^{2+} entry and hence reduce contraction. Thus the SR is playing a role in controlling membrane excitability and hence contractility. The SR also plays a role in the relaxation of force. This is not primarily due to a direct sequestering of large amounts of Ca^{2+}, but rather that the SR directs Ca^{2+} to the surface membrane extrusion mechanisms, i.e. Ca-ATPase and Na$^+$/Ca^{2+} exchanger. This enables them to act more efficiently, and therefore aids relaxation. Recent direct measurements of SR luminal content show decreases with agonist application but not during spontaneous activity; confirming the results described above. This technique will be used to better characterize the uterine SR, its control and relevance to normal and abnormal labours.

2002 Role of the sarcoplasmic reticulum in smooth muscle. Wiley, Chichester (Novartis Foundation Symposium 246) p 6–25

Control of uterine activity is vitally important for successful pregnancy and parturition. Unfortunately, there are still far too many pre-term births or term emergency caesarean sections due to failure of the control and regulation of uterine contractility and our subsequent inability to intervene pharmaceutically in useful ways. The cost in human and financial terms of these failures is high, and drives both clinical and basic science research endeavours to gain a better understanding of contractility and its control. Studies on the sarcoplasmic reticulum (SR) of uterus aim at improving our understanding of uterine contractility, as well as investigating fundamental aspects of the role of the SR in cells.

In this paper we briefly describe uterine force production and relaxation, discuss the role of the SR in these pathways and then present data obtained in single cells measuring SR Ca^{2+} directly.

Force production in the uterus

The uterus is a spontaneously active, phasic smooth muscle. Pacemaker activity within its membrane causes depolarization and trains of action potentials (Wray 1993). The pacemaker region is not anatomically fixed, nor do there appear to be specific pacemaker cells. Simultaneous force and electrical recordings show pacemaker potentials, although these are so far uncharacterized in terms of the currents contributing to them. More is known, however, about the relationship between membrane potential and intracellular Ca^{2+} concentration ($[Ca^{2+}]_i$) (Shmigol et al 1998a). Both L- and T-type Ca^{2+} channels have been reported in uterine smooth muscle cells (myometrium), and will respond to the depolarization produced by the action potential trains (Inoue & Sperelakis 1991). We (Shmigol et al 1998a) have characterized the properties of voltage-activated Ca^{2+} transients in pregnant rat myometrium and found that: (i) a rapid increase in $[Ca^{2+}]_i$ occurs at the same time as inward Ca^{2+} current (I_{Ca}) upon membrane depolarization, with a threshold of $-50\,mV$; (ii) nifedipine (10 M) abolishes both I_{Ca} and the increase in $[Ca^{2+}]_i$, suggesting that L-type Ca^{2+} current is the major source of Ca^{2+} ions entering the cell to produce the Ca^{2+} transient; and (iii) when depolarizing pulses are applied rapidly (3 Hz) to mimic the normal action potential trains, then a more tetanic rise of $[Ca^{2+}]_i$ occurred.

The rise in Ca^{2+} causes the formation of Ca^{2+}–calmodulin and the subsequent activation of myosin light chain kinase (MLCK). MLCK phosphorylates the myosin light chains on Ser19 and thereby stimulates the myosin ATPase and cross-bridge cycling and contraction. It is clear therefore that depolarization produces a rise in $[Ca^{2+}]_i$ and that voltage-gated Ca^{2+} entry is an important part of that process. The question arises as to whether, under physiological conditions, Ca^{2+}-induced Ca^{2+} release (CICR) from the uterine SR occurs. In addition, the influence of agonists and inositol-1,4,5-trisphosphate ($InsP_3$) production on the SR and the Ca^{2+} signal needs to be explored. We will therefore now consider the SR in uterine myocytes.

Uterine SR

The uterus contains an extensive SR, occurring both close to the plasmalemma (peripheral SR) and throughout the cytoplasm (central SR), and this is reported

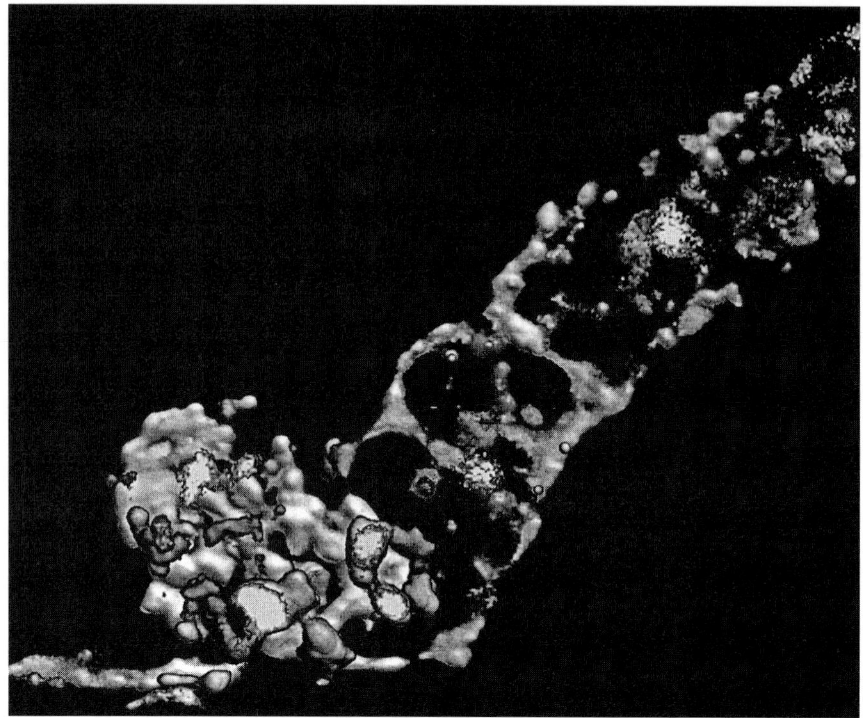

FIG. 1.   A confocal image from a freshly isolated uterine myocyte from pregnant rat. The SR was loaded with mag-fluo-4 to show its structure and distribution. The stack of images was acquired using a Perkin Elmer Ultra *VIEW* LCI confocal imaging system. Images were combined to visualize the 3D distribution of the SR (A.V. Shmigol & S. Wray, unpublished data).

to increase in pregnancy (Broderick & Broderick 1990, Somlyo 1985). There have, however, been no recent studies examining the control or mechanism of this gestational alteration. Using confocal microscopy and fluorescent Ca^{2+}-binding indicators we find a rich reticular network over the entire cell, with certain regions fluorescing more brightly, indicating Ca^{2+} 'hot spots' within the SR (Fig. 1).

The uterine SR possesses a Ca-ATPase and both ryanodine (RyR) and $InsP_3$ ($InsP_3R$) receptors. All three types of RyRs are found in the uterus (Martin et al 1999). The single-channel properties of the three subtypes of RyR appear to be rather similar but there are reported differences between them in their caffeine sensitivity. There is much current interest in assessing the physiological contribution of the RyR subtypes. All three $InsP_3Rs$ have also been reported in the uterus (Martin et al 1999), with no data available on whether changes occur in

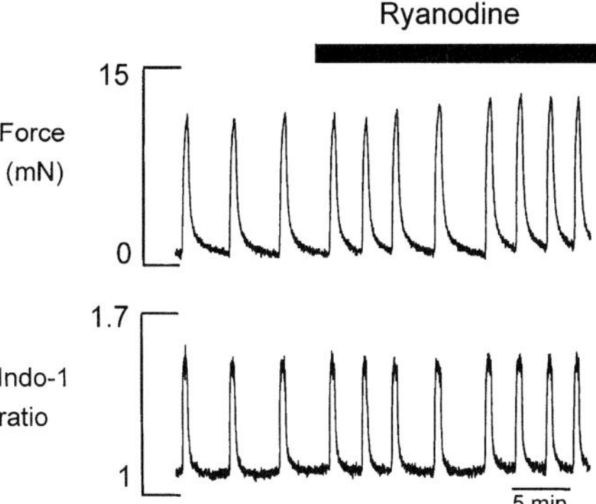

FIG. 2.   The effect of modulating CICR on uterine contraction. Spontaneous Ca^{2+} transients (Indo-1 ratio) and contractions in term pregnant, non-labouring human myometrium, before and after 50 μM ryanodine application. (Taken from Kupittayanant et al 2002.)

pregnancy or labour. Ca^{2+} also modulates the opening of InsP$_3$Rs and in turn, Ca^{2+} release from InsP$_3$Rs may stimulate RyR and CICR. These interactions and the spatial distribution of InsP$_3$R and RyR may account for differences in the responses of smooth muscle to stimulation.

We will now address the importance of each of these release mechanisms, and then return to their role in the rise of Ca^{2+} with depolarization.

CICR in the uterus

This process, and its contribution to the Ca^{2+} transient in the uterus, has proved difficult to elucidate. The simplest conclusion is that CICR does not play a role in intact myometrium, despite the presence of RyRs. The experimental basis for this can be summarized as follows:

(1)   In all species examined (rat, human and mouse) ryanodine application does not inhibit spontaneous contraction (Taggart & Wray 1998, Kupittayanant et al 2002). Figure 2 shows the effect of ryanodine on human myometrium. There is no decrement in either force or Ca^{2+}, nor was their course greatly affected, in preparations from pregnant or non-pregnant animals, as well as in single cells (Arnaudeau et al 1994a, Holda et al 1996).

(2)   Caffeine (10 mM), an agonist for CICR, is unable to produce any rise of Ca^{2+} or force in myometrium (Taggart & Wray 1998, Kupittayanant et al 2001a). These data are also in agreement with earlier force measurements reporting no rise in force (Savineau & Mironneau 1990); indeed caffeine relaxes the uterus, due to its inhibitory action on the phosphodiesterase that breaks down cAMP (Arnaudeau et al 1994a).

(3)   There have been no reports of sparks occurring in uterine cells. Ca^{2+} sparks are considered to be due to spontaneous releases of Ca^{2+} from RyR.

There are indications at the single cell level, that there may be a subtle role for CICR, e.g. when pairs of depolarizing pulses were applied to single cells, the increase in $[Ca^{2+}]_i$ produced by the second pulse was larger than that produced by the first pulse (Shmigol et al 1998a). This facilitation was abolished by ryanodine. It may be therefore that the threshold for CICR activation is high in uterine myocytes, and this is why no CICR could be seen during a single depolarizing pulse. Others have also suggested that the $[Ca^{2+}]_i$ necessary to initiate CICR in smooth muscle is high ($>1\,\mu M$) and indeed therefore higher than that level reached during the Ca^{2+} transient (Sanders 2001). Caffeine has been reported to be unable to release SR Ca^{2+} in isolated cells (Lynn et al 1993), although in a more recent study in pregnant rats, around 30% of the cells were caffeine sensitive (Martin et al 1999). Interestingly, the RyR3, isolated from rabbit uterus and expressed in a non-muscle cell line, did express caffeine sensitivity. Thus it may be that *in vivo* or intact myometrial preparations, there is modulation of the RyR altering its caffeine sensitivity. It appears that coupling between L-type Ca^{2+} entry and Ca^{2+} release from RyR in smooth muscle is much looser than that occurring with depolarization in striated muscle (Collier et al 2000).

In summary, in intact myometrium there is no clear physiological role for CICR despite the presence of RyR. This is the conclusion reached by others for a variety of smooth muscles e.g. taeniae coli (Iino 1989), stomach (Guerrero et al 1994), tracheal myocytes (Fleischmann et al 1996) and portal vein (Kamishima & McCarron 1996).

InsP$_3$-induced Ca^{2+} release in the uterus

A major component of agonist stimulation of smooth muscle involves their activation of phospholipase C (PLC), subsequent to them binding to G-protein coupled receptors. PLC in turn leads to the production of InsP$_3$ and diacylglycerol. Unlike CICR, InsP$_3$-induced Ca^{2+} release (IICR) is clearly present in the uterus. In both rat and human myometrium, if agonists are added in zero external Ca^{2+}, a small, transient release of $[Ca^{2+}]_i$ and accompanying contraction can be seen (Taggart & Wray 1997, Luckas et al 1999). That these Ca^{2+} releases were arising from the SR could be demonstrated by emptying the SR, e.g. with cyclopiazonic acid (CPA; $20\,\mu M$), a specific inhibitor of the myometrial SR Ca-ATPase. Application of agonist now fails to release Ca^{2+} and there is no rise in force. In contrast, incubation with ryanodine to block CICR had no effect on the agonist-induced Ca^{2+} and force increases (Kupittayanant et al 2001a). This is consistent with the data discussed above, i.e. that CICR does not play a role in intact myometrium. The agonist-induced SR Ca^{2+} release and force in zero external $[Ca^{2+}]$ were transient, indicating either the limited store of Ca^{2+} or ability to release Ca^{2+}. The magnitude of the agonist-induced Ca^{2+} and force changes were significantly enhanced in myometrium from pregnant compared with non-pregnant rats, suggesting a gestational change in the SR (Taggart & Wray 1998).

Thus a clear release of Ca^{2+} and a functional effect can be seen in the myometrium, attributable to $InsP_3R$. This released Ca^{2+} is however only small compared with that occurring when Ca^{2+} entry is taking place, which of course is the usual physiological condition. We will now discuss the effects of emptying and refilling the SR.

The uterine SR Ca-ATPase

The SR is able to take up Ca^{2+} against the electrochemical gradient between its lumen and the cytoplasm due to presence of active Ca^{2+} pumps in the SR membrane, known as SERCA. The uterus appears to express SERCA2a and SERCA2b (Tribe et al 2000), and both are increased in uterine samples taken from women in labour, compared to those not in labour. Ca^{2+} binding proteins such as calreticulin and calsequestrin buffer the Ca^{2+} within the SR, so that total Ca^{2+} within the SR may reach millimolar amounts. The activity of SERCA is under physiological regulation as well as being influenced by the gradient of Ca^{2+} across the SR membrane.

Emptying the SR by using cyclopiazonic acid (CPA) produced no inhibition of force or Ca^{2+}, in rat or human uterus (Taggart & Wray 1998, Kupittayanant et al 2002), but rather both were significantly increased (Fig. 3). That is, inhibition of the SR potentiates force and Ca^{2+}, especially in pregnant myometrium. This leads us to suggest that in the uterus the SR Ca^{2+} store acts to feedback and limit contractions, as is thought for vascular smooth muscle (Herrera & Nelson 2002, this volume). Briefly Ca^{2+} release from SR RyR produces Ca^{2+} sparks, which activate K^+ channels, producing spontaneous transient outward currents (STOCs) and cell hyperpolarization. The uterus possesses Ca^{2+}-activated K^+ channels (K_{Ca}), and their modulation and expression may alter with pregnancy and labour (Khan et al 2001). Thus it is tempting to suggest a similar mechanism exists in the uterus and that it plays a role in governing contractility changes during labour.

Another way the SR might modulate excitability in the uterus is via Ca^{2+}-activated Cl^- (Cl_{Ca}) channels. The evidence for Cl_{Ca} in the uterus is limited but Arnaudeau et al (1994b) showed that oxytocin in rat cells appeared able to stimulate these channels, and we also find their activation via voltage-gated Ca^{2+} entry (K. Jones, A.V. Shmigol & S. Wray, unpublished observation).

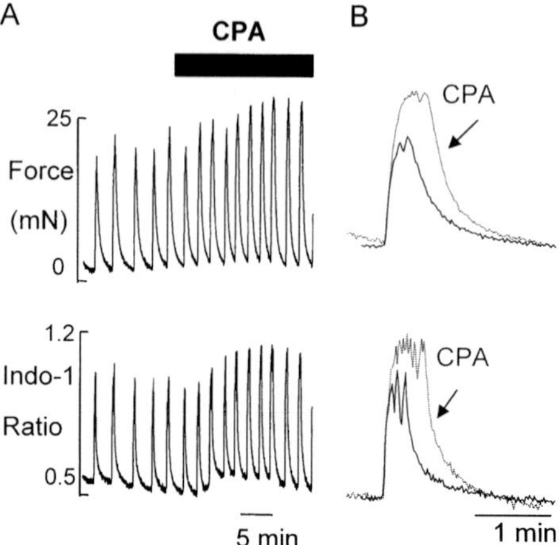

FIG. 3.   The effect of inhibiting the SR on contraction. (A) Spontaneous Ca transients and force, in human pregnant myometrium, before and during 20 μM cyclopiazonic acid (CPA) application. (B) Superimposed force and Ca^{2+} records taken from A, under control conditions and in the presence of CPA (dotted trace). (Taken from Kupittayanant et al 2002.)

Within the uterus there is close association between the peripheral SR and specialized regions of the plasma membrane containing elements important to Ca^{2+} signalling and homeostasis. One such element is the Na^+/Ca^{2+} exchanger (Taggart & Wray 1997). Interestingly this coupling was found to be specific to the pregnant myometrium. Much work remains to be done to further our understanding of how Ca^{2+} signalling events are integrated and the role of microdomains (Lee et al 2001).

Thus the role of the SR in the uterus can be seen to incorporate excitability via Ca^{2+}-activated channels, as well as releasing some Ca^{2+} to support agonist-induced contractions. As with other smooth muscles it is necessary to consider the spatiotemporal nature of the Ca^{2+} signals produced, if we are to fully understand the ultimate functional effects on contraction. So far the discussion has focussed on the elevation of the Ca^{2+} signal with stimulation. We will now consider the equally important process of lowering Ca^{2+} and its role in relaxation in the uterus.

Relaxation and Ca^{2+} uptake by the SR

The rise in $[Ca^{2+}]_i$ and MLCK phosphorylation of myosin need to be reversed for relaxation. It is very important to the correct functioning of the uterus that its contractions are phasic and not maintained (tonic), as blood flow falls in the myometrium at the peak of contraction, due to the occlusion of its blood vessels (Larcombe-McDouall et al 1999). Relaxation involves a cessation of voltage-gated

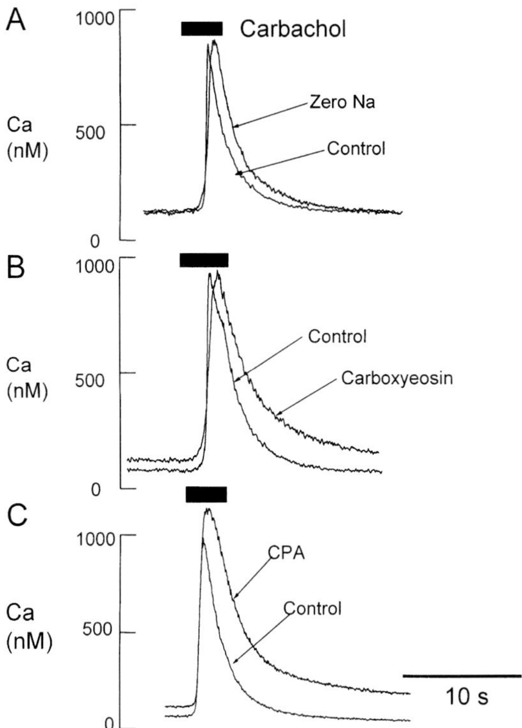

FIG. 4. Mechanisms to lower [Ca] in rat uterine myocytes. The cells were stimulated with carbachol and then the rate of decay of the Ca^{2+} transient determined. The traces show the effects of inhibiting; Na/Ca exchange (top traces), surface membrane Ca-ATPase (middle traces) and SR Ca-ATPase (bottom traces). (Taken from Shmigol et al 1999.)

Ca^{2+} entry, Ca^{2+}-uptake into the SR by SERCA, Ca^{2+} extrusion from the cell and dephosphorylation of the myosin light chains. The type 1 phosphatase, myosin light chain phosphatase (MLCP) dephosphorylates myosin. As with MLCK its activity is physiologically regulated, e.g. its activity is decreased following phosphorylation via Rho-associated kinase (Somlyo & Somlyo 2000). In the uterus we have found a small but significant reduction of force, but not Ca^{2+} when Rho-associated kinase is inhibited (Kupittayanant et al 2001b).

We have examined Ca^{2+} extrusion mechanisms and the SR in relaxation (Fig. 4). These data can be summarized as follows (Shmigol et al 1998b, 1999):

(1) In single uterine cells measuring $[Ca^{2+}]_i$ simultaneously with I_{Ca} and eliciting Ca^{2+} transients by repetitive membrane depolarizations, blocking the surface membrane Ca-ATPase with carboxyeosin decreased the rate of the decay of the Ca^{2+} transient by around 30–50%.

(2) Removal of external Na^+ decreased the rate of $[Ca^{2+}]_i$ decay by around 50%, indicating an important role for the Na^+/Ca^{2+} exchanger, as mentioned above.

(3) When both the Na^+/Ca^{2+} exchanger and plasma membrane Ca-ATPase were inhibited the cells failed to restore Ca^{2+} after stimulation.

(4) The rate of decay of the $[Ca^{2+}]_i$ following carbachol stimulation was slowed when SERCA was inhibited by CPA, by $>50\%$. However, the decay of the Ca^{2+} transient was abolished if plasmalemmal extrusion mechanisms were inhibited.

Thus in the uterus the SR contributes to the decay of the $[Ca^{2+}]_i$ transients but, acting alone, it cannot significantly reduce Ca^{2+}. The conclusion reached is that the SR takes up Ca^{2+} and then releases it close to the plasmalemmal Ca^{2+} extrusion sites (i.e. the SR is acting in series with the surface membrane extrusion mechanisms). This translocation is thought to be mediated by the SR forming a superficial barrier in the narrow space between the SR and plasmalemma and the SR releasing the Ca^{2+} in a vectorial manner (van Breemen 1977).

In summary, in the uterus where L-type Ca entry represents the most significant mechanism for elevating $[Ca^{2+}]_i$, then it is necessary that surface membrane extrusion mechanisms exist to remove the Ca^{2+} gained by the cell. In the uterus this task is performed equally by the Na^+/Ca^{2+} exchanger and the Ca-ATPase. However the SR contributes to this process by taking up Ca^{2+} and vectorially releasing it to the extrusion sites, to maximize their efficiency.

In the concluding section we will discuss recent work where SR luminal Ca ($[Ca^{2+}]_L$) has been directly measured in uterine cells.

Simultaneous measurements of SR and cytosolic $[Ca^{2+}]$

If we are to better understand the functioning of the SR then we require direct measurements of the SR $[Ca^{2+}]$ ($[Ca^{2+}]_L$). There have been few studies, in any cell type where this has been achieved (ZhuGe et al 1999, Chatton et al 1995, Golovina & Blaustein 1997). We therefore developed a methodology to record simultaneous, time-resolved, direct SR and cytosolic Ca^{2+} changes (Shmigol et al 2001). We combined a low-affinity Ca^{2+} sensitive indicator, mag-fluo-4 (K_d 22 μM) with a high-affinity indicator, Fura-2 (K_d 0.14 μM) (Shmigol et al 2001). As shown in Fig. 5, this gave clear and reproducible signals from $[Ca^{2+}]_i$ and $[Ca^{2+}]_L$. We detected reductions in $[Ca^{2+}]_L$ and increases in $[Ca^{2+}]_i$ with agonist applications to the uterine cells. There was some replenishment of the SR, in the continued presence of agonist and absence of external Ca^{2+}, but extracellular Ca^{2+} was required for full replenishment. Agonist-induced oscillation of $[Ca^{2+}]_i$,

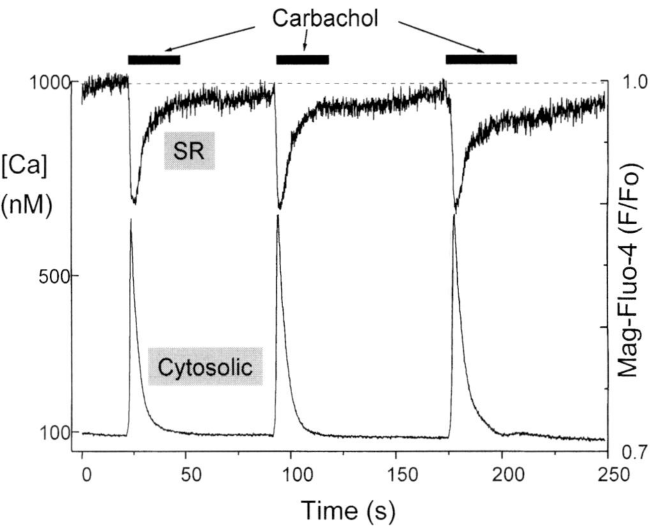

FIG. 5. Simultaneous and reproducible measurements of changes in SR and cytosolic Ca^{2+} in response to agonists. Pregnant rat myometrial cells were superfused with Krebs (2 mM Ca^{2+}) and Carbachol (in Ca-free solution) applied as indicated. Repetitive applications of Carbachol evoked reproducible $[Ca^{2+}]_i$ elevations (bottom trace) and SR $[Ca^{2+}]_L$ decreases (upper traces). Left-hand axis: cytosolic Ca^{2+}. Right-hand axis: intraluminal (SR) Ca^{2+} changes, measured with mag-fluo-4. (Taken from Shmigol et al 2001.)

accompanied by successive release and re-uptake of SR Ca^{2+}, occurred and were abolished by inhibition of the SR Ca-ATPase.

We could detect no changes in $[Ca^{2+}]_L$ during spontaneous transients, consistent with the conclusions drawn above. Indeed, we found that if SR Ca^{2+} was elevated, then spontaneous activity was suppressed until SR Ca^{2+} levels returned to their initial value. This is also consistent with SR activity inhibiting rather than potentiating $[Ca^{2+}]_i$ and force.

Thus it can be seen that direct monitoring of the SR Ca^{2+} is providing new information about the relationship between $[Ca^{2+}]_i$, $[Ca^{2+}]_L$ and Ca^{2+} signalling in the cell. It should now also be possible to look in detail at the kinetics of the release process and the effects of Ca^{2+} load on this.

Conclusions

In the uterus the SR is a pronounced and complex organelle, which appears to be under some degree of hormonal control. For example, its size, SERCA expression and release mechanisms show alteration with pregnancy. As yet there has been little study of the molecular and mechanistic processes that lead to these changes in the

SR. Their elucidation may be of relevance to our understanding of alterations in SR function in pathophysiological processes, such as hypertension and pre-term labour.

There is a general similarity between species in the role of SR in the uterus. For example, the SR is acting to limit force and Ca^{2+} transients in the spontaneously active uterus. The mechanism for this is at least in part via the control of surface membrane excitability. Such activity in a temporal manner may aid the phasic nature of uterine contractions. That is, stimulation leads to L-type Ca^{2+} entry and contraction but also increases SR Ca^{2+} content. This leads to release of SR Ca^{2+} vectorially to the surface membrane and the K_{Ca}, which then act to reduce further Ca^{2+} entry, and hence contribute to relaxation. The SR contributes to relaxation by taking up Ca^{2+}. However, the capacity of the SR to take up this Ca^{2+} is very limited. Rather, the SR passes the small amount of Ca^{2+} it takes up in a vectorial fashion to the Na^+/Ca^{2+} exchanger and Ca-ATPase on the surface membrane, and Ca^{2+} entry will be balanced by Ca^{2+} extrusion. That agonists can produce $InsP_3$ and stimulate Ca^{2+} release and force in the uterus has been demonstrated. Our work using mag-fluo-4 shows the decreases in SR $[Ca^{2+}]_L$ accompanying the rises in $[Ca^{2+}]_i$. For the uterus we are still lacking, however, a quantitative assessment of this contribution of agonist action compared to their other effects, such as on Ca^{2+} entry and sensitization of the contractile machinery. When we have that information we will be able to examine how these elements may be altered in labours which fail to progress or which are characterized by hyperactivity, threatening fetal oxygenation and uterine rupture, or in processes leading to pre-term birth.

Acknowledgements

We are grateful to the MRC and Wellcome Trust for funding this work. SK is supported by the Royal Thai Government.

References

Arnaudeau S, Lepretre N, Mironneau J 1994a Oxytocin mobilizes calcium from a unique heparin-sensitive and thapsigargin-sensitive store in single myometrial cells from pregnant rats. Pflugers Arch 428:51–59

Arnaudeau S, Lepretre N, Mironneau J 1994b Chloride and monovalent ion-selective cation currents activated by oxytocin in pregnant rat myometrial cells. Am J Obstet Gynecol 171:491–501

Broderick R, Broderick KA 1990 Ultrastructure and calcium stores in the myometrium. In: Carsten ME, Miller JD (eds) Uterine function: molecular and cellular aspects. Plenum Press, New York, p 1–70

Chatton JY, Liu H, Stucki JW 1995 Simultaneous measurements of Ca^{2+} in the intracellular stores and the cytosol of hepatocytes during hormone-induced Ca^{2+} oscillations. FEBS Lett 368:165–168

Collier ML, Ji G, Wang Y, Kotlikoff MI 2000 Calcium-induced calcium release in smooth muscle: loose coupling between the action potential and calcium release. J Gen Physiol 115:653–662

Fleischmann BK, Wang YX, Pring M, Kotlikoff MI 1996 Voltage-dependent calcium currents and cytosolic calcium in equine airway myocytes. J Physiol 492:347–358

Golovina VA, Blaustein MP 1997 Spatially and functionally distinct Ca^{2+} stores in sarcoplasmic and endoplasmic reticulum. Science 275:1643–1648

Guerrero A, Fay FS, Singer JJ 1994 Caffeine activates a Ca^{2+}-permeable nonselective cation channel in smooth muscle cells. J Gen Physiol 104:375–394

Holda JR, Oberti C, Perez-Reyes E, Blatter LA 1996 Characterization of an oxytocin-induced rise in $[Ca^{2+}]_i$ in single human myometrium smooth muscle cells. Cell Calcium 20:43–51

Iino M 1989 Calcium-induced calcium release mechanism in guinea pig taenia caeci. J Gen Physiol 94:363–383

Inoue Y, Sperelakis N 1991 Gestational change in Na^+ and Ca^{2+} channel current densities in rat myometrial smooth muscle cells. Am J Physiol 260:C658–C663

Kamishima T, McCarron JG 1996 Depolarization-evoked increases in cytosolic calcium concentration in isolated smooth muscle cells in rat portal vein. J Physiol 492:61–74

Khan R, Mathroo-Ball B, Arulkumaran S, Ashford ML 2001 Potassium channels in the human myometrium. Exp Physiol 86:255–264

Kupittayanant S, Burdyga TV, Wray S 2001 The effects of inhibiting Rho-associated kinase with Y-27632 on force and intracellular calcium in human myometrium. Pflüger's Arch 443:112–114

Kupittayanant S, Luckas MJM, Wray S 2002 Effects of inhibiting the sarcoplasmic reticulum on spontaneous and oxytocin-induced contractions of human myometrium. Br J Obstet Gynaecol 108:1–8

Larcombe-McDouall JB, Buttell N, Harrison N, Wray S 1999 In vivo pH and metabolite changes during a single contraction in rat uterine smooth muscle. J Physiol 518:783–790

Lee Y-H, Hwang M-K, Morgan KG, Taggart MJ 2001 Receptor-coupled contractility of uterine smooth muscle: from membrane to myofilaments. Exp Physiol 86:283–288

Luckas MJ, Taggart MJ, Wray S 1999 Intracellular calcium stores and agonist induced contractions in isolated human myometrium. Am J Obstet Gynecol 181:468–476

Lynn S, Morgan JP, Gillespie JI, Greenwell JR 1993 A novel ryanodine-sensitive calcium release mechanism in cultured human myometrial smooth muscle cells. FEBS Lett 330:227–230

Martin C, Hyvelin JM, Chapman KE, Marthan R, Ashley RH, Savineau JP 1999 Pregnant rat myometrial cells show heterogeneous ryanodine- and caffeine-sensitive calcium stores. Am J Physiol 277:C243–C252

Herrera GM, Nelson MT 2002 SR and membrane currents. In: Role of the sarcoplasmic reticulum in smooth muscle. Wiley, Chichester (Novartis Found Symp 246) p 189–207

Sanders KM 2001 Mechanisms of calcium handling in smooth muscles. J Appl Physiol 91:1438–1449

Savineau JP, Mironneau J 1990 Caffeine acting on pregnant rat myometrium: analysis of its relaxant action and its failure to release Ca^{2+} from intracellular stores. Br J Pharmacol 99:261–266

Shmigol A, Eisner DA, Wray S 1998b Carboxyeosin decreases the rate of decay of the $[Ca^{2+}]_i$ transient in uterine smooth muscle cells isolated from pregnant rats. Pflüger's Arch 437:158–160

Shmigol A, Eisner DA, Wray S 1998a Properties of voltage-activated $[Ca^{2+}]_i$ transients in single smooth muscle cells isolated from pregnant rat uterus. J Physiol 511:803–811

Shmigol AV, Eisner DA, Wray S 1999 The role of the sarcoplasmic reticulum as a Ca^{2+} sink in uterine smooth muscle cells. J Physiol 520:153–163

Shmigol AV, Eisner DA, Wray S 2001 Simultaneous measurements of changes in sarcoplasmic reticulum and cytosolic [Ca^{2+}] in rat uterine smooth muscle cells. J Physiol 531:707–713

Somlyo AP 1985 Excitation-contraction coupling and the ultrastructure of smooth muscle. Circ Res 57:497–507

Somlyo AP, Somlyo AV 2000 Signal transduction by G-proteins, rho-kinase and protein phosphatase to smooth muscle and non-muscle myosin II. J Physiol 522:177–185

Taggart MJ, Wray S 1997 Agonist mobilization of sarcoplasmic reticular calcium in smooth muscle: functional coupling to the plasmalemmal Na^+/Ca^{2+} exchanger. Cell Calcium 22:333–341

Taggart MJ, Wray S 1998 Contribution of sarcoplasmic reticular calcium to smooth muscle contractile activation: gestational dependence in isolated rat uterus. J Physiol 511:133–144

Tribe RM, Moriarty P, Poston L 2000 Calcium homeostatic pathways change with gestation in human myometrium. Biol Reprod 63:748–755

van Breemen C 1977 Calcium requirement for activation of intact aortic smooth muscle. J Physiol 272:317–329

Wray S 1993 Uterine contraction and physiological mechanisms of modulation. Am J Physiol 264:C1–C18

ZhuGe R, Tuft RA, Fogarty KE, Bellve K, Fay FS, Walsh JV 1999 The influence of sarcoplasmic reticulum Ca^{2+} concentration on Ca^{2+} sparks and spontaneous transient outward currents in single smooth muscle cells. J Gen Physiol 113:215–228

DISCUSSION

Nelson: We have done similar experiments with ryanodine on urinary bladder smooth muscle. We observe a small increase in contraction frequency in response to ryanodine. We have done experiments in which we first blocked either Ca^{2+}-activated K^+ (BK) channels or small-conductance Ca^{2+}-activated K^+ (SK) channels, and then added ryanodine. This led to an enormous increase in contractility. I am proposing that the effects of CPA would be larger if BK or SK channels were first inhibited.

Wray: If we add iberiotoxin and do the CPA experiment, it still has effects, although they are not as large. This is another reason why I think that the K_{Ca} channels aren't the full story. We have not yet tried apamin to block SK channels. But I agree: there are many inputs into this system. However, if functionally we are getting increases in force and Ca^{2+} with manoeuvres such as emptying the SR, this is telling us that it is playing an important role. Or is this too simplistic?

Nelson: I'd also point out that when you say that there is no change in SR Ca^{2+}, you probably wouldn't pick up Ca^{2+} sparks in these averaged measurements, so you would miss these events if they are occurring. And when you add CPA and measure relaxation, the cells are not voltage clamped. You argue that membrane excitability is affected. Then the addition of CPA in a non-voltage-clamp situation could affect the apparent changes in decay that might be related to changes in Ca^{2+} or excitability.

Wray: We have tried to look at the decay currents both when agonists are present and also under voltage clamp conditions. The results are pretty similar; the SR facilitates decay of the Ca^{2+} transient, but not by acting directly. With respect to the SR Ca^{2+} not changing, it would be more accurate to say that it doesn't produce a detectable change. If we do confocal experiments and look for sparks, it's extremely difficult to find them in the rat. We have had more luck with the human myocytes. I think there's something — another element — that we are missing.

Sanders: What is ryanodine doing pharmacologically?

Wray: We have used a range of concentrations, where we can either have the RyRs in the subconductance state or the blocked state. It doesn't make a difference to the conclusions — it has little or no effect in the uterus.

Sanders: If ryanodine is working in this blocked open state, its effects ought to look a lot like those of thapsigargin. Do you have a positive control for your ryanodine studies to show that this will work? In other words, does caffeine release Ca^{2+} in the presence of ryanodine?

Wray: Caffeine is not a releaser in uterine myocytes. A paper by Martin et al (1999) showed that in a subset of cells with ryanodine receptors, caffeine does not release from them. We speculate that there is an uncoupling between the receptor and the channels. Caffeine relaxes the uterus, presumably because of its effect in elevating cAMP. There are some similarities, though, between ryanodine and CPA, because if the ryanodine was doing anything it was slightly potentiating force and slightly increasing frequency, rather similarly to CPA. The ryanodine Fig. 1 showed was the one where it was most neutral, because I was trying to show that it was not diminishing force.

Sanders: I was looking at the rises in basal Ca^{2+} from thapsigargin and CPA.

Wray: These are more pronounced than what we get with ryanodine. Whether this is to do with the time course, I'm not sure. If extrusion mechanisms are more stimulated to a greater extent by faster or slower changes in Ca^{2+}, then it may be difficult to compare.

Sanders: Have you ever put caffeine on and then added ryanodine?

Wray: No.

Blaustein: I have a question that concerns the state of the uterine smooth muscle. The non-pregnant state is likely to be different from the pregnant state, and certainly the pre-term or term state, where the contractile mechanism may be different. Is it? Do some of these things change dramatically at that time?

Wray: We have compared the pregnant and non-pregnant state. It doesn't make a great deal of difference. The indications are that if we put CPA or ryanodine on a pregnant uterus, we can get slightly larger potentiations of force. Whatever this pathway is, it is probably expressed more in the pregnant uterus. But these are non-labouring pregnant women or rats.

Blaustein: In the pregnant state the uterus is trying to suppress contraction except at the time of labour.

Wray: Yes, I agree there's evidence that expression of K^+ channels, for example, does change in labour. For human myometrium the picture is far from complete, partly because it is difficult to get samples from women in labour. I agree that we would expect changes.

Blaustein: Would you expect the SR to play more of a role and SR Ca^{2+} to be more important in labour?

Wray: If I wanted to be conventional I'd say yes, but actually I don't think so. It is interesting that the neonatal uterus seems to have a massive store and excitability is not as important in the neonate, from our preliminary data.

Nelson: We are looking at SK channels, and we have SK3-overexpressing mice. These mice have difficulty giving birth because these SK3 channels are overexpressed in the uterus, and the smooth muscle can't contract properly. We have to suppress this expression (it's a conditional knockout) in order to keep our colony going.

Wray: Control of excitability is going to be vital to these processes.

Young: With regard to the differentiation between the non-labouring and labouring human uterus, I was struck by the differences across the species in the thapsigargin experiment that you mentioned. In your human tracing, the frequency of the uterine contractions was largely unchanged, whereas this was exactly the opposite in the mouse. The strength of the contraction was really not that much greater than in the control segment. This speaks not only to the differences between the rat and the human, but also potentially to any major changes between the spontaneous labour and whether or not this was oxytocin-stimulated in order to get the contraction. The factors that enter into the initiation of labour could be affecting the SR and its role. It is slightly dangerous to draw conclusions comparing the normal processes in human labour with clinically useful methods to modulate functional contractility in order to implement delivery. The SR may be switching roles completely.

Wray: I can't disagree about it being dangerous to extrapolate to women in labour. I'm saying that it would be nice to know a lot more about it. However, in all species and gestational states, the SR is acting to limit contraction, but we have not yet examined labouring tissues.

Young: This is clearly a major change from what we have normally thought about the role of Ca^{2+} metabolism. It is also coupled with a unique situation, where the uterus, against all odds, is trying to remain quiescent. This is the first trigger that I have seen that can explain that.

van Breemen: Does the manner by which Ca^{2+} is released have an effect on the subsequent contraction? For example, in your experiments where ATP releases a small amount of Ca^{2+} it gives a large contraction, and where ionomycin releases a huge amount of Ca^{2+} it gives only a minor contraction.

Wray: The speed of the Ca^{2+} release seems to be important: faster rises will stimulate these mechanisms better than slow ones. For example, if we put on CPA or thapsigargin, often we will see a substantial elevation of basal Ca^{2+}, but it is quite slow. We don't see a corresponding rise of the basal level of force production in many of the tissues. It seems that there is a lot of Ca^{2+} there but the contractile machinery is not interested in it. It will presumably also be the case that extrusion mechanisms handle small, slow releases, and hence little or no contraction.

Somlyo: You showed that there are other mechanisms that come into play, such as Ca^{2+} sensitization. I would expect an agonist such as a purinergic agonist to activate the Rho/Rho kinase system, which ionomycin does not.

Taylor: That is certainly what we see with oxytocin on spontaneous contractions in human myometrium.

Hirst: It might also suggest that the site of Ca^{2+} release is important, even before the mechanisms altering sensitivity come in. As far as I'm concerned, Ca^{2+} going through L-type Ca^{2+} channels is ineffective at triggering contraction, whereas store-released Ca^{2+} is much more effective. Even at the level you are talking about, it would be the site of release that is important.

Taylor: In our experiments, the oxytocin itself caused no change in Ca^{2+} concentration — that occurs at much higher concentrations than are required to bring about the change in Ca^{2+} sensitivity. But still oxytocin improved the Ca^{2+} tension relationships through what looks like a Rho kinase pathway.

Hirst: How long does it take to activate a sensitization pathway?

Taylor: It is difficult for us to get a time course, because we are relying on spontaneous contractions.

Paul: We have been talking about the SR as a site of release, as if it were one thing. There is superficial SR, deep SR and ER, with three different ryanodine receptor isoforms and at least two different SERCAs. Are these on the same vesicles, or different vesicles? What's the role of the ER? Do they communicate with one another? There are a lot of questions about this compartmentalization and vectorial release. The mitochondria also intrigue me: the SR can still function quite well after CCCP (carbonyl cyanide *m*-chlorophenylhydrazone), which inhibits mitochondrial Ca^{2+} uptake.

Somlyo: When we look at ryanodine and $InsP_3$ receptors, they are distributed wherever there is SR.

Nixon: There may be some variation in the subtype distribution. If there is localization of one particular subtype, this will change the profile of Ca^{2+} release.

Somlyo: At the gross level the distribution looks even. However, even though there is continuity within the lumen, there is calreticulin and calsequestrin. When we are dealing with different regions we have to deal with the tortuosity factor that biophysicists have been concerned with for many years. There is adsorption-hindered diffusion, which gives a different calculation for how long it takes. We haven't sorted out factors such as this.

Bolton: Electron microscopy depends heavily on fixing the tissue in glutaraldehyde or other similar substances. I'm not sure that such fixed tissues give an accurate picture of the SR, such as that given by using vital stains on living cells. The SR seems to be much more extensive in living cells than it does in electron photomicrographs, and one also has to bear in mind that Ca^{2+} waves pass along smooth muscle cells quite quickly (we've seen speeds of 100 μm/s). I understand that this is not compatible with simple diffusion, so there is CICR. If this is taking place there has to be quite a reasonable density, and not a discontinuity, of whatever the release sites are (presumably RyRs) along the cell. The picture of the SR we get from electron photomicrographs is a little misleading, and the newer stains we are using on living cells give us a more accurate picture.

Somlyo: It all depends on the fixation method. We use osmium ferricyanide to selectively infiltrate the SR. If we use intermediate high-voltage electron microscopy, we can look at thicker specimens, and this technique provides more extensive views than obtained from the usual thin sections. This is the same information we get when we infiltrate the SR with Fluo-3. These pictures are pretty reliable. Furthermore, if you want to confirm without chemical fixation, methods such as rapid freezing are available. All these techniques give the same pictures, which vary according to the smooth muscle type.

Blaustein: This is an important point: is everyone talking about the same smooth muscle? It varies from cell to cell.

Lompré: I'd also add that within the same smooth muscle tissue, there is variation between cells in different conditions. This is true for the uterus, and also vascular smooth muscle, depending on whether the cells are proliferating or not.

Blaustein: Earlier, Rick Paul raised the issue of the release sites. One thing we have seen in cultured cells is that there are frequent release sites. We also saw this when we were looking at sparks. There are some special areas within the cells. The SR is not uniform in this sense. The question is where those release sites are. There is another issue concerning Ca^{2+}-induced Ca^{2+} release. If there is freely diffusible Ca^{2+} within the SR, Ca^{2+} is released into the cytosol and this triggers release next door. Then if there is diffusion away from the next door site simply because Ca^{2+} has been vacated from the place it is released from, there is a problem. As one part of the SR is emptied of Ca^{2+}, if Ca^{2+} is freely diffusible it will not be there at the next sites that haven't released yet. This is something that hasn't been addressed. There are some real problems in terms of having a freely communicating intraluminal SR.

Eisner: There are experiments in pancreatic acinar cells showing that Ca^{2+} can 'tunnel' through the ER from one end of the cell to the other.

Blaustein: There is a problem with such experiments: there is no spatial resolution. It is as if Ca^{2+} were disappearing in one part of the cell and appearing in another part, but if you look at Petersen's pictures (Park et al 2000) they are very low resolution, and you cannot tell what has happened inside the cell.

Taylor: You mentioned hotspots of release, and I wasn't sure what the stimulus was for that. Do different agonists do subtly different things in the way that they bring about a Ca^{2+} signal? Do different agonists share the same hotspots? There is some precedent in other cells for different extracellular stimuli using the same intracellular messenger and nonetheless talking differently to the ER with it.

Blaustein: I've not looked at that issue. We have looked at one agonist, and have not tried to compare agonists. It's a good question though. You would expect that the different receptors might be at different places and therefore produce different hotspots, if this were to be the case.

Fry: I want to return to a topic that Sue Wray was addressing about whether the SR causes relaxation or contraction of the tissue. One of the features of many smooth muscles is that they hypertrophy. Very often, when smooth muscle cells get bigger they become more spontaneously active (at least in the urinary and genital tract). Is the relationship lost between the SR and the membrane as the cell gets bigger, going from a holding down state to a rather more releasing state? Is there a change in activity or distribution of SR as cells hypertrophy?

Wray: I don't have any quantitative data on this, nor do I think there are any available.

Brading: Someone has got to start looking at this issue. At the moment everyone is looking at normal smooth muscle cells. In any dysfunction there are likely to be changes in 3D relationships.

Somlyo: There's a problem here. In hypertrophied muscle there is an increase in rough ER because of the increase in protein production. The next problem is going to be sorting out how much of the rough ER takes up Ca^{2+}.

Kotlikoff: Susan Wray, I was impressed by your comment that you have no caffeine responses. One of the things that differs between different smooth muscle tissues is the state of expression of RyR2, which is the cardiac isoform that is also found in some smooth muscles, although not all. RyR2 is more sensitive to caffeine than RyR3, which is more ubiquitously expressed and has lower sensitivity. This may be a telling statement about the functional role of the RyR in the tissue that you are looking at.

Wray The uterus is reported to express all three types of RyRs.

Brading: I keep hearing about the effects of ryanodine on RyRs when it is applied to smooth muscles. Do we actually have evidence as to what ryanodine is doing at that level in smooth muscle cells? It is a widely used tool.

Nelson: There have been a couple of bilayer studies on the effects of ryanodine on RyR channels.

A general statement. We have been discussing the differences between smooth muscle types, but what about looking at the common themes throughout smooth muscle? One of these is CICR. Could we generate a list of things that we agree on that might be different in cardiac muscle? The one I'd put up first is CICR: does the Ca^{2+} channel signal the RyRs in the same way in smooth muscle as it does in cardiac muscle. I would say maybe not.

Sanders: There's some possibility of that occurring in bladder.

Nelson: Gil Wier has done some seminal work on local control of heart muscle. I don't think there is local control of ryanodine receptors by Ca^{2+} channels in any smooth muscle.

Wier: I think you showed that voltage depolarization in smooth muscle increased the frequency of Ca^{2+} sparks. What you are asserting is that this was not local control; that it was just due to an elevation of cytoplasmic and SR Ca^{2+}.

Nelson: But we don't see what you do: the Ca^{2+} channel opening and then a local blast of Ca^{2+} turning on the RyRs in a few milliseconds. In smooth muscle this process is slower and is quite different. This may reflect a fundamentally different arrangement of the Ca^{2+} channels with respect to the RyR and this may be common in smooth muscle.

Blaustein: We have seen the same thing in cultured cells: we see no evidence for CICR. There are two possible explanations. First, the smooth muscle cells have a different ryanodine receptor isoform. Second, the rate of rise of Ca^{2+} is much slower in smooth muscle. In cardiac muscle there is a blast of Ca^{2+}, with a rapid rise. Perhaps the slower effect means that smooth muscle cells don't get to threshold at the ryanodine receptor. The question is whether it is the channel or the process in general.

Somlyo: The answer would have to come from the kind of experiment that Gerrit Isenberg and others have done in other tissues, figuring out how much Ca^{2+} goes in, then measuring how much cytoplasmic Ca^{2+} is increased and the buffering power of the cell. If you find that the amount that goes in is not enough to account for the total increase in cellular Ca^{2+} levels, you know there must be CICR. Otherwise you have to assume that enough Ca^{2+} goes in during an action potential to account for all the Ca^{2+} buffering in the cytosol plus giving you enough Ca^{2+} to cause a contraction. The arithmetic is simple, even if the experiments aren't.

Eisner: The problem with this sort of experiment is that with the right sort of voltage clamp experiment you can show evidence that CICR, under some circumstances, can amplify the signal. The question is, does it do it normally.

Iino: The critical experiment would be to look at the luminal Ca^{2+} in the SR. If you depolarize the cell and the luminal Ca^{2+} is increased, there should be mechanisms of overloading the SR. If there is CICR, there will be a decrease in the SR Ca^{2+}.

Bolton: All the processes that are going on in the cell are going on all the time. Whatever level of Ca^{2+} you have in whatever compartment at steady state, there is essentially an equilibrium between those processes storing it and those processes letting the gradients run down. If you open channels in the SR, the gradients will run down, but at the same time the SR Ca^{2+} pump will start to increase its activity. The idea that there are fixed stores, or that a particular agonist can release one store and another can't, is quite a wrong way of looking at it. What one has to think about is the relationship between the number of pores that can be opened in the SR, and how fast or how active the SR Ca^{2+} current is. Everything is in balance, and we are looking at a fluid system that we are perturbing at certain times in ways that we don't understand in detail. It will be very difficult to understand exactly what is going on.

Paul: At 24 h in organ culture we find that RyR2 has been up-regulated by a factor of two or three.

Kotlikoff: It is very rare to find a cultured cell that responds to caffeine, and you almost never find Ryr2 in a cultured cell. It would be nice when we talk about RyRs to know which one we are talking about (if we have the knowledge).

Hellstrand: With regard to the question of what ryanodine is doing, if you put it onto a muscle it eliminates most SR Ca^{2+} release, because it opens up the SR Ca^{2+} channels and there is some level of communication between the stores. If you put ryanodine there you essentially kill InsP$_3$-induced responses as well. In organ culture we have found that if you culture in the presence of ryanodine for a couple of days, the InsP$_3$-induced release reappears, whereas there is still no ryanodine-induced release, and Ca^{2+} waves occur (Dreja et al 2001). We know that mitochondrial activity affects the properties of Ca^{2+} waves, so there is a Ca^{2+}-dependent modulation of SR release on this level which thus presumably involves InsP$_3$ receptors and not RyRs. It is very hard to distinguish these two receptor populations in the way they interact on the SR.

Brading: No one has mentioned yet that RyR function might be modified by other molecules such as the cADP ribose receptor.

References

Dreja K, Nordström I, Hellstrand P 2001 Rat arterial smooth muscle devoid of ryanodine receptor function: effects on cellular Ca^{2+} handling. Br J Pharmacol 132:1957–1966

Martin C, Hyvelin JM, Chapman KE, Marthan R, Ashley RH, Savineau JP 1999 Pregnant rat myometrial cells show heterogeneous ryanodine- and caffeine-sensitive calcium stores. Am J Physiol 277:C243–C252

Park MK, Petersen OH, Tepikin AV 2000 The endoplasmic reticulum as one continuous Ca^{2+} pool: visualization of rapid Ca^{2+} movements and equilibration. EMBO J 19:5729–5739

Relationship between the sarcoplasmic reticulum and the plasma membrane

Cheng-Han Lee, Damon Poburko, Kuo-Hsing Kuo, Chun Seow,
Cornelis van Breemen[1]

The iCAPTUR⁴E Center, University of British Columbia, St. Paul's Hospital, 1081 Burrard
Street, Vancouver, BC, V6Z 1Y6 and Cardiovascular Sciences, Children's and Women's Health
Centre of British Columbia, 4500 Oak Street, Vancouver, BC, Canada V6H 3N1

Abstract. Ionic interactions between the plasma membrane (PM) and the sarcoplasmic
reticulum (SR) play a crucial role in smooth muscle activation and homeostasis. The
most common form of Ca^{2+} signalling seen in vascular smooth muscle of conduit
arteries and capacitance veins consists of repetitive asynchronous Ca^{2+} waves. In the
inferior vena cava of the rabbit these waves are initiated by Ca^{2+} release via $InsP_3$
receptors ($InsP_3R$) and propagated by regenerative Ca^{2+} release. Maintenance of the
$[Ca^{2+}]_i$ oscillations is dependent on Ca^{2+} entry through the Na^+/Ca^{2+}-exchanger (NCX)
which is driven in the reverse mode by Na^+ entry through non-specific cation channels.
The latter are also responsible for depolarization and activation of voltage-gated Ca^{2+}
channels. The sarcoplasmic–endoplasmic reticulum Ca^{2+}-ATPase (SERCA) in the sheet-
like junctional SR is responsible for refilling and completing the cycle. Under resting
conditions the interaction between the superficial SR and the NCX is reversed with
Ca^{2+} release channels supplying Ca^{2+} to the NCX in the PM to be extruded in exchange
of extracellular Na^+. It is proposed that the above Ca^{2+} transport between the SR lumen
and the extracellular space takes place at PM–SR junctions across a narrow junctional
space.

*2002 Role of the sarcoplasmic reticulum in smooth muscle. Wiley, Chichester (Novartis Foundation
Symposium 246) p 26–47*

Ca^{2+} is the main intracellular signalling molecule in smooth muscle. Fluctuation in
local cytoplasmic $[Ca^{2+}]$ near Ca^{2+}-sensitive effector molecules allows for specific
regulation of multiple functions. These temporal fluctuations and spatial
variations of cytoplasmic $[Ca^{2+}]$ are dependent on the interactions of ion
transport proteins located in the plasma membrane (PM) and membranes of the
sacoplasmic reticulum (SR), nuclear envelope and mitochondria. These

---

[1]This chapter was presented at the symposium by Dr van Breemen, to whom correspondence
should be addressed

interactions are either transient or steady-state processes, which prevent ionic concentrations from reaching equilibrium. In general, a cytoplasmic Ca^{2+} gradient results from the separation of a Ca^{2+} source (channel supplying Ca^{2+} from the extracellular space or organelle) from a sink (Ca^{2+} pump or exchanger). As will be discussed in more detail below, the degree of this separation varies from close alignment, less than 10 nm, causing the two transport molecules to function as a unit, also called a duo-porter, to distances of several hundred nanometres. Recent evidence indicates that the cytoskeleton and specialized junctional proteins play an essential role in maintaining the functional alignments between the cell membrane and SR (Takeshima et al 2000). The resulting microstructural arrangements of apposing membranes create diffusional barriers defining different types of restricted spaces within the cytoplasm. The combined effects of the separation of Ca^{2+} sources and sinks added to the diffusional barriers discussed above cause the Ca^{2+} concentration in the restricted spaces to be different from the bulk $[Ca^{2+}]_i$. The functional importance of this is that Ca^{2+}-sensitive ion channels selective for K^+, Cl^- and Ca^{2+} and enzymes such as protein kinase C (PKC) and phospholipase C (PLC), which are located in membranes bordering a restricted space, can be regulated separately from the myofilaments occupying the bulk of the cytoplasm. In addition temporal variations in the Ca^{2+} signal, such as waves and oscillations, may encode messages for the nucleus (Dolmetsch et al 1998). In principle it is therefore possible for Ca^{2+} to regulate the processes of contraction, relaxation, growth, migration and matrix synthesis independent of each other. This presentation focuses on the interactions between the SR and plasma membrane (PM) during smooth muscle activation and rest, which may be responsible for enhancing the informational content of the smooth muscle Ca^{2+} signal.

$[Ca^{2+}]_i$ oscillations

Agonist-induced $[Ca^{2+}]_i$ oscillations constitute a ubiquitous intracellular signalling mechanism in both excitable and non-excitable cells. Although all smooth muscle $[Ca^{2+}]_i$ oscillations depend on PM–SR interactions there are two fundamentally different types of $[Ca^{2+}]_i$ oscillations depending on their immediate source of Ca^{2+}. When Ca^{2+} used to generate each spike is derived directly from the extracellular space, it typically appears as a uniform rise in $[Ca^{2+}]_i$ throughout the cell. Even though intracellular Ca^{2+} release through a Ca^{2+}-induced Ca^{2+} release (CICR) mechanism may be involved in amplifying the Ca^{2+} signal, it occurs secondarily to $[Ca^{2+}]_i$ elevation from stimulated Ca^{2+} entry. In this case where $[Ca^{2+}]_i$ rises more or less evenly across the entire cell no apparent Ca^{2+} waves are observed. In contrast, when the endoplasmic reticulum (ER)/SR is the immediate Ca^{2+} source for each Ca^{2+} spike, $[Ca^{2+}]_i$ initially rises in specific cellular loci and the regional elevation in $[Ca^{2+}]_i$ propagates in a wave-like fashion throughout the length of the cell. In vascular smooth muscle cells, both non-wave-like and wave-like $[Ca^{2+}]_i$ oscillations are observed.

Synchronized non-wave-like $[Ca^{2+}]_i$ oscillations

The non-wave-like $[Ca^{2+}]_i$ oscillations are commonly observed in vascular smooth muscle cells (VSMCs) of small resistance arteries and arterioles, and typically display a frequency of ~ 0.05 Hz and ~ 0.1 Hz respectively (Yip & Marsh 1996, Bartlett et al 2000, Mauban et al 2001, Peng et al 2001). The Ca^{2+} spikes are driven by periodic depolarizations stimulating the opening of voltage-gated Ca^{2+} channels (VGCCs). Since the smooth muscle cells are electrically coupled via gap junctions these non-wave-like oscillations in $[Ca^{2+}]_i$ are synchronized between neighbouring VSMCs and have the functional role of signalling force oscillations or vasomotion in the small arteries and arterioles. The oscillations in force follow the oscillations in $[Ca^{2+}]_i$ with a delay of approximately 300 ms, the time required for phosphorylation of myosin light chain (MLC) and the initiation of cross-bridge cycling. Unfortunately, the mechanism underlying the generation of the synchronized oscillations of VSMC membrane potential has not been definitively elucidated. Thus far, K_{Ca}, Cl_{Ca}, endothelium and SR have all been implicated. Recently, Peng et al (2001) forwarded the intriguing hypothesis that synchronized $[Ca^{2+}]_i$ oscillations may be initiated by asynchronous Ca^{2+} waves, which are discussed below.

Asynchronous wave-like $[Ca^{2+}]_i$ oscillations

Since the first report by Iino and co-workers in 1994 (Iino & Tsukioka 1994), asynchronous wave-like $[Ca^{2+}]_i$ oscillations have emerged as an universal mode of Ca^{2+} signalling in *in situ* VSMCs (Table 1). Confocal microscopy of intracellular Ca^{2+}-sensitive dyes in intact blood vessels, reveals recurrent intracellular Ca^{2+} waves travelling through the longitudinal axis of the ribbon-shaped VSMCs. These waves, which are usually but not always initiated by agonists, do not propagate between cells.

Ca^{2+} waves and contraction

The main function of vascular smooth muscle is to distribute blood flow through selective vasoconstriction and vasomotion. The latter is clearly associated with oscillations in $[Ca^{2+}]_i$, but it was long thought that tonic contraction was initiated by SR Ca^{2+} release and maintained by a steady state elevation of $[Ca^{2+}]_i$ dependent on influx. However, confocal microscopy of intact blood vessels has

TABLE 1 Summary of reported asynchronous wave-like $[Ca^{2+}]_i$ oscillations in *in situ* VSMCs

Tissue (Reference)	Stimulus	Average frequency	Average wave velocity	Inhibited by
Rat tail artery (isometric) (Iino et al 1994, Kasai et al 1997, Asada et al 1999)	Noradrenaline 0.1 μM 0.3 μM 1.0 μM	NA (increased with increasing [drug])	NA $\sim$20 m/s NA	30 μM ryanodine, 50 mM caffeine
	Angiotensin Local release 100 nM	$\sim$0.13 Hz $\sim$0.19 Hz	18.3+1.0 m/s	100 nM U-73122, 5 μM CPA
Rat mesenteric artery (isobaric, at 70 mmHg) (Mauban et al 2001, Miriel et al 1999)	Phenylephrine 300 nM 1 μM 5 μM	0.051 + 0.002 Hz 0.067 $\sim$ 0.37 Hz NA	NA NA 32.4+1.7 μm/s	20 μM CPA
Rat Aorta (isometric) (Asada et al 1999)	Spontaneous (possibly local RAS)	NA	NA	
Rabbit inferior vena cava (isometric) (Ruehlmann et al 2000, Lee et al 2001)	Phenylephrine 0.15 μM 1.5 μM 15 μM 150 μM	0.044+0.01 Hz 0.186+0.027 Hz 0.442+0.044 Hz 0.511+0.025 Hz	16.8+1.3 m/s 27.0+3.5 m/s 70.9+5.5 m/s 89.6+7.4 m/s	100 μM ryanodine, 25 mM caffeine, 75 μM 2-APB, 10 μM CPA, 2 μM thapsigargin
Rat mesenteric artery (isometric) (Peng et al 2001)	Noradrenaline 0.1 $\sim$0.5 μM 10 μM	0.05 Hz NA	$\sim$36 m/s NA	10 mM caffeine, 10 μM ryanodine, 1 μM thapsigargin
Rat cerebral artery (isobaric) (Jaggar & Nelson 2000, Jaggar 2001)	UTP 10 μM 30 μM	NA (increased with increasing [drug])	NA 30+20 m/s	10 μM ryanodine
	Luminal pressure 10 mmHg 60 mmHg	0.15+0.0 3Hz 0.29+0.02 Hz	NA NA	10 μM ryanodine, 100 nM thapsigargin, 25 μM diltiazem
	High K$^+$ 6 mM $[K^+]_{ext}$ 30 mM $[K^+]_{ext}$	0.14+0.02 Hz 0.33 + 0.02 Hz	NA NA	25 μM diltiazem

NA, not available.

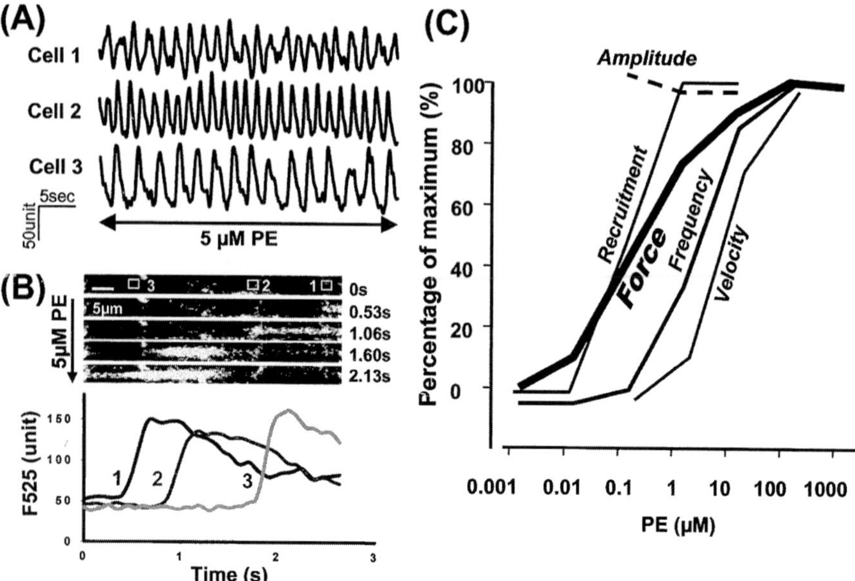

FIG. 1.   Phenylephrine (PE)-mediated asynchronous wave-like $[Ca^{2+}]_i$ oscillations and contraction in the rabbit inferior vena cava. (A) $[Ca^{2+}]_i$ oscillations recorded in neighbouring smooth muscle cells within the intact vessel are not synchronized between cells as they each display different frequency of oscillations. (B) Individual Ca^{2+} spike in PE-mediated $[Ca^{2+}]_i$ oscillations are wave-like as different regions (1, 2 and 3) in the same ribbon-shaped VSMC experience sequential rise of $[Ca^{2+}]$ in time. (C) The [PE]-dependence in force generation is compared to the [PE]-dependence in the percentage recruitment of cells, the amplitude of the $[Ca^{2+}]_i$ oscillations, the frequency of the $[Ca^{2+}]_i$ oscillations and the apparent velocity of the recurring Ca^{2+} waves. (Experimental traces reproduced, with permission from Lee et al 2001.)

shown that agonist-induced contractions are maintained at a steady state level by asynchronous wave-like $[Ca^{2+}]_i$ oscillations in single smooth muscle cells (Fig. 1A and 1B), which summate to give a steady state elevation in $[Ca^{2+}]_i$ for the whole tissue (Ruehlmann et al 2000). Figure 1 relates the concentration of the α adrenergic agonist phenylephrine (PE) to force developed by the inferior vena cava (IVC) of the rabbit and the various parameters of the $[Ca^{2+}]_i$ oscillations. It is clear that at the lower concentration range force depends on the number of cells recruited to display fixed amplitude $[Ca^{2+}]_i$ oscillations, and at higher concentrations of the agonist force is regulated by the frequency of the oscillations and perhaps also by the velocity of the recurring Ca^{2+} waves. It is clear that the additive effect of asynchronous Ca^{2+} waves is a steady state rise in the average tissue $[Ca^{2+}]_i$, but is there a specific functional advantage to the oscillatory pattern of Ca^{2+} release? A possibility considered in this presentation is that under these conditions the SR periodically releases Ca^{2+} in the vicinity of calmodulin, tethered to the thin filaments, and thus enhances the specificity of the Ca^{2+} signal for stimulation of contraction.

In addition to causing tonic contraction, low frequency ($<0.05\,Hz$) asynchronous wave-like $[Ca^{2+}]_i$ oscillations, which themselves are associated with only minimal development of tone, appear to be instrumental in the initiation of vasomotion in the rat mesenteric artery (Peng et al 2001). In this proposed scenario asynchronous Ca^{2+} waves can activate an unknown depolarizing current through the plasma membrane of each VSMC. The activation of such a depolarizing current can by chance occur synchronously in a sufficiently high fraction of VSMCs within the artery to entrain cycles of membrane depolarization and repolarization in electrically coupled VSMCs. The resulting synchronized but intermittent activation of L-type VGCCs in all VSMCs subsequently produces synchronized non-wave-like $[Ca^{2+}]_i$ oscillations which underlie vasomotion.

In contrast to the above, Ca^{2+} waves have also been associated with the induction of dilatation of resistance arteries. In this case the wave-like Ca^{2+} release is thought to achieve higher concentrations near the K_{Ca} on the PM, with the relaxing effect of the resulting hyperpolarization outweighing the local stimulation of contraction (Jaggar 2001). In accordance with this finding, openings of K_{Ca} have been reported following Ca^{2+} waves in smooth muscle cells (Young et al 2001).

Mechanism of wave-like $[Ca^{2+}]_i$ oscillations

Although rhythmic events have been observed in VSM for many decades their molecular mechanisms remain to be fully elucidated. One reason for this is that blood vessels display different types of rhythmic activity. The molecular basis for PE-induced $[Ca^{2+}]_i$ oscillations has been resolved in some detail for smooth muscle of the inferior vena cava of the rabbit, which may therefore serve as a basis for comparison with other types of blood vessels (Lee et al 2001). As shown in Table 1 and Fig. 2A, SR store depletion or sarcoplasmic–endoplasmic reticulum Ca^{2+} ATPase (SERCA) blockade can inhibit the wave-like $[Ca^{2+}]_i$ oscillations, indicating the importance of the SR-mediated Ca^{2+} release. In the case of the rabbit inferior vena cava, waves of SR-mediated Ca^{2+} release begin with the opening of inositol-1,4,5-trisphosphate (InsP$_3$)-sensitive Ca^{2+} release channels (InsP$_3$R), since they are prevented or instantly blocked by concentrations of 2-aminoethoxydiphenyl borate (2-APB), which block InsP$_3$R (Fig. 2B). As some of the released Ca^{2+} is extruded to the extracellular space recurrence of the Ca^{2+} waves depends on replenishment of the SR by stimulated Ca^{2+} influx. This occurs through three different pathways: VGCCs, non-selective cation channels (NSCCs) and the Na^+/Ca^{2+} exchanger (NCX) in its reverse-mode. In this large vein the VGCCs play a relatively minor role since nifedipine will only decrease

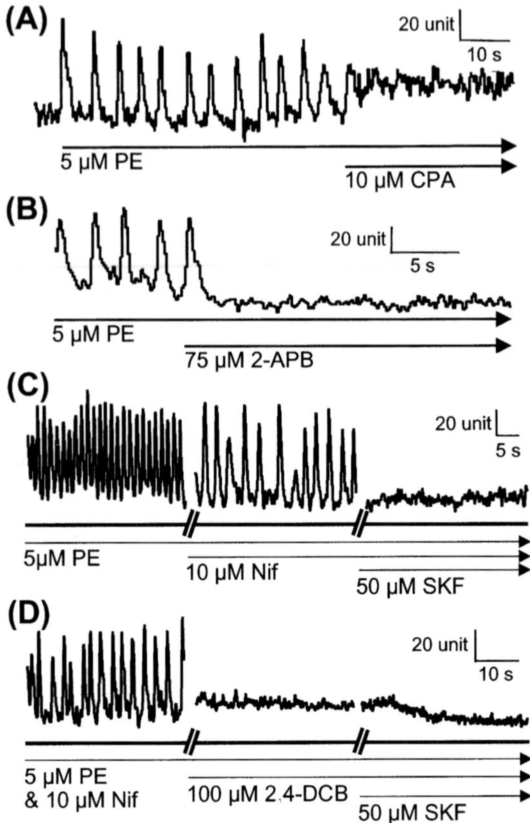

FIG. 2. Mechanism of phenylephrine (PE)-mediated wave-like $[Ca^{2+}]_i$ oscillations in the rabbit inferior vena cava. (A) PE-mediated $[Ca^{2+}]_i$ oscillations are completely inhibited by $10\,\mu M$ cyclopiazonic acid (CPA), but the average $[Ca^{2+}]_i$ remains elevated. (B) PE-mediated $[Ca^{2+}]_i$ oscillations are abolished by $75\,\mu M$ 2-aminoethoxydiphenyl borate (2-APB). (C) Application of $10\,\mu M$ nifedipine (Nif) reduced the frequency of PE-mediated $[Ca^{2+}]_i$ oscillations while additional application of SKF96365 (SKF) completely abolished the remaining $[Ca^{2+}]_i$ oscillations. (D) Application of $100\,\mu M$ 2,4-dichlorobenzamil (2,4-DCB) completely inhibited nifedipine-resistant PE-induced $[Ca^{2+}]_i$ oscillations and lowered the $[Ca^{2+}]_i$ to a level that is slightly higher than baseline. Additional application of SKF96365 returned the $[Ca^{2+}]_i$ level to baseline. (Experimental traces reproduced with permission from Lee et al 2001.)

the frequency of the oscillations (Fig. 2C) and inhibit contraction by 27%. Blockade of the NCX eliminates the oscillations (Fig. 2D), leaving a slightly elevated $[Ca^{2+}]_i$, which can be reduced to baseline value by the receptor-operated channel/store-operated channel (ROC/SOC) blocker SKF96365 (Fig. 2D) and also by 2-APB (Fig. 2B). These observations led to the postulation of the following

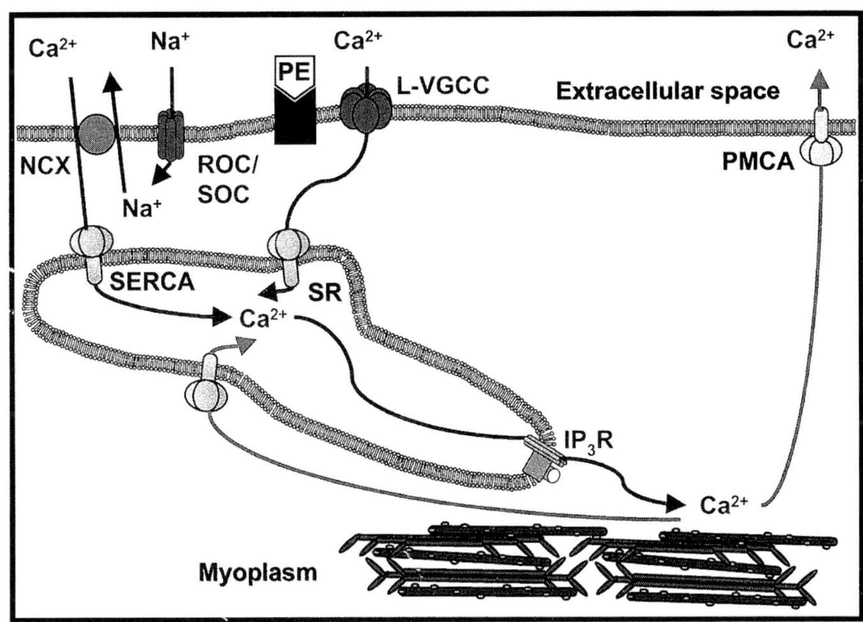

FIG. 3.   Model for Ca^{2+} movements during phenylephrine (PE)-mediated $[Ca^{2+}]_i$ oscillations in VSMC from rabbit IVC. IP_3 generated as a consequence of α adrenergic activation releases Ca^{2+} via $InsP_3R$ toward the myoplasm to activate the myofilaments. Subsequent Ca^{2+} removal occurs partially through the PMCA, while the remainder is taken up into the SR by the SERCA. ROC/ SOC, activated by receptor activation and SR depletion, allows entry of mainly Na^+ and some Ca^{2+}. This Na^+ entry raises $[Na^+]_i$ in the PM–SR junctional space between the plasma membrane and the SR membrane and drives the NCX in the reverse-mode to supply Ca^{2+} to SERCA to refill the SR lumen, thereby completing the cycle. PE, phenylephrine; NCX, sodium/calcium exchanger; ROC/SOC, receptor-operated channel/store-operated channel; SERCA, sarcoplasmic endoplasmic reticulum Ca^{2+}-ATPase; L-type VGCC, L-type voltage-gated Ca^{2+} channel; PMCA, plasma membrane Ca^{2+}-ATPase; IP_3R, InP_3-sensitive Ca^{2+} release channel; SR, sarcoplasmic reticulum. (Figure reproduced with permission from Lee et al 2001.)

sequence of events illustrated in Fig. 3. (1) PE activates PLC which catalyses the synthesis of $InsP_3$; (2) activation of $InsP_3R$ and Ca^{2+} release from the SR near calmodulin tethered to the thin filaments; (3) opening of NSCCs in the plasma membrane and influx of mainly Na^+ and some Ca^{2+} into the PM–SR junctional space; (4) depolarization, opening of VGCCs and reversal of NCX resulting in Ca^{2+} influx; and (5) Ca^{2+} uptake into the SR by SERCA. The identity and the mode of activation of the NSCC are not yet resolved. The rabbit vena cava does express Trp1 (T. Szado & C. van Breemen, unpublished results), which has been shown to constitute a NSCC and to be activated by SR Ca^{2+} release, but direct evidence for its participation in the above events requires further experimentation with knockout or antisense techniques. It is of interest to note that oscillatory inward non-selective cationic current has been described in endothelin-stimulated rat aorta, a large vessel that exhibits asynchronous wave-like $[Ca^{2+}]_i$ oscillations as well (Salter & Kozlowski 1998).

The constant amplitude of the propagating waves indicates regenerative transient Ca^{2+} release from the SR. The regenerative nature depends on the positive feedback of increasing $[Ca^{2+}]_i$ on the $InsP_3$-sensitivity of $InsP_3R$ and the recruitment of Ca^{2+}-sensitive ryanodine-sensitive Ca^{2+} release channel (ryanodine receptor; RyR). The propagation is due to elevation of $[Ca^{2+}]_i$ to the threshold for activation of clusters of release channels in adjacent portions of the SR. In some smooth muscle the threshold value has been observed as the inflection point between a 'foot' segment and the steep portion of the upstroke of $[Ca^{2+}]_i$ elevation. The relative involvement of $InsP_3R$ and RyR appears to vary between different smooth muscle preparations. The delayed negative feedback on release, which is essential for oscillatory behaviour, has been ascribed to a number of mechanisms: (1) inhibition of $InsP_3R$ by high $[Ca^{2+}]_i$ (Iino & Tsukioka 1994) or inhibition of RyR by adaptation/inactivation mechanisms (Sitsapesan & Williams 2000, Lamb et al 2000); (2) inhibition of $InsP_3R$ by low luminal SR Ca^{2+} (Missiaen et al 1992); and (3) time-dependent inactivation of both IP_3R (Hajinoczky & Thomas 1997). The latter mechanism would imply a maximal limit to the frequency of the wave-like $[Ca^{2+}]_i$ oscillations as it requires time for the channels in the inactivated state to return to the closed resting state. This is supported by the observation that $[Ca^{2+}]_i$ oscillations in the rabbit IVC appear to peak at a frequency of $\sim 0.5\,Hz$ irrespective of further increases in the agonist concentration (Ruehlmann et al 2000).

As the SR takes up Ca^{2+} from the surrounding cytoplasm the NSCC may close and as a result of repolarization the VGCC would also close. Stimulation of SERCA, PM Ca^{2+}-ATPase (PMCA) and possibly NCX by elevated $[Ca^{2+}]_i$, while release terminates, is responsible for the down-stroke of the $[Ca^{2+}]_i$ oscillation. In this context it would be interesting to obtain evidence for NCX reversal during each cycle. The next Ca^{2+} wave may start at the frequent discharge sites (Gordienko et al 1998, 2001) when the SR luminal Ca^{2+} has been recharged and spillover from the SR raises the local $[Ca^{2+}]_i$ to threshold once again. The fact that waves tend to originate from the same area within the cell, has been explained by the observation that such sites posses a higher local density of SR near PM Ca^{2+} channels and are devoid of buffering mitochondria (Gordienko et al 2001).

A conceptually simpler model of $[Ca^{2+}]_i$ oscillations is one where $[InsP_3]$ oscillates (Hirose et al 1999). The delayed negative feedback in this case is Ca^{2+} activation of PKC, which then inhibits PLC. However, it is generally believed that $[InsP_3]$ oscillations can only give rise to low frequency $[Ca^{2+}]_i$ oscillations and cannot produce the high frequency ($\sim 0.5\,\mathrm{Hz}$) $[Ca^{2+}]_i$ oscillations observed in these VSMCs.

The apparent complexity of the mechanism of repetitive Ca^{2+} waves leaves many questions unanswered. For example if the $[InsP_3]_i$ sets the Ca^{2+} sensitivity of the $InsP_3R$ and if this is one of the determinants of the threshold for Ca^{2+} release, why are the amplitudes constant with increasing [PE]? It would be expected that the lower threshold would allow Ca^{2+} release at a lower luminal $[Ca^{2+}]$. In certain cases the amplitude even increases with increasing agonist concentration, which is a change opposite to that predicted by the above model. Perhaps RyRs are mainly responsible for wave propagation. Many aspects of RyR regulation during $[Ca^{2+}]_i$ oscillations are as yet unknown. For example Ca^{2+} is not the only stimulus as cADP ribose and nicotinic acid adenine dinucleotide phosphate (NAADP) have also been shown to activate these channels (Li et al 2001, Yusufi et al 2001). Thus far we have no information on whether the concentrations of these mediators are elevated or oscillate.

Ultrastructure of VSMCs

The SR is composed of an interconnected tubular and sheet-like network whose membranes surround the SR lumen, which has an elevated Ca^{2+} content. It extends throughout the spindle-shaped VSMCs and is contiguous with the nuclear envelope (Somlyo 1985). It contributes to Ca^{2+} signalling by virtue of active Ca^{2+} transport via SERCA from the cytoplasm to the SR lumen and Ca^{2+} release from the SR into the cytoplasm via $InsP_3R$ and RyR. The SR has been classified according to its location as superficial or deep, with distinct functions being ascribed to the superficial SR. In domains where the SR apposes the PM it creates a narrow space, which extends in two dimensions for about 0.5–$1\,\mu m$ and has a depth of between 20 and 40 nm. The structures responsible for this spacing have not yet been identified, although in some instances 'feet' similar to those seen in triadic junctions in skeletal muscle have been reported (Somlyo 1985) and proteins called 'junctophillins' have been isolated from the diads of cardiac muscle (Takeshima et al 2000). The narrow cytoplasmic space between the junctional SR and PM is referred to as the PM–SR junctional space and is thought to present an imperfect barrier to diffusion of Ca^{2+} and Na^+. As can be seen from the electron micrograph serial sections and schematic 3D interpretations in Fig. 4, the caveolae are able to perforate the junctional SR sheet, such that their apices frequently remain in contact with the bulk cytoplasm. As shown in Fig. 4 (panels E, F and G), the apices of the caveolae are frequently close to the perpendicular SR sheets that appear to arise from the superficial SR sheets. Because of these varying geometric arrangements it is plausible that different membrane domains perform specialized functions. The PM–SR junctions are likely the sites for interactions

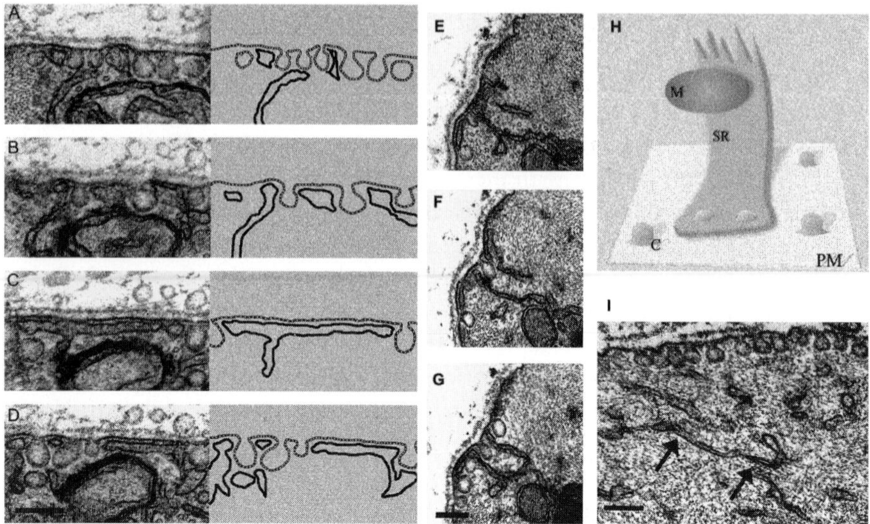

FIG. 4.   Ultrastructure of vascular smooth muscle of the rabbit inferior vena cava revealed with electron microscopy. Serial cross-sections of VSMCs are shown in series 1 (panel A–D) and series 2 (panel E–G). Series 1 illustrates the close spatial apposition between the superficial SR sheet and the PM with the apices of the caveolae perforating through the superficial SR sheets to come into contact with the bulk cytoplasm. The membranes of the PM (dotted line) and the SR (solid line) in panel A–D are outlined to the right of the respective panels. The close apposition between the superficial SR sheet, the PM and the neck region of the caveolae creates a narrow and expansive restricted space. Series 2 illustrates the perpendicular sheets of SR, which appear to arise from the superficial SR sheets. Mitochondria also come into close contact with the perpendicular SR sheets. Panel H contains a stylized illustration of the close association between the superficial SR sheet, which is continuous with the perpendicular sheet, the perforating caveolae (C), the PM and a mitochondrion (M). Panel I shows calyculin-A mediated dissociation of the superficial SR sheets from the PM (see arrows). The black scale bar indicated represents 200 nm of distance.

between NSCC, NCX and SERCA during SR refilling, and NCX and InsP$_3$R and/ or RyR during SR unloading. In support of this postulate, we found that dissociation of the superficial SR sheets from the PM (Fig. 4, arrows) by calyculin A inhibits the agonist-induced wave-like $[Ca^{2+}]_i$ oscillations (C. H. Lee, K. H. Kuo, C. Y. Seow and C. van Breemen, unpublished results). On the other hand the non-junctional PM, including the apices of many of the caveolae, which contain the VGCCs, may introduce Ca^{2+} into a peripheral region of the cytoplasm from which SERCA located on the perpendicular SR sheets may buffer Ca^{2+} before it reaches the deeper myoplasm to attenuate contraction. On the other hand it is also possible that RyRs in this location mediate CICR to potentiate the signal. In resistance vessels where the SR is strongly coupled to K_{Ca} it is conceivable that both K^+ channels and RyRs border the PM–SR junctional space. The same may apply to PM enzymes which require rather high $[Ca^{2+}]$ for activation such as PKC and PLC.

As mentioned above, the junctional SR is connected to sheets of perpendicular SR (Fig. 4), which extend from the PM through a peripheral cytoplasmic region with lower myofilament density into the myoplasm. It is proposed that during the active state of wave-like $[Ca^{2+}]_i$ oscillations, Ca^{2+} taken up by the junctional SR is released by these perpendicular sheets near the calmodulin, which is tethered to the myosin light chain kinase (MLCK) of the thin filaments (M. Walsh, personal communication, 2001). This process would enhance the specificity and efficiency of Ca^{2+} regulation of contraction.

In addition to forming close contacts with the PM, the SR network also comes into close contact with the mitochondria (Nixon et al 1994, Rizzuto et al 1998), forming yet another diffusionally restricted space (Fig. 4, panel E–G). This space, approximately 60–80 nm wide and sandwiched between the SR and mitochondrial membranes, also appears to be functionally important. As the SR network penetrates deeper into the cell, it inserts into the nuclear membrane such that the lumen of the perinuclear SR network is continuous with the lumen of the nuclear envelope (Somlyo 1985).

Ca^{2+} cycling at resting state

Even in the resting state, Ca^{2+} enters the smooth muscle cells and cycles through intracellular organelles. The mechanisms involved in the resting cycle differ from those seen during stimulated Ca^{2+} oscillations as they are functioning to sustain a constant low $[Ca^{2+}]_i$. The basal rate of Ca^{2+} influx into smooth muscle has been estimated to be approximately $16\,\mu$moles/litre of cells per min, which is equivalent to more than a hundred times the resting $[Ca^{2+}]_i$ per minute (Meisheri et al 1980). This raises the questions: (1) how does so much Ca^{2+} enter unstimulated cells; (2) how is this Ca^{2+} distributed within the cell; and (3) how is it extruded in order to maintain resting levels of $[Ca^{2+}]_i$ and vascular tone? It is also unknown whether basal Ca^{2+} influx via the leak pathway exhibits variations in magnitude in different regions of the PM. It is becoming clear that the PM is made up of a patchwork of domains, some covered by dense bodies where the myofilaments are attached, some apposed by the superficial SR and the remainder facing the bulk cytoplasm. It would be of considerable interest to determine the relative sizes of these domains for various types of smooth muscle and relate them to specific functional properties.

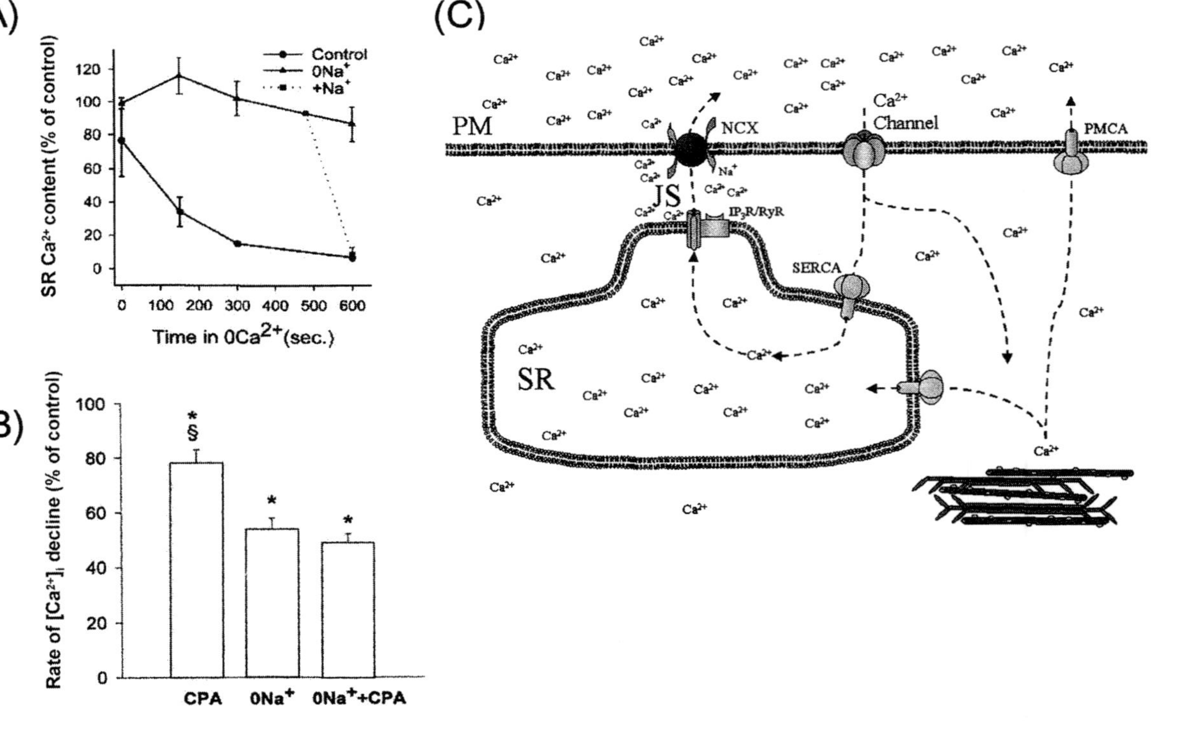

FIG. 5.   Maintenance of the 'superficial buffer barrier' depends on NCX-assisted Ca^{2+} transport from the SR lumen to the extracellular space. (A) Rate of loss of SR Ca^{2+} content, measured as a caffeine transient, into Ca^{2+} free perfusate at room temperature. (B) Rate of decline in $[Ca^{2+}]_i$ from an elevated level, measured as fura 2 fluorescence ratio, into Ca^{2+} free superfusate, which is either Na^+ free or contains 10 μM CPA or is Na^+ free and contains CPA. (C) This cartoon represents a model for maintained buffering by the superficial SR of Ca^{2+} entry. Ca^{2+} taken up by SERCA is subsequently released into the SR–PM junctional space from where it is extruded by the NCX.

Resting Ca^{2+} influx is attributed partly to a certain level of the open probability of excitable channels, including ROCs, SOCs and VGCCs. Some SOCs may stay open due to the basal level of $InsP_3$, while the resting membrane potential of vascular smooth muscle allows for a certain degree of activation of the Ca^{2+} window current (Nelson et al 1990). Additionally a non-specific influx of Ca^{2+}, referred to as the Ca^{2+} leak, substantially contributes to basal influx; however, its precise mechanism remains elusive (van Breemen et al 1972).

The resting $[Ca^{2+}]_i$ is approximately 80 nM and is maintained at that low level in spite of the fact that every half second the equivalent of the total free cytosolic Ca^{2+} enters the cells. Two mechanisms protect the cell from drastic changes in $[Ca^{2+}]_i$ as a result of Ca^{2+} entry: (1) the presence of fixed and diffusible Ca^{2+} binding sites in the cytoplasm and (2) sequestration by SERCA in the SR. It has been shown in several types of VSM that stimulated influx is more effective in raising Ca^{2+} concentration near the myofilaments when Ca^{2+} uptake in the SR is inhibited by either blockade of SERCA or opening of release channels (van Breemen et al 1986). Thus in the resting smooth muscle the peripheral SR takes up Ca^{2+} entering the cells before it can equilibrate with the deeper myoplasm. Continuous unloading of the superficial SR to the extracellular space, involving coupling of SR release channels to NCX, permits buffering of Ca^{2+} influx to continue (Nazer & van Breemen 1998). In support of this postulate, Fig. 5 shows that Na^+ removal from the extracellular space inhibits the loss of Ca^{2+} from the SR, while returning Na^+ leads to a rapid transfer of Ca^{2+} from the SR to the extra-cellular space (ECS). This cyclical process of Ca^{2+} buffering and unloading has been referred to as the 'superficial buffer barrier' (van Breemen et al 1995). The definition of the superficial SR remains somewhat vague including the junctional SR and sheets of SR within about 300 nm of the PM.

In conclusion the ionic interactions between the SR and PM are responsible for fine-tuning the smooth muscle Ca^{2+} signal in terms of fluctuating local cytoplasmic Ca^{2+} gradients and repetitive Ca^{2+} waves.

References

Asada Y, Yamazawa T, Hirose K, Takasaka T, Iino M 1999 Dynamic Ca^{2+} signalling in rat arterial smooth muscle cells under the control of local renin-angiotensin system. J Physiol 521:497–505

Bartlett IS, Crane GJ, Neild TO, Segal SS 2000 Electrophysiological basis of arteriolar vasomotion in vivo. J Vasc Res 37:568–575

Dolmetsch RE, Xu K, Lewis RS 1998 Calcium oscillations increase the efficiency and specificity of gene expression. Nature 392:933–936

Gordienko DV, Bolton TB, Cannell MB 1998 Variability in spontaneous subcellular calcium release in guinea-pig ileum smooth muscle cells. J Physiol 507:707–720

Gordienko DV, Greenwood IA, Bolton TB 2001 Direct visualization of sarcoplasmic reticulum regions discharging Ca^{2+} sparks in vascular myocytes. Cell Calcium 29:13–28

Hajnoczky G, Thomas AP 1997 Minimal requirements for calcium oscillations driven by the IP_3 receptor. EMBO J 16:3533–3543

Hirose K, Kadowaki S, Tanabe M, Takeshima H, Iino M 1999 Spatiotemporal dynamics of inositol 1,4,5-trisphosphate that underlies complex Ca^{2+} mobilization patterns. Science 284:1527–1530

Iino M, Tsukioka M 1994 Feedback control of inositol trisphosphate signalling by calcium. Mol Cell Endocrinol 98:141–146

Iino M, Kasai H, Yamazawa T 1994 Visualization of neural control of intracellular Ca^{2+} concentration in single vascular smooth muscle cells in situ. EMBO J 13:5026–5031

Jaggar JH 2001 Intravascular pressure regulates local and global Ca^{2+} signaling in cerebral artery smooth muscle cells. Am J Physiol 281:C439–C448

Jaggar JH, Nelson MT 2000 Differential regulation of Ca^{2+} sparks and Ca^{2+} waves by UTP in rat cerebral artery smooth muscle cells. Am J Physiol 279:C1528–C1539

Kasai Y, Yamazawa T, Sakurai T, Taketani Y, Iino M 1997 Endothelium-dependent frequency modulation of Ca^{2+} signalling in individual vascular smooth muscle cells of the rat. J Physiol 504:349–357

Lamb GD, Laver DR, Stephenson DG 2000 Questions about adaptation in ryanodine receptors. J Gen Physiol 116:883–890

Lee CH, Poburko D, Sahota P, Sandhu J, Ruehlmann DO, van Breemen C 2001 The mechanism of phenylephrine-mediated $[Ca^{2+}]_i$ oscillations underlying tonic contraction in the rabbit inferior vena cava. J Physiol 534:641–650

Li PL, Tang WX, Valdivia HH, Zou AP, Campbell WB 2001 cADP-ribose activates reconstitutes ryanodine receptors from coronary arterial smooth muscle. Am J Physiol 280:H208–H215

Mauban JRH, Lamont C, Balke CW, Wier WG 2001 Adrenergic stimulation of rat resistance arteries affects Ca^{2+} sparks, Ca^{2+} waves, and Ca^{2+} oscillations. Am J Physiol 280:H2399–H2405

Meisheri K D, Palmer RF, van Breemen C 1980 The effects of amrinone on contractility, Ca^{2+} uptake and cAMP in smooth muscle. Eur J Pharmacol 61:159–165

Miriel VA, Mauban JR, Blaustein MO, Wier WG 1999 Local and cellular Ca^{2+} transients in smooth muscle of pressurized rat resistance arteries during myogenic and agonist stimulation. J Physiol 518:815–824

Missiaen L, Taylor CW, Berridge MJ 1992 Luminal Ca^{2+} promoting spontaneous Ca^{2+} release from inositol trisphosphate-sensitive stores in rat hepatocytes. J Physiol 455:623–640

Nazer MA, van Breemen C 1998 Functional linkage of Na^+-Ca^{2+} exchange and sarcoplasmic reticulum Ca^{2+} release mediates Ca^{2+} cycling in vascular smooth muscle. Cell Calcium 24:275–283

Nelson MT, Patlak JB, Worley JF, Standen NB 1990 Calcium channels, potassium channels, and voltage dependence of arterial smooth muscle tone. Am J Physiol 259:C3–C18

Nixon GF, Mignery GA, Somlyo AV 1994 Immunogold localization of inositol 1,4,5-trisphosphate receptors and characterization of ultrastructural features of the sarcoplasmic reticulum in phasic and tonic smooth muscle. J Muscle Res Cell Motil 15:682–700

Peng H, Matchkov V, Ivarsen A, Aalkjaer C, Nilsson H 2001 Hypothesis for the initiation of vasomotion. Circ Res 88:810–815

Rizzuto R, Pinton P, Carrington W et al 1998 Close contacts with the endoplasmic reticulum as determinants of mitochondrial Ca^{2+} responses. Science 280:1763–1766

Ruehlmann DO, Lee CH, Poburko D, van Breemen C 2000 Asynchronous Ca^{2+} waves in intact venous smooth muscle. Circ Res 86:E72–E79

Salter K J, Kozlowski RZ 1998 Differential electrophysiological actions of endothelin-1 on Cl^- and K^+ currents in myocytes isolated from aorta, basilar and pulmonary artery. J Pharmacol Exp Ther 284:1122–1131

Sitsapesan R, Williams A J 2000 Do inactivation mechanisms rather than adaptation hold the key
    to understanding ryanodine receptor channel gating? J Gen Physiol 116:867–872
Somlyo AP 1985 Excitation–contraction coupling and the ultrastructure of smooth muscle. Circ
    Res 57:497–507
Takeshima H, Komazaki S, Nishi M, Iino M, Kangawa K 2000 Junctophilins: A novel family of
    junctional membrane complex proteins. Mol Cell 6:11–22
van Breemen C, Farinas BR, Gerba P, McNaughton ED 1972 Excitation–contraction coupling
    in rabbit aorta studied by the lanthanum method for measuring cellular calcium influx. Circ
    Res 30:44–54
van Breemen C, Cauvin C, Johns A, Leijten P, Yamamoto H 1986 Ca^{2+} regulation of vascular
    smooth muscle. Fed Proc 45:2746–2751
van Breemen C, Chen Q, Laher I 1995 Superficial buffer barrier function of smooth muscle
    sarcoplasmic reticulum. Trends Pharmacol Sci 16:98–105
Yip KP, Marsh DJ 1996 $[Ca^{2+}]_i$ in rat afferent arteriole during constriction measured with
    confocal fluorescence microscopy. Am J Physiol 271:F1004–F1011
Young RC, Schumann R, Zhang P 2001 Intracellular calcium gradients in cultured human
    uterine smooth muscle: a functionally important subplasmalemmal space. Cell Calcium
    29:183–189
Yusufi ANK, Cheng J, Thompson MA, Chini EN, Grande JP 2001 Nicotinic acid-adenine
    dinucleotide phosphate (NAADP) elicits specific microsomal Ca^{2+} release from mammalian
    cells. Biochem J 353:531–536

DISCUSSION

Eisner: Your model of how the oscillation is maintained requires that the NCX
in the surface membrane be very close to the SERCA in the SR. But presumably
you can't have the $InsP_3$ and RyRs in this place, because if they were they would
release Ca^{2+} into the space and this would make it impossible for backwards NCX
to bring Ca^{2+} in. Later on, towards the end of your paper, the SERCA has moved
away from the Na^+/Ca^{2+} exchanger and the RyR is close to the Na^+/Ca^{2+} exchange
in order that Ca^{2+} could be pumped out of the cell. Do you require two different
models to explain the two different sorts of experiments, and do either of them fit
with what is known about the localization of the various receptors?

van Breemen: You are talking about the next set of experiments we intend to do, so
I can only hypothesize. We don't need two different models. We cannot have a lot
of RyRs facing a narrow junctional space, but it is probably OK to have $InsP_3$
receptors there. These are more regulated by the filling state of the SR; when it is
empty they presumably close down again. The space has very great diffusion
limitations. It is only 20 nm wide, and as much as 500 nm in diameter. Any ion
that you put in this space will probably have great problems diffusing
immediately out into the bulk of the cytoplasm. This reminds me of an
experiment that Karaki did where he was labelling his cells with aequorin
(Karaki et al 1997). During one stimulation with PE, he saw the aequorin light
up, but then he had to wait some 8 h for the aequorin to diffuse back into this space.

Sanders: When you talk about store-operated channels, are you considering TRP channels as being part of this?

van Breemen: We have used RT-PCR to look in this tissue, and find strong expression of TRP1, but not of the other TRP channels. The main one in this tissue is probably TRP1, which fits with a non-specific Na^+-permeable channel.

Sanders: I thought there was a report of TRP channels being blocked by 2-APB.

van Breemen: In this case it wasn't. If we prevent the refilling, we can't block with 2-APB. There was only one kind of source that was blocked by the SKF, and this one could be blocked by SKF but not with the 2-APB if we prevented refilling.

Sanders: Did you look for all the TRPs?

van Breemen: Yes, but there's a slight problem in that we were looking in rat but didn't have all the rat primers; we had to use mouse and human. There's a lot of analogy, though.

Blaustein: In my paper I'll show some data on TRP1 in culture, indicating that TRP1 is localized to the microdomains opposite the reticulum (Blaustein et al 2002, this volume). We have some data showing that if we use an antisense knockout of *Trp1*, this takes out store-operated channel activity. The classical measurement we use involves emptying the store and then demonstrating a re-entry of Ca^{2+} after we add back external Ca^{2+}: this is abolished if we knockout *Trp1*. I am aware of other data where this occurs with TRP4. We have seen several TRPs expressed using immunoblotting. There are several TRPs and they may well be expressed as heteromultimers. Whether it is SOCs or ROCs, or both, some of them clearly behave like a SOC.

David Eisner asked earlier about where the SERCA is located. I think it is distributed. There must be some SERCA at the periphery, in the SR at the junctions between the SR and the plasma membrane, but there must be some elsewhere as well. I don't have any direct evidence for this — we are in the process of looking at this — but I think the SERCA is distributed among the SR and has two different functions. One of these is to take up Ca^{2+} at the junctional region and the other is to take up Ca^{2+} when it is released in the more peripheral regions.

Eisner: Presumably, there needs to be an appreciable fraction of the SERCA at that junctional region, otherwise no matter how high Ca^{2+} gets there won't be more than a certain incremental increase of Ca^{2+} back into the SR.

Blaustein: It depends on the region. If you are talking about the space between the plasma membrane and the SR, that volume is very small. There will only be a few Ca^{2+} ions in such a space, even if it is a couple of hundred nanometres wide. This is with a free Ca^{2+} concentration of 100 nM.

Paul: There are two SERCA isoforms in smooth muscle, 2A and 2B, in roughly equal proportions. We suspect that there is a reason for this: for example, one may be the housekeeping isoform found all over the place.

Taylor: Coming back to the PE-regulated Ca^{2+} entry pathway, this component was blocked by the SKF compound, and not 2-APB. You suggested it was a store-operated Ca^{2+} entry. What is the evidence for this being a store-regulated Ca^{2+} entry pathway? Many people are finding 2-APB to be a selective blocker of capacitative Ca^{2+} entry. This would sit comfortably with it not being store-regulated at all, but another receptor-regulated pathway.

van Breemen: The only evidence is that when CPA is used to prevent filling of the store, it is not possible to block with 2-APB. If PE is added to CPA, there is no further increase in the Ca^{2+} concentration. There are no additive effects; it seems like they are activating the same channel. I don't want to claim that I know how this particular SOC is activated. Perhaps it is through local Ca^{2+}, or through depletion of the SR. So far, it seems to be related to the SR being full or not.

Taylor: We have shown that in A7r5 cells, when one activates a receptor, there is a shutdown of the capacitative pathway together with an activation of another pathway. Merely not seeing another effect when PE is added on top of empty stores is not a conclusive demonstration that the pathway that was activated by PE was a SOC entry pathway.

Hellstrand: You have shown the coding of the response to PE in terms of frequency of oscillations, but not amplitude modulation. Have you come across any example where the modulation of amplitude also changes the force of contraction, or do you think the amplitude is constant in this system?

van Breemen: In the literature I've seen cases where the amplitude increases, and I've seen many cases where the baseline goes way up. In terms of this tissue, it seems to be mainly the amplitude and not the baseline.

Hellstrand: Is the baseline high enough to activate contraction by itself?

Somlyo: Isn't the contractile system a damped system, and it is damped by calmodulin?

Hellstrand: That is what I am getting at. There are a lot of phase shifts in this system. One observation we have made is that under hypoxia we see a decrease in amplitude but an increase in frequency of the waves. We are trying to model a case where this would account for reduction of force simply on the basis of non-linearity of the $[Ca^{2+}]_i$ versus myosin phosphorylation versus force reactions. It seems intuitively that this could explain why there can be a reduction in force although there is no reduction in the overall level of global Ca^{2+}. Is amplitude modulation something that people have seen?

van Breemen: So you are thinking in terms of why you would get a reduction in the force when you get rid of the specific release of Ca^{2+}, even though the average Ca^{2+} stays up. Michael Walsh has some evidence for why releasing Ca^{2+} close to the activation site may be required for efficient stimulation of contraction.

Walsh: I will present some of this evidence in the general discussion session (see p 48).

Wier: Along these lines, I was fascinated by your results showing the effect of CPA in reducing force although the average Ca^{2+} remained high. What do you think might be the mechanism behind this? Are Ca^{2+} waves somehow more efficacious in activating force? Many of us have thought that they might actually be less efficacious.

van Breemen: I'm not the first person to observe this. Many people have seen that when SERCA is blocked, the force:Ca^{2+} ratio tends to go down. The general explanation for this is that Ca^{2+} goes into a non-contractile compartment. Mordy Blaustein has said that the junctional space may be too small to account for this. But it is possible that in a peripheral region of the cytoplasm extending to 200–300 nm from the cell membrane, the myofilaments are sparse and may not contribute a lot to the contraction. This is a novel idea, which we plan to test more rigorously.

Paul: I would like people to tell me about the data they haven't published, where they have seen Ca^{2+} go up without a concomitant change in force. We have seen this with CPA, and I just came back from Igor Wendt's group at Monash where he has used peroxide to get Ca^{2+} to rise to fairly high levels without triggering a contraction. These of course are whole tissue measurements: whether this has anything to do with frequency addition, I don't know. I know people have seen Ca^{2+} go up to appreciable levels without seeing a contraction.

Sanders: The same thing happened with thapsigargin. It's as if CPA were doing something besides blocking SERCA pumps. It doesn't always match with thapsigargin.

van Breemen: In our case we added thapsigargin and it had exactly the same effect as CPA. But I agree: there can definitely be differences.

Blaustein: The diffusion out of this restricted space is very slow. If it isn't, there are some effects that we simply can't explain. It is not releasing enough Ca^{2+} from this space to make a difference. I think Andrew Somlyo's point about calmodulin acting as a damper on these oscillations is extremely important: it means that if you raised the baseline Ca^{2+} a little bit, this will not have nearly as important an effect as large peak oscillations. If Ca^{2+} is bound to calmodulin it comes off relatively slowly during oscillations, so you don't really get down to the trough. Even if you abolish the oscillations so that the average level of Ca^{2+} in the trough goes up a little bit, this isn't enough to do the trick. This is the difference; I don't see any problem with these observations.

Isenberg: I would like to address the question of the diffusional barrier once more. There is one argument that this barrier is flat sheet. On the other hand, Casey van Breemen has shown that in the region of the caveolae, there are interruptions of these sheets. If we have large holes like this, the barrier shouldn't be as tight as you suggested and should empty much faster.

Blaustein: The question is whether those holes are holes in the barrier or not. If you look at the reticulum where the caveolae stick up, this is not a hole in the barrier because the diffusion is restricted within the barrier itself. These holes are places where the plasma membrane at the top of the caveolae can actually communicate with bulk cytoplasm. This is a different region than that between the plasma membrane and the reticulum.

van Breemen: I like to think of there being two restricted diffusional spaces. One is the junctional space that is very narrow and for which it is clear that there is tremendous diffusional restriction. The other one is a space that is much wider, but still has a lower myofilament density. It may not contribute as much to the contractility of the whole cell. You can still have buffering of Ca^{2+} from this second restricted diffusional space which is then recycled out. The evidence for the 20 nm wide junctional space is clear, but I don't think it explains everything.

Brading: We are talking about these spaces as though they were constant. What worries me about this is that when one starts moving ions, it is likely that there will be changes to volumes. All sorts of things may be going on here. I also wanted to introduce what is going on with the Na^+, because I think this is going to be quite important. I also don't know what exactly is happening when Ca^{2+} moves in and out of the SR, and whether there is any shift in osmolarity or changes in volume.

Sanders: With traditional electron microscopy the space looks as if it might be as short as 20 nm. Is this the same sort of volume that might be predicted by freeze fixation?

Somlyo: It is within the measurement errors. There is a fenestration of the SR sheet, and sticking out come the caveolae. No one has really measured accurately this distance, or the distance between the caveolae and SR on top. The surface coupling space is pretty consistent. With regard to what is different in smooth muscle, if you are talking about the SR at the junction having Ca-ATPase or not, we don't know. What we do know from freeze–fracture studies of striated muscle is that the Ca-ATPase does not seem to be at the SR terminal cisternae. We don't know the answer in smooth muscle, but if there is Ca-ATPase at the junctional surface itself, this is different from what one sees in striated muscle.

Bolton: My understanding is that the vena cava is a tonic smooth muscle, and therefore it is right at one end of a spectrum of smooth muscle contractile types. This means that, it generates tension slowly, it maintains tension when it is activated and it doesn't show action potentials. Its organization may be very different from a phasic smooth muscle that normally operates by action potentials and cannot maintain tension.

van Breemen: I completely agree. For example, we have looked at the effect of the SR on the K_{Ca} channels, and it is zero in the vena cava. This is completely opposite to what is found in the resistance vessel. There is a tremendous variability in the nature of local Ca^{2+} regulation.

Nelson: Relating to the proximity, we have probed this area by looking at sparks and the communication in the large conductance Ca^{2+}-activated K^+ (BK) channels. The measurements are consistent with close apposition of the RyRs and spark sites to the BK channels. This would be consistent with a proximity of 10–20 nm seen by electron microscopy. Also, mobile buffers are unable to compete with Ca^{2+} at the BK channel, which is also consistent with this idea. As a probe of what happens to local Ca^{2+} we have looked at the decay of a spark. Nothing we could do, such as zero Na^+ or lanthanum, had any effect on the decay, suggesting that diffusion was responsible.

Sanders: Decay suggests that diffusion out of that space is fairly rapid.

Nelson: It is a little slower than heart, but it is still pretty quick.

van Breemen: Is it possibly regulated by SERCA?

Nelson: We have the phospholamban knockout mouse model, and John Lederer was able to peel out a component of the decay that was due to reuptake. We didn't see any difference between the control and the knockouts. Presumably it is happening, but we couldn't see any difference in the decay in the phospholamban knockouts, as was seen in heart muscle, nor could we see any effect of lanthanum or zero Na^+. Examining the decay of the spark would be a good indicator of local Ca^{2+} removal, though. It would also be worth examining the decay of the BK current.

Blaustein: As you say, you can't get the Ca^{2+} chelators in there to have a big effect, either. It must therefore be pretty local. Could there be local reuptake and does it have to be phospholamban-sensitive?

Nelson: We were unable to do manoeuvres such as have been done in heart. I would say that the BK channels are much better local Ca^{2+} centres, and their decay was also unaffected.

Blaustein: But you couldn't really see Fluo-3 or Fluo-4 in that area, because it is such a small volume. You can only see this when you are looking at sparks and the Ca^{2+} concentration changes are large.

Nelson: Admittedly, we are looking at some reflection of what is going on. We are not looking at the Ca^{2+} right under the BK channel.

Somlyo: Could some of the BK be used for binding to low affinity cytosolic buffers?

Nelson: That's possible.

Somlyo: Keep in mind that if you measure total cytoplasmic Ca^{2+}, it greatly exceeds the concentration of known cytoplasmic buffers, even if you take into account the 30–40 μM calmodulin that no one can completely account for in smooth muscle cells.

Nelson: One other point. We are examining Ca^{2+} waves. Blocking Ca^{2+} channels dilates cerebral arteries but not Ca^{2+} waves. Gil Wier, when you blocked Ca^{2+} channels in mesenteric arteries, didn't you observe waves?

Wier: Yes, we reported this in 1999 in pressurized arteries. Nifedipine produced a complete relaxation but the synchronous Ca^{2+} waves continued.

van Breemen: I think that the relationship between the K_{Ca} channels and the SR in the cerebral resistance arteries is such as to cause relaxation, so even the waves cause relaxation.

Paul: Do you worry about using non-ratiometric dyes here when these things are supposed to be contracting and oscillating? I know people have shown waves with Fura-2, but it always makes me a little nervous when non-ratiometric dyes are used, because some of this could be artefactual.

van Breemen: We stretch the tissue a lot, so there is hardly any shape change.

Somlyo: Does the stretch affect the oscillations?

Blaustein: The key question is whether the volume changes. The volume inside the cell doesn't really change much. The cell gets fatter and shorter, but because the volume doesn't change much, the concentration of dye stays the same.

Somlyo: If the images are confocal, you really should not have a volume change, because you are measuring a very constant slice.

Bolton: It might be a different slice.

Burdyga: You showed that high K^+ produced an almost instantaneous rise in Ca^{2+} all over the place. Is the SR acting as a sink or a source of Ca^{2+}?

van Breemen: In this tissue the SR tends to act as a sink for the $[Ca^{2+}]_i$ rise in high K^+. The rise is slower when the sink is more effective.

Nelson: Another issue that you are touching on is the on-rate of Ca^{2+} onto the target sites. If the on-rate is slow, this will impact on the frequency of the Ca^{2+} signal.

Eisner: You could address this by putting in one of the fluorescent calmodulins, which change fluorescence when they bind Ca^{2+}.

References

Blaustein MP, Golovina VA, Song H et al 2002 Organization of Ca^{2+} stores in vascular smooth muscle: functional implications. In: Role of the sarcoplasmic reticulum in smooth muscle. Wiley, Chichester (Novartis Found Symp 246) p 125–141

Karaki H, Ozaki H, Hori M et al 1997 Calcium movements, distribution and function in smooth muscle. Pharmacol Rev 49:157–230

General discussion I

The role of calmodulin in smooth muscle contraction

Walsh: I would like to describe an experiment to indicate that there's a specific pool of calmodulin involved in regulating contraction, and that the actual mechanism of Ca^{2+} regulation of smooth muscle contraction might be significantly different from what we had previously envisaged. Following the increase in cytosolic $[Ca^{2+}]$, most people think that it binds to calmodulin and the complex then diffuses to the myofilaments, where it interacts with myosin light chain kinase (MLCK) and activates the enzyme, phosphorylating myosin and leading to contraction. But this mechanism of freely diffusible cytosolic calmodulin is not consistent with many of the data in the literature. We went back and looked at this issue, and this is the experiment I'd like to describe. The preparation we use is de-endothelialized rat tail artery vascular smooth muscle. We begin with an intact de-endothelialized preparation showing a K^+-induced contraction, which is reversible on removal of K^+. Then we skin the preparation with Triton X-100, which removes all the membranes including the SR. Then, as many people have shown before with various smooth muscle preparations, we can induce contraction with Ca^{2+} alone. This is consistent with at least some calmodulin being retained in the skinned preparation, which would argue against simple cytosolic calmodulin being responsible for contraction. This Ca^{2+}-induced contraction is reversible on removal of Ca^{2+}. Then we induce another Ca^{2+} contraction, and in the continued presence of Ca^{2+} we add the calmodulin antagonist, trifluoperazine (TFP). This competitively displaces the $Ca^{2+}/$ calmodulin from the MLCK, allowing us to wash it out of the system. This correlates with relaxation. We then remove all the calmodulin by several washes in zero $Ca^{2+}/5$ nM EGTA solution. Now, when Ca^{2+} is added there is no contractile response. We have shown by Western blotting that there is no calmodulin left. The novel finding here was that we could add back purified calmodulin in the absence of Ca^{2+}, then wash out any unbound calmodulin, and recover the Ca^{2+}-induced contraction. We showed by Western blotting that this exogenous calmodulin binds to the preparation and accounts for the Ca^{2+}-induced contraction, which again is reversible on removal of Ca^{2+}. We are suggesting that there is a discrete pool of calmodulin that represents about 10% of the 40 μM total tissue content that Andrew Somlyo referred to earlier. We suggest that this is permanently anchored to the MLCK, either at resting $[Ca^{2+}]$, where it has zero or two Ca^{2+} ions bound to it, or even at zero $[Ca^{2+}]$ under experimental conditions. Then it is the diffusion of Ca^{2+} to the myofilaments, where it interacts with this permanently bound calmodulin, which is responsible for activating the kinase and triggering contraction. This is relevant to the spatial issue of Ca^{2+} release from the SR. You could have Ca^{2+} release in areas remote from the myofilaments that will never trigger a contractile response. A local increase in $[Ca^{2+}]$ in the vicinity of the myofilaments would be required to activate contraction in the smooth muscle.

Paul: How does that stoichiometry work out? I always thought there was 10-fold less kinase than myosin filaments.

Walsh: That's correct.

Paul: Where do you think the kinase lives?

Walsh: There's strong evidence that it is anchored to the actin filament through an N-terminal domain that Jim Stull's group has defined very precisely. The age-old question remains: how does an anchored MLCK molecule gain access to an adequate number of myosin heads to account for the phosphorylation stoichiometry that can be achieved in muscle?

Brading: With actin in isolation, can you find out anything about the kinetics of binding of kinase onto actin molecules?

Walsh: The kinase binds to actin in the isolated state with a K_d of about $0.8\,\mu\mathrm{M}$ and to myofilaments with a K_d of about $0.1\,\mu\mathrm{M}$. No measurements of the on- and off-rates have been made, however. *In situ*, most importantly, MLCK appears to be permanently bound since it does not dissociate from detergent-treated smooth muscle tissues, implying that the off-rate *in situ* is extremely slow or zero.

Brading: So you would need something happening *in situ*.

Nelson: There is an emerging theme of calmodulin being tightly bound to almost all its targets. For example, the SK channel is gated by calmodulin, and is extremely tightly bound. Another example is the voltage-dependent Ca^{2+} channel, where calmodulin is also bound and involved in inactivation.

Walsh: In many of these cases the binding is through so-called IQ motifs, which is sometimes a Ca^{2+}-independent interaction. The example I have just presented is clearly different.

Nelson: Yes, it is different, because you remove calmodulin. Calmodulin cannot be readily dissociated from the SK channel.

Walsh: I would say, however, that, while you can dissociate calmodulin from MLCK with TFP, you probably can't get it off under physiological conditions.

Kotlikoff: What do you know about the binding motif here?

Walsh: All we know about it at the moment is that the Ca^{2+}-independent interaction of calmodulin with the myofilaments is different from the interaction of Ca^{2+}/calmodulin with the isolated kinase. The question is: is this calmodulin bound to MLCK or something else? The evidence we have so far is indicative that it is bound to MLCK. We have some preliminary cross-linking evidence to support this, but more work is needed to nail this down. Clearly the interaction is different *in situ* than with purified proteins.

Brading: Phosphorylation of all these peculiar things that are floating around might lead to alterations in the rate of decay.

Walsh: That's another issue that needs to be looked at and it is known that both calmodulin and MLCK can be phosphorylated.

Paul: One thing still puzzles me. I thought the activation of MLCK by Ca^{2+} was in the nanomolar range biochemically. Then in the skinned fibres it has always been sort of a puzzle, because you are now having to use micromolar/submicromolar levels. There may be some real differences in the binding constants.

Walsh: Definitely. With the purified kinase, you can't detect any binding of calmodulin in the absence of Ca^{2+}, so there is no affinity at all. *In situ*, on the other hand, it binds very tightly in zero $[Ca^{2+}]$ and does not appear to dissociate. It is misleading to use binding constants that one measures in solution with the purified proteins in the context of the muscle. As we have heard, nothing is free or cytosolic. What do these binding constants really mean? We have looked at the simplest situation where you have MLCK sitting on an actin filament, either with or without tropomyosin. In this case, the binding of calmodulin is still Ca^{2+} dependent. So something else is required in the skinned muscle to account for the Ca^{2+}-independent interaction of calmodulin.

Taylor: A naïve question: on the one hand your inability to wash out the calmodulin made it look like it was held by high-affinity binding, yet presumably the TFP, which seemed to work rather quickly, depends on calmodulin dissociating to prevent it from rebinding. Does this imply it is dissociating quite quickly?

Walsh: We have gone through several repetitive wash cycles and quantified the calmodulin release. It is barely detectable. The TFP is actually competing directly with MLCK for the same site on calmodulin, so I think it is simply a question of concentration. The TFP concentration is 0.4 mM.

Taylor: Unless it is exerting some allosteric effect, the calmodulin still has to come off.

Walsh: Yes, it does. I think, though, that it is exerting some steric effect and causing the displacement.

Somlyo: When you add back calmodulin, does the TFP come off? I ask because there are a couple of papers showing that TFP does interact with myosin. We found the same thing, and we are desperately trying to remove the TFP. Have you been able to remove the TFP that might be bound to the myosin?

Walsh: We haven't looked at this directly. We did quantify myosin light chain levels throughout by Western blotting, and there is no loss at all. We were hoping that the level of TFP was low enough to affect calmodulin but not myosin light chains, and this appears to be the case.

Fry: But the force was diminished in your second contraction.

Walsh: There's always a little bit of run down in those long experiments, so even in the controls you would see a 10% drop in maximal force.

Sanders: I wanted to raise another issue about calmodulin: its regulation of TRP channels. There's some really nice work from Lutz Birnbaumer's group showing binding sites for calmodulin and negative regulation of TRP channels. Presumably when $[Ca^{2+}]$ goes up in this space, Ca^{2+} entry through certain TRP channels is modulated. Casey van Breemen, you said that you only found TRP1. I'm surprised about that, because most smooth muscles seem to express 4 and 6, and possibly 7. TRP6 appears to be part of the explanation of receptor-operated Ca^{2+} entry: the non-selective conductance that is present in a lot of smooth muscles. I was wondering how these channels might fit into your model.

van Breemen: Perhaps the problem with the rabbit is that we didn't have primers from the rabbit. I can definitely say that we have a lot of TRP1, but I can't say that we don't have some of the other channels.

Kotlikoff: Returning to the issue of receptor-operated or store-operated channels, one of the things I noticed in your traces was that your caffeine responses came back to baseline very smoothly. There was not the sustained elevation of $[Ca^{2+}]$ that you might predict. One of the things that is very different about voltage-clamped smooth muscle between a caffeine response and a phospholipase receptor-linked response is this additional current that is seen with the phospholipase receptor. I didn't see any functional evidence of a sustained response in your caffeine experiments. Have you voltage clamped these cells and looked to see whether there is any current there? If not, do you think this is because the current is just too small to be able to measure?

van Breemen: The comparison with the receptor activation would be unequal in that receptor activation does many things. But if you compare it with SERCA blockade, then there is still a difference between caffeine and SERCA blockade. It is just a little more puzzling to explain. One way you could explain it is if ryanodine channels were facing the junctional space.

Kotlikoff: Do you know whether there is a current? If you just release the store do you see a current?

van Breemen: I have not done this experiment.

Bolton: SERCA blockade is different. If you are just blocking uptake, the Ca^{2+} is going to come out rather slowly if there are no channels open.

van Breemen: SERCA blockers also hold the SERCA in the open position, so some Ca^{2+} also leaks through the pump itself.

Ca^{2+} signalling and Ca^{2+}-activated K$^+$ channels in smooth muscle

John G. McCarron, Karen N. Bradley and Thomas C. Muir

Neuroscience and Biomedical Systems, Institute of Biomedical and Life Sciences, West Medical Building, University of Glasgow, Glasgow G12 8QQ, UK

Abstract. In smooth muscle, transient subsarcolemma increases in Ca^{2+} of ~ 200 nM from the sarcoplasmic reticulum activate Ca^{2+}-activated K$^+$ channels (K$_{Ca}$) in the sarcolemma giving rise to spontaneous transient outward currents (STOCs). In the present study we have examined whether (1) STOCs are spatially restricted membrane currents, (2) single K$_{Ca}$ channel activity is regulated by changes in *bulk average* cytosolic Ca^{2+} concentrations ([Ca^{2+}]$_c$) without concomitant *local subsarcolemma* Ca^{2+} changes, and (3) a relationship exists between the voltage-dependent Ca^{2+} current (I$_{Ca}$) and K$_{Ca}$ channel activity. Guinea-pig single colonic myocytes were voltage clamped in the whole cell configuration (to measure macroscopic currents) and bulk average [Ca^{2+}]$_c$ measured simultaneously using the dye Fura-2. Single channel activity was also recorded with a second electrode, on the same cell, in the cell-attached mode. K$_{Ca}$ activity was identified by reversal potential and conductance measurements. If STOCs are not spatially restricted events but reflect increased K$_{Ca}$ channel activity throughout the sarcolemma, the voltage-dependence of single K$_{Ca}$ channels should be similar to that of STOCs. Prolonged depolarization (-60 mV to $+50$ mV) increased [Ca^{2+}]$_c$, the amplitude and frequency of STOCs, and single K$_{Ca}$ channel activity. [Ca^{2+}]$_c$ peaked around -20 mV. Between -50 and -20 mV, the increase in STOC frequency was markedly voltage-dependent (e-fold for 5 mV depolarization); beyond -20 mV less so. Single K$_{Ca}$ channel activity increased about e-fold for a 20 mV depolarization and thus was demonstrably different in this respect from that of STOC activity, evidence consistent with the proposed spatially restricted nature of STOCs. Simultaneous depolarization (3 s) of both the whole cell and the membrane patch, from -70 to 0 mV, elevated [Ca^{2+}]$_c$ to about 800 nM and evoked single K$_{Ca}$ channel activity, the latter began after about 10 ms, peaked around 100 ms, then declined. On repolarization to -70 mV K$_{Ca}$ channel activity ceased abruptly. Depolarization (to 0 mV) of the whole cell only, with the patch transmembrane potential maintained at -70 mV, increased [Ca^{2+}]$_c$ to about 800 nM but, importantly, did not increase K$_{Ca}$ channel activity. Conversely depolarization (to 0 mV) of the patch alone, the whole cell being maintained at -70 mV, did not alter the bulk [Ca^{2+}]$_c$ but evoked single K$_{Ca}$ channel activity. The time course of K$_{Ca}$ channel activity was remarkably similar to that of I$_{Ca}$ suggesting that Ca^{2+} influx through voltage-dependent Ca^{2+} channels may serve as a trigger for K$_{Ca}$ channel activation. Together these results suggest that STOCs are spatially restricted membrane currents and that K$_{Ca}$ channels are sensitive to both depolarization and local subsarcolemma Ca^{2+} increases but not to alterations in [Ca^{2+}]$_c$.

2002 Role of the sarcoplasmic reticulum in smooth muscle. Wiley, Chichester (Novartis Foundation Symposium 246) p 52–70

Alterations in the cytosolic free Ca^{2+} concentration ([Ca^{2+}]$_c$) regulate several smooth muscle activities, such as metabolism and gene expression, as well as being the major trigger for contraction. Early measurements of $[Ca^{2+}]_c$ indicated that Ca^{2+} signals existed in different forms which varied in amplitude and duration, and showed oscillatory behaviour at different frequencies (Woods et al 1986, Chatton et al 1997). From this evidence, speculation arose that such differences in duration and frequency could encode biological information and even determine the qualitative nature of the cellular response. Indeed, recent evidence suggests that not only the temporal features but in particular the spatial distribution of Ca^{2+} influence cellular activity (Simon & Llinas 1985, Smith & Augustine 1988, Adler et al 1991, Landolfi et al 1998).

Scattered compartments of high $[Ca^{2+}]$ — even the maintenance of standing gradients of the ion within regions of the cell — are facilitated by the latter's subcellular architecture. For example, the sarcoplasmic reticulum (SR) is not randomly distributed but makes numerous close connections with mitochondria and especially with plasma membranes (12–20 nM; Devine et al 1972, Rizzuto et al 1998) throughout the cell. Together these structures restrict diffusion of Ca^{2+} and allow high concentrations to be maintained within the cell. The high concentrations within these areas may vary from each other and from the bulk average $[Ca^{2+}]_c$ and may regulate important aspects of the cellular response. For example, one such region of the cytoplasm between the sarcolemma and SR (variously described as subsarcolemmal 'fuzzy space' in heart cells, 'active zone' in presynaptic neurons, or 'superficial buffer barrier' in smooth muscle), contributes to the signals enabling the selective targeting of effectors and increasing the speed of response — independently of bulk average $[Ca^{2+}]_c$ (Simon & Llinas 1985, Smith & Augustine 1988, Adler et al 1991, Landolfi et al 1998, van Breemen et al 1995). Moreover, Ca^{2+} signals generated in the subsarcolemma space may lead to responses (relaxation) opposite to that (contraction) produced by increases in bulk average $[Ca^{2+}]_c$ (Nelson et al 1995). Among the most important signals derived from these locally high concentrations of Ca^{2+} are those which activate sarcolemma membrane channels giving rise to periodic inward currents (STICs; Large & Wang 1996), regulate inositol-1,4,5-trisphosphate (InsP$_3$)-sensitive Ca^{2+} store refilling (McCarron et al 2000) and, particularly relevant to the present study, produce spontaneous transient outward currents (STOCs; Bolton & Imaizumi 1996) in the sarcolemma.

STOCs arise from the concerted activation of up to 100 Ca^{2+}-activated K$^+$ channels (K$_{Ca}$) in the sarcolemma as a consequence of Ca^{2+} release from the SR. This release may take the form either of non-propagating focal events such as 'sparks' or 'puffs', or of more regenerative Ca^{2+} waves. Since a temporal correlation between them exists, STOCs have been attributed to focal non-propagating Ca^{2+} release events. However, this view requires that each STOC is a spatially restricted membrane current that occurs at selected areas of membrane closely apposed to the SR. While essential to the prevalent hypothesis for STOCs, the spatially restricted nature of the current has hitherto remained an untested assumption; this has been assessed in the present study. To do this the voltage-dependence of STOCs and of single K_{Ca} channels have been compared. If a STOC is a spatially restricted membrane current, arising from the activity of a closely clustered group of up to 100 K_{Ca} channels, rather than from the activity of a similar number of channels scattered throughout the sarcolemma, then the shape of the accumulated single-channel currents would be expected to resemble that of a STOC as in single Na^+ channel activity and its relationship to the macroscopic current (Sigworth & Neher 1980). Implicit in this view is that the voltage-dependence of STOCs should be similar to that of single K_{Ca} channels.

K_{Ca} channel activity is regulated not only by its sensitivity to the sarcolemma voltage but also by the concentration of Ca^{2+} at the cytoplasmic face of the channel. Yet sensitivity of the K_{Ca} channels to Ca^{2+} (over normal physiological membrane potentials) is sufficiently low (Barrett et al 1982, Markwardt & Isenberg 1992) to preclude their activation by physiological bulk average $[Ca^{2+}]_c$ (Becker et al 1989). Local Ca^{2+} increases from the SR, however, are effective in channel activation even though they reach only some 200 nM. The effectiveness of such low Ca^{2+} concentrations suggests that K_{Ca} channels sensitivity *in vivo* may exceed that proposed for *in vitro* conditions and as a result physiological increases in bulk average $[Ca^{2+}]_c$ could activate K_{Ca} channels. Alternatively, the reported low sensitivity of the K_{Ca} channels to Ca^{2+} is correct and the magnitude of local $[Ca^{2+}]$ has been substantially underestimated. If the latter proposal is correct, physiological increases in bulk average $[Ca^{2+}]_c$ could not activate the channels. Accordingly, a second objective of the present study was to determine if increases in bulk average $[Ca^{2+}]_c$ to similar or even larger values than those of local $[Ca^{2+}]$ might also activate K_{Ca} channels.

In addition to the influence of Ca^{2+} released from the SR on K_{Ca} channel activity, the latter may also be affected by the influx of Ca^{2+} through adjacent voltage-dependent Ca^{2+} channels. A third objective of the present study was therefore to examine the relationship between Ca^{2+} influx across the sarcolemma and K_{Ca} channel activity.

Methods

From largely circular colonic smooth muscle of male guinea-pigs (500–700 g killed by cervical dislocation and bleeding) single smooth muscle cells were enzymatically dissociated as previously described (McCarron & Muir 1999).

Current recordings

Membrane currents were measured using conventional, whole-cell, tight seal recording. The composition of the extracellular solution was (mM): Na glutamate 80, NaCl 60, $MgCl_2$ 1.1, $CaCl_2$ 3, HEPES 10, and glucose 10 (pH 7.4 with NaOH).

Unless otherwise stated the whole cell pipette solution contained (mM): KCl 130, NaATP 2, MgCl₂ 3, HEPES 10, Fura-2 0.05, pH 7.2 with KOH. The single channel pipette solution contained (mM): KCl 40, NaCl 75, HEPES 10, pH 7.4 with KOH.

Whole-cell currents were amplified by an Axopatch 1D (Axon Instruments), filtered at 500 Hz (four-pole bessel filter), and sampled at 5 kHz using a Digidata interface and Axotape software. Single channel currents were amplified by an Axopatch 200B filtered at 1 kHz (8-pole bessel filter), and sampled at 5 kHz using a Digidata interface and Axotape software. In experiments examining K_{Ca} activity with local and bulk average changes in $[Ca^{2+}]_c$ the patch electrode voltage was adjusted after consideration of the whole cell pipette voltage with the patch transmembrane potential $(P_{VM}) = V_{whole\ cell} - V_{patch}$ as described in the text.

Ca²⁺ Measurements

$[Ca^{2+}]_c$ was measured using the membrane-impermeable dye Fura-2 (potassium salt; 50 μM) introduced into the cell from the patch pipette. Fluorescence was measured by a microfluorimeter which consisted of an inverted fluorescence microscope (Nikon diaphot) and a photomultiplier tube with a bialkali photocathode. The excitation wavelengths (340 and 380 nm, 7 nm bandpass) were provided by a PTI Deltascan (Photon Technology International Inc, London UK). The cell was illuminated every 10 ms for 8.5 ms with each wavelength. A complete 340/380 ratio and $[Ca^{2+}]_c$ measurements were made therefore at a frequency of 50 Hz. The excitation light passed through a 425 nm short pass filter (76% transmission at 340 nm and 80% transmission at 380 nm) and a field stop diaphragm used to reduce background fluorescence. A 400 nm long pass dichroic mirror (94% transmission at 510 nm) reflected the excitation wavelengths onto the cell. A 570 nm short pass dichroic mirror (82% transmission at 510 nm) passed the emission light through a 505 nm barrier filter (60 nm bandpass, 88% transmission at 510 nm) onto the photomultiplier for photon counting. Longer wavelengths from bright field illumination with a 610 nm Shott glass filter (90% transmission) were reflected onto a CCD camera mounted on to the viewing port of the delta scan allowing the cell to be monitored during the course of the experiments. Background fluorescence was measured with the pipette attached to the cell but before rupturing its membrane. This background was subtracted from the fluorescence counts obtained during the experiments. The K_D for Fura-2 was determined as 280 nM from an *in vitro* calibration. R_{MIN} and R_{MAX} were also determined from *in vitro* calibrations and decreased by 15% to adjust for cell viscosity.

Results

Voltage-dependence of STOCs and of single K_{Ca} channels

Application of the two-electrode patch clamp arrangement in whole cell and on-cell configurations in a single colonic myocyte is shown in Fig. 1. In the experiments

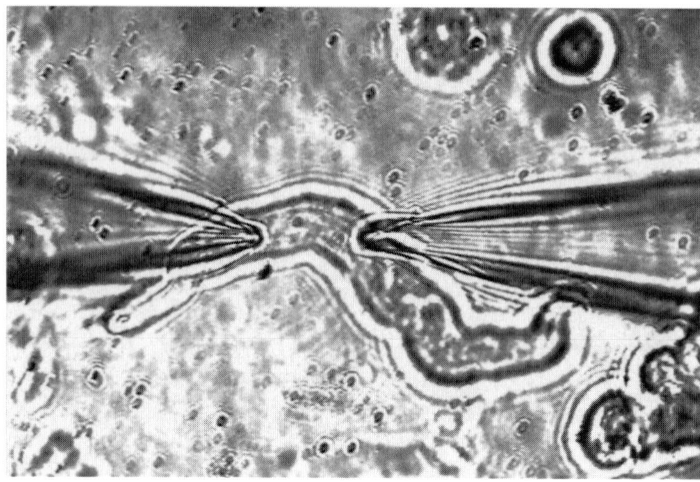

FIG. 1. Two-electrode patch clamp arrangement. Single colonic myocytes were voltage clamped simultaneously in whole cell and on-cell configurations.

in Fig. 2 the patch electrode voltage was 0 mV, therefore the patch membrane potential (P_{VM}) followed the whole cell electrode membrane potential. Whole cell depolarization (-60 to $+50$ mV, Fig. 2) altered $[Ca^{2+}]_c$, STOC frequency and amplitude, as represented by the spiking activity and single K_{Ca} channel activity. $[Ca^{2+}]_c$ increased over the range of -50 mV to -20 mV then declined. Over the range of -30 mV to 10 mV STOC activity substantially increased; single K_{Ca} channel activity also increased but less so. Over the range -50 mV to -20 mV, the probability of STOC occurrence was markedly voltage-dependent, increasing e-fold for a 5 mV depolarization but less so at membrane potentials positive to -20 mV (Fig. 3). Single channel activity (Fig. 3) was distinct from STOC activity in its voltage-dependence at potentials up to $\sim +10$ mV increasing e-fold for a 20 mV depolarization, a value similar to that reported elsewhere (Carl et al 1996). Together these results support the proposal that STOCs are spatially restricted currents which arise from focally released Ca^{2+} from the SR.

K_{Ca} channel activity and bulk average $[Ca^{2+}]_c$

With the patch electrode voltage at 0 mV, depolarization of the cell via the whole cell electrode (-70 mV to 0 mV) changed the potentials of both the P_{VM} and the whole cell from -70 mV to 0 mV (Fig. 4 left panel; i.e. $P_{VM} = V_{\text{whole cell}} - V_{\text{patch}}$), elevated $[Ca^{2+}]_c$ to about 800 nM, and evoked single K_{Ca} channel activity (bottom two panels). On repolarization to -70 mV, K_{Ca} channel activity ceased abruptly. In Fig. 4 (middle panel) depolarization of the whole cell from -70 mV to 0 mV, with the P_{VM} being held at -70 mV (by applying a simultaneous pulse to the patch

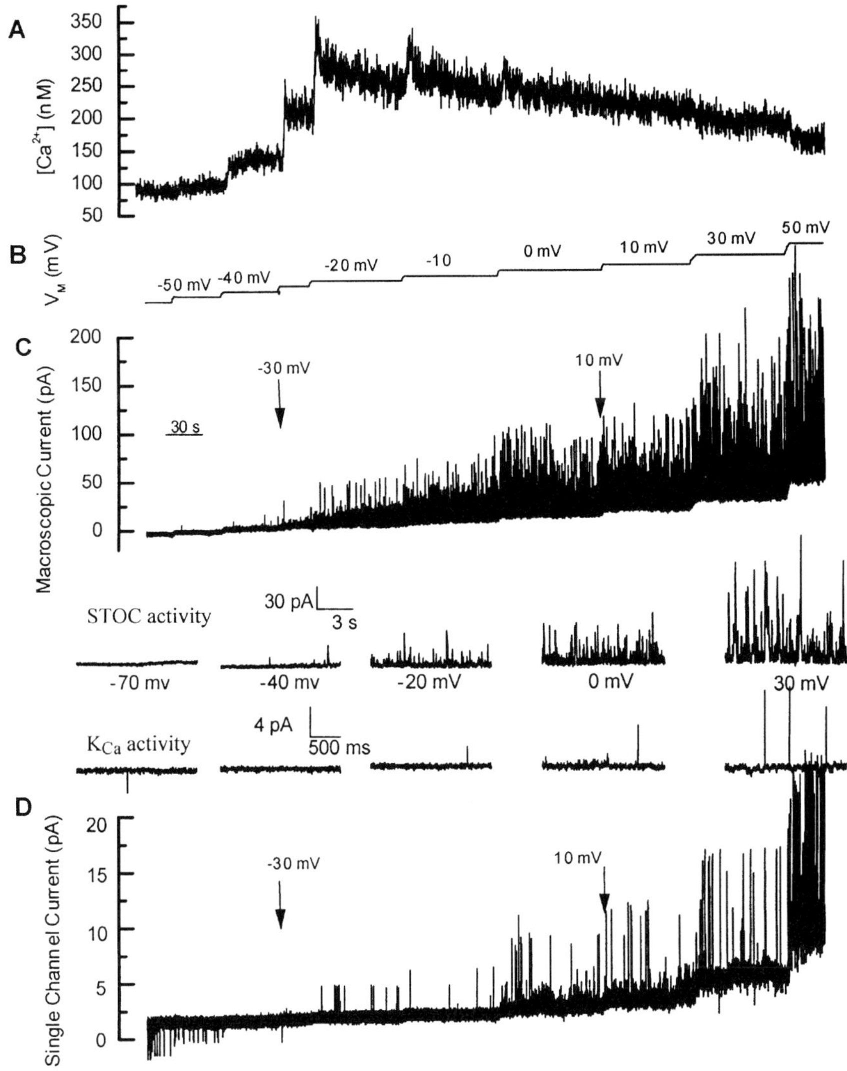

FIG. 2. [Ca^{2+}]$_c$, whole cell currents and single K$_{Ca}$ channel activity in response to depolarization in single colonic myocytes. Cells were depolarized over the membrane potential range -60 to $+50$ mV (B) so altering [Ca^{2+}]$_c$ (A), STOC frequency and amplitude, as represented by the spiking activity in C, and single K$_{Ca}$ channel activity (D). [Ca^{2+}]$_c$ increased over the voltage range of -50 mV to -20 mV then declined. Over the range of -30 mV to $+10$ mV, STOC activity was substantially increased while the increase in single K$_{Ca}$ channel activity was less pronounced (D). The insets between C and D show STOC activity (upper panel) and K$_{Ca}$ channel activity on an expanded time base.

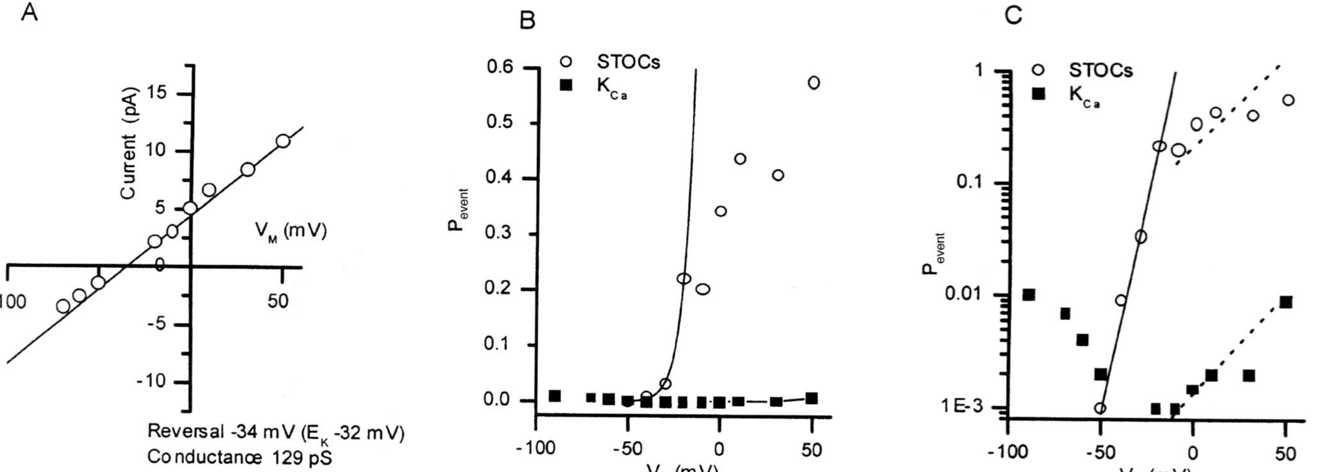

FIG. 3. Single K_{Ca} channel activity and STOCs. (A) Single channels had a conductance of 129 pS (data from Fig. 1) and reversed close to E_K suggesting that the channels were indeed of the large conductance K_{Ca} type. (B and C) Over the range -50 mV to -20 mV the probability of a STOC occurrence was markedly voltage-dependent, increasing e-fold for a 5 mV depolarization. Between -90 mV and -50 mV, K_{Ca} open probability decreased perhaps reflecting a diminishing driving force for Ca^{2+}. Positive to -20 mV STOC activity was less voltage-dependent (C dotted line open circles). Single-channel activity was distinct from STOC activity to $\sim +10$ mV. C is a semi-log plot of the data presented in B.

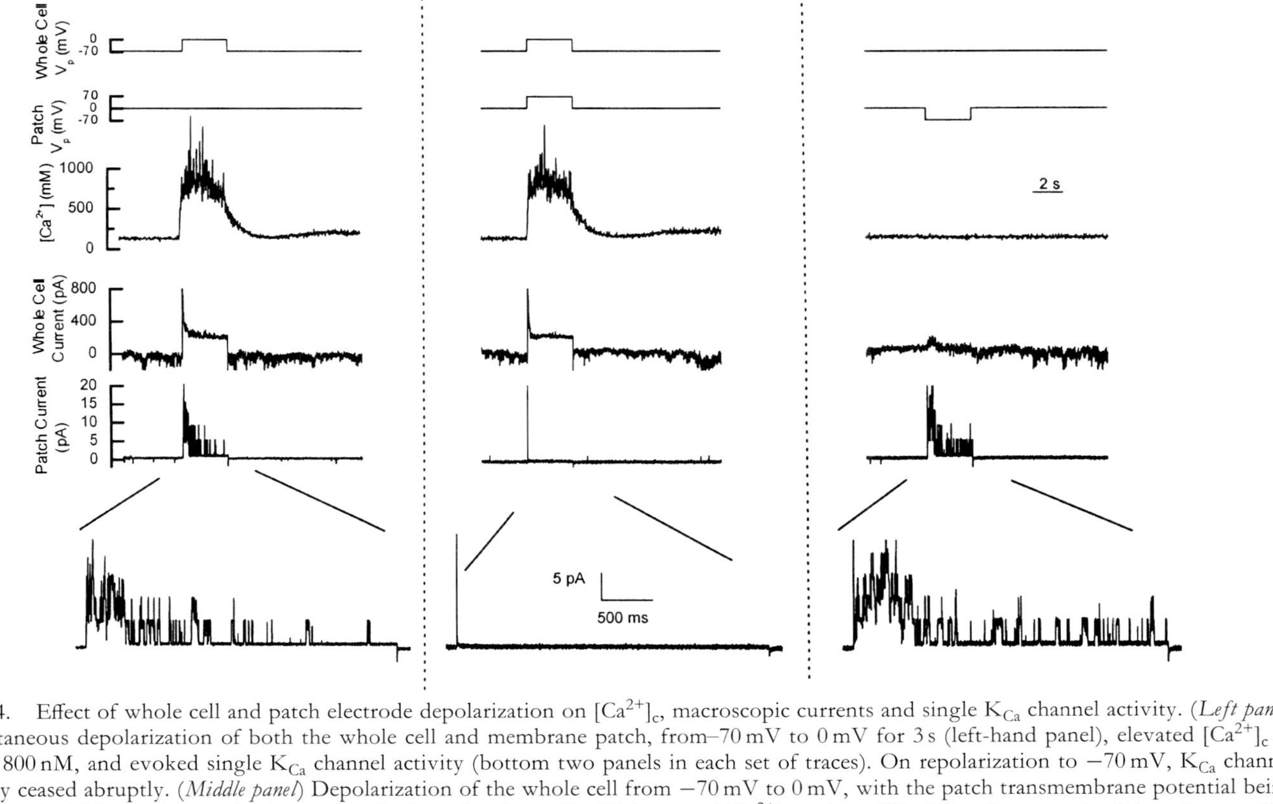

FIG. 4.   Effect of whole cell and patch electrode depolarization on $[Ca^{2+}]_c$, macroscopic currents and single K_{Ca} channel activity. (*Left panel*) Simultaneous depolarization of both the whole cell and membrane patch, from–70 mV to 0 mV for 3 s (left-hand panel), elevated $[Ca^{2+}]_c$ to about 800 nM, and evoked single K_{Ca} channel activity (bottom two panels in each set of traces). On repolarization to −70 mV, K_{Ca} channel activity ceased abruptly. (*Middle panel*) Depolarization of the whole cell from −70 mV to 0 mV, with the patch transmembrane potential being held at −70 mV (by applying a simultaneous pulse from 0 mV to +70 mV), increased $[Ca^{2+}]_c$ to about 800 nM but importantly, did not increase K_{Ca} channel activity (bottom two panels). (*Right panel*) Maintaining the whole cell voltage at −70 mV, while depolarizing the patch membrane potential alone to 0 mV (by applying a voltage pulse to the patch from 0 mV to −70 mV) did not alter the bulk $[Ca^{2+}]_c$ but evoked high single K_{Ca} channel activity (bottom two panels).

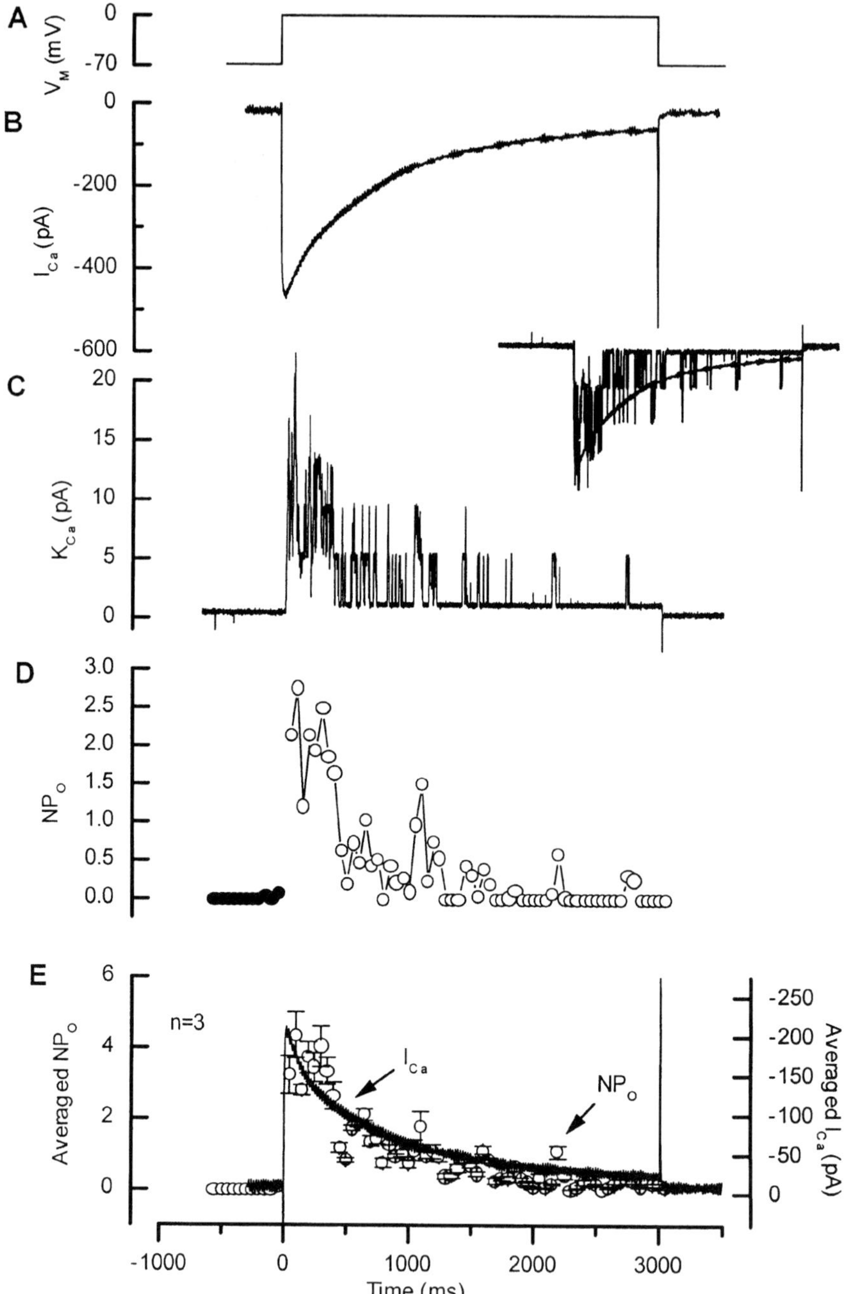
A
V_M (mV)
0
-70
B
I_Ca (pA)
0
-200
-400
-600
C
K_Ca (pA)
20
15
10
5
0
D
NP_O
3.0
2.5
2.0
1.5
1.0
0.5
0.0
E
Averaged NP_O
6
4
2
0
n=3
I_Ca
NP_O
Averaged I_Ca (pA)
-250
-200
-150
-100
-50
0
-1000
0
1000
2000
3000
Time (ms)

electrode from $0\,mV$ to $+70\,mV$), increased $[Ca^{2+}]_c$ to about $800\,nM$ but importantly, did not increase K_{Ca} channel activity. In Fig. 4 (right panel) when the whole cell membrane voltage was held at $-70\,mV$ and the P_{VM} depolarized to $0\,mV$ (by applying a voltage pulse to the patch electrode from $0\,mV$ to $-70\,mV$) the bulk $[Ca^{2+}]_c$ did not change but high single K_{Ca} channel activity was evoked.

Ca^{2+} influx and K$_{Ca}$ activity

To explore the interplay between Ca^{2+} influx and K_{Ca} channel activity, we examined the relationship between the latter and I_{Ca}. Depolarization (Fig. 5) of both the whole cell and membrane patch to $0\,mV$ (from $-70\,mV$) evoked single K_{Ca} channel activity which began after about $10\,ms$, peaked usually within $500\,ms$ then declined. Thus after the initial peak in activity, the frequency of channel opening decreased during depolarization, even though these channels do not exhibit voltage-dependent inactivation at $0\,mV$ (see Blatz & Magleby 1987). This observation suggests that changes in subsarcolemma Ca^{2+} occurred during the depolarization. To examine the relationship with voltage-dependent Ca^{2+} influx, we compared the time course of I_{Ca} and the kinetics of single K_{Ca} activity. In these experiments, to enable examination of inward I_{Ca}, outward currents were blocked by dialysing the cells with CsCl ($140\,mM$, KCl substitution) and TEA ($20\,mM$) in the bathing solution (NaCl substitution). Depolarization ($-70\,mV$ to $0\,mV$, Fig. 5)-activated I_{Ca} had a similar time course to that of single channel activity evoked by depolarization of the whole cell. A plot of the summarized open probability (NPo) and I_{Ca} (Fig. 5) demonstrated a considerable similarity during the inactivation of I_{Ca}, though I_{Ca} peaked before K_{Ca} activity. These results suggested that, since the time course of inactivation of I_{Ca} and the decline in K_{Ca} activity was similar, that Ca^{2+} influx may serve as a trigger for activation of the K_{Ca} channels.

Discussion

Focal Ca^{2+} release from the SR may accumulate at certain sites in the subsarcolemma (Gordienko et al 1998) to activate several (up to 100) K_{Ca}

FIG. 5. (*Opposite*) Relationship between I_{Ca} and K_{Ca} channel activity in single colonic myocytes. To examine the relationship between I_{Ca} and K_{Ca} channel activity, we blocked K_{Ca} by dialysing the cells with $140\,mM$ CsCl (KCl substitution) and $20\,mM$ TEA in the bathing solution (NaCl substitution). In these cells I_{Ca} was examined following depolarization from $-70\,mV$ to $0\,mV$ (A). It had a similar time course to that of single channel activity after the whole cell depolarization (B and inset). K_{Ca} channel activity was summarized by plotting the open probability (NPo calculated from a sliding $50\,ms$ window) against time (C). Indeed a plot of the summarized NPo and summarized I_{Ca} (E) demonstrates a considerable similarity during the inactivation of I_{Ca} though I_{Ca} peaked before K_{Ca} activity. Error bars have been omitted from I_{Ca} for clarity ($n = 3$). These results suggest that since the time course of inactivation of I_{Ca} and the decline in K_{Ca} activity are similar, Ca^{2+} influx may serve as a trigger for activation of the K_{Ca} channels.

channels so generating STOCs (Benham & Bolton 1986, Nelson et al 1995). Hitherto, evidence for this view has been drawn largely from the temporal correlation which occurs between focal Ca^{2+} release and STOCs (Nelson et al 1995, Perez et al 1999, Bayguinov et al 2001). If true, then each STOC should be an event localized to a special region of the sarcolemma closely apposed to released Ca^{2+} i.e. STOCs are spatially restricted to certain membrane areas depending upon the proximity of high Ca^{2+} concentrations. If STOCs were derived not from a spatially restricted K_{Ca} channel activity, but from a generalized increase in K_{Ca} channel activity throughout the sarcolemma, then the voltage dependence of the two events should be similar. Conversely, a dissimilar voltage dependence between the two would suggest that STOCs were not produced by a generalized increase in K_{Ca} channel activity. The present investigation has clarified the position. Substantial differences were found between the voltage-dependence of STOCs (e-fold increase for 5 mV depolarization) and that of single K_{Ca} channels (e-fold for 20 mV depolarization) providing support for the view that STOCs are spatially restricted to certain areas of the sarcolemma.

Interestingly, over the membrane potential range of -50 mV to -20 mV the increase in STOC activity exceeded that of single K_{Ca} channel opening. One interpretation of this observation is that those channels involved in the generation of STOCs are modulated by other influences beside their intrinsic voltage-dependence. One such influence could be the influx of Ca^{2+} via voltage-dependent Ca^{2+} channels. Over the same voltage range (-50 mV to -20 mV) where STOC frequency exceeded that of single K_{Ca} channel opening, bulk average $[Ca^{2+}]_c$ had also increased significantly (present study) because of entry via voltage-dependent Ca^{2+} channels (McCarron et al 2000). Indeed the voltage-dependence of STOCs (e-fold increase for 5 mV depolarization), over this voltage range, is similar to that of the voltage-dependent Ca^{2+} channel (e-fold increase for 4.5 mV depolarization, Quayle et al 1993) suggesting that the activity of the Ca^{2+} channel may directly or indirectly act as a trigger for STOCs. Several possible mechanisms may link Ca^{2+} channel activity and STOCs and include: (1) the influx of Ca^{2+} via the Ca^{2+} channel (*vide infra*) may activate a Ca^{2+}-induced Ca^{2+} release localized to certain areas of the cell; (2) the elevation in bulk average $[Ca^{2+}]_c$ may trigger increased Ca^{2+} release from the SR; and (3) an elevation in bulk average $[Ca^{2+}]_c$ may increase SR lumenal $[Ca^{2+}]$ which in turn may elevate Ca^{2+} release (Cheng et al 1996).

Further support for a link between voltage-dependent Ca^{2+} channel activity and K_{Ca} channel activity was the present observation that during a depolarization (after ~ 200 ms) the time course of change in the open probability of K_{Ca} channels was similar to that of I_{Ca} as it inactivated. Since K_{Ca} channels themselves do not exhibit voltage-dependent inactivation at 0 mV, the reduction in the open probability of the channel presumably reflects a decline in the subsarcolemma Ca^{2+} concentration as a consequence of the inactivation of I$_{Ca}$. Ca^{2+} entry via channels may generate locally high subsarcolemma concentrations of the element (Chad & Eckert 1984). However, whether or not Ca^{2+} entry via the voltage-dependent Ca^{2+} channels itself activates the K$_{Ca}$ channel or if the ion triggers additional release from the store to activate the channel remains to be determined.

The importance of subsarcolemma Ca^{2+} concentration, rather than bulk average [Ca^{2+}]$_c$, in the generation of STOCs is highlighted by the findings reported here that single K$_{Ca}$ channels may not be activated by increases in bulk average [Ca^{2+}]$_c$ (up to $\sim$800 nM) in the absence of local subsarcolemma Ca^{2+} increases. The reported sensitivity of the K$_{Ca}$ channel to Ca^{2+} should prevent their activation by normal physiological [Ca^{2+}]$_c$ (Barrett et al 1982, Markwardt & Isenberg 1992). Yet the channels are clearly active during STOCs. Thus the [Ca^{2+}] occurring during a Ca^{2+} spark (an increase of $\sim$200 nM) may be a substantial underestimate of the concentration of Ca^{2+} to which K$_{Ca}$ may be exposed. Indeed, using the K$_{Ca}$ channel itself as a Ca^{2+} indicator, estimates of the local [Ca^{2+}] giving rise to a STOC have been obtained. Thus the increased open probability of the channel during a STOC suggested that Ca^{2+} concentrations of up to 100 μM may have been reached (Perez et al 1999).

In this study, features of the control of both K$_{Ca}$ and STOCs have been examined. The results suggest that STOCs may be spatially restricted membrane currents and that K$_{Ca}$ channels are sensitive to both depolarization and local subsarcolemma Ca^{2+} increases but not to alterations in bulk average [Ca^{2+}]$_c$.

Acknowledgements

This work was funded by the Wellcome Trust (054328/Z/98/Z) and British Heart Foundation (PG/2001079). The authors would like to acknowledge the highly skilled technical assistance of Mr J.W. Craig.

References

Adler EM, Augustine GJ, Duffy SN, Charlton MP 1991 Alien intracellular calcium chelators attenuate neurotransmiter release at the squid giant synapse. J Neurosci 11:1496–1507

Bayguinov O, Hagen B, Sanders KM 2001 Muscarinic stimulation increases basal Ca^{2+} and inhibits spontaneous Ca^{2+} transients in murine colonic myocytes. Am J Physiol 280:C689–C700

Barrett JN, Magleby KL, Pallotta BS 1982 Properties of single calcium-activated potassium channels in cultured rat muscle. J Physiol 331:211–230

Becker PL, Singer JJ, Walsh JV Jr, Fay FS 1989 Regulation of calcium concentration in voltage-clamped smooth-muscle cells. Science 244:211–214

Benham CD, Bolton TB 1986 Spontaneous transient outward currents in single visceral and vascular smooth muscle cells of the rabbit. J Physiol 381:385–406

Blatz AL, Magleby KL 1987 Calcium-activated potassium channels. Trends Neurosci 10:463–467

Bolton TB, Imaizumi Y 1996 Spontaneous transient outward currents in smooth muscle cells. Cell Calcium 20:141–152

Carl A, Lee HK, Sanders KM 1996 Regulation of ion channels in smooth muscle by calcium. Am J Physiol 271:C9–C34

Chad JE, Eckert R 1984 Calcium domains associated with individual channels can account for anomalous voltage relations of Ca-dependent responses. Biophys J 45:993–999

Chatton JY, Liu HY, Stucki JW 1997 Modulation of hormone-induced calcium oscillations by intracellular pH in rat hepatocytes. Am J Physiol 272:G954–G961

Cheng H, Lederer MR, Lederer WJ, Cannell MB 1996 Calcium sparks and $[Ca^{2+}]_i$ waves in cardiac myocytes. Am J Physiol 270:C148–C159

Devine CE, Somlyo AV, Somlyo AP 1972 Sarcoplasmic reticulum and excitation-contraction coupling in mammalian smooth muscles. J Cell Biol 52:690–718

Gordienko DV, Bolton TB, Cannell MB 1998 Variability in spontaneous subcellular calcium release in guinea-pig ileum smooth muscle cells. J Physiol 507:707–720

Landolfi B, Curci S, Debellis L, Pozzan T, Hofer AM 1998 Ca^{2+} homeostasis in the agonist-sensitive internal store: functional interactions between mitochondria and the ER measured in situ in intact cells. J Cell Biol 142:1235–1243

Large WA, Wang Q 1996 Characteristics and physiological role of the Ca^{2+}-activated Cl^- conductance in smooth muscle. Am J Physiol 271:C435–C454

McCarron JG, Muir TC 1999 Mitochondrial regulation of the cytosolic Ca^{2+} concentration and the $InsP_3$-sensitive Ca^{2+} store in guinea-pig colonic smooth muscle. J Physiol 516:149–161

McCarron JG, Flynn ERM, Bradley KN, Muir TC 2000 Two Ca^{2+} entry pathways mediate $InsP_3$-sensitive store refilling in guinea-pig colonic smooth muscle. J Physiol 525:113–124

Markwardt F, Isenberg G 1992 Gating of maxi K^+-channels studied by Ca^{2+} concentration jumps in excised inside-out multi-channel patches (myocytes from guinea pig urinary bladder). J Gen Physiol 99:841–862

Nelson MT, Cheng H, Rubart M et al 1995 Relaxation of arterial smooth muscle by calcium sparks. Science 270:633–637

Quayle JM, McCarron JG, Asbury JR, Nelson MT 1993 Single calcium channels in resistance-sized cerebral arteries from rats. Am J Physiol 264:H470–H478

Perez GJ, Bonev AD, Patlack JB, Nelson MT 1999 Functional coupling of ryanodine receptors to K_{Ca} channels in smooth muscle cells from rat cerebral arteries. J Gen Physiol 113:229–237

Rizzuto R, Pinton P, Carrington W et al 1998 Close contacts with the endoplasmic reticulum as determinants of mitochondrial Ca^{2+} responses. Science 280:1763–1766

Simon SM, Llinas RR 1985 Compartmentalization of submembrane calcium activity during calcium influx and its significance in transmitter release. Biophys J 48:485–498

Sigworth FJ, Neher E 1980 Single Na^+ channel currents observed in cultured rat muscle cells. Nature 287:447–449

Smith SJ, Augustine GJ 1988 Calcium ions, active zones and synaptic transmitter release. Trends Neurosci 11:458–464

van Breemen C, Chen Q, Laher I 1995 Superficial buffer barrier function of smooth muscle sarcoplasmic reticulum. Trends Pharmacol Sci 16:98–105

Woods NM, Cuthbertson KSR, Cobbold PH 1986 Repetitive transient rises in cytoplasmic free calcium in hormone-treated hepatocytes. Nature 319:600–602

DISCUSSION

Nelson: We have several papers on simultaneous measurements of sparks and large conductance K_{Ca} channel (BK) currents (Perez et al 1999, 2001). These studies point to Ca^{2+} being in the order of 10–100 μM when it activates the channels. We have no evidence that argues in favour of a cluster of BK channels. Given the density of channels, a uniform distribution would be sufficient to provide the transient BK currents that are seen. There is no need to invoke spatially restricted channels. We have papers on vasoconstriction and protein kinase C (PKC) inhibition of sparks (Bonev et al 1997, Jaggar et al 2000). Vasoconstrictors such as UTP decreased spark frequency. This inhibition could be explained by activation of PKC, because they are blocked by PKC blockers and mimicked by PKC activation. SR Ca^{2+} level is unchanged by UTP or PKC activation, measured with caffeine pulses (Bonev et al 1997, Jaggar et al 2000).

McCarron: I disagree. I don't think that your data do show that the SR Ca^{2+} content remained unchanged. I think it changed quite a lot. It was down by about 60% from the numbers in the paper. When you applied caffeine to maximally discharge the SR Ca^{2+} content, the increase in Ca^{2+} produced by this was 402 nM. The increase in Ca^{2+} produced by InsP$_3$ was 249 nM. This is 61% of the overall Ca^{2+} content (Jaggar et al 2000).

Nelson: The store content plus and minus UTP was unchanged (Jaggar et al 2000).

McCarron: UTP would presumably have produced InsP$_3$, which would have released Ca^{2+} from the store — precisely the opposite of what you are saying.

Nelson: It was a long-term treatment, taking place over minutes (Bonev et al 1997, Jaggar et al 2000).

McCarron: The time course of the suppression of STOCs with UTP was within a minute.

Nelson: Yes, but it was sustained over tens of minutes. Clearly, in the first few minutes if the SR is depleted this may contribute to the initial suppression but the tonic effect appeared to reflect PKC activation.

McCarron: We found the tonic effect unaltered by PKC antagonists but blocked by InsP$_3$ receptor antagonists (McCarron et al 2002).

Sanders: There has been a theme in smooth muscle for a long time that to get STOCs you have to depolarize the cells. In your data here you showed a clear transition as the membrane potential was moved positive to -30: the STOC activity went way up. You are getting into the window current range for Ca^{2+} influx to be activated tonically. Moreover, Bolton's group showed that depolarization also regulates sparking frequency. When you go to very negative potentials, you don't see sparks. Then, when you go to higher potentials, corresponding with the development of STOCs, you start to see sparks. How important is Ca^{2+} entry to both sparking and STOC development?

McCarron: The voltage dependence of STOCs over that range (-50 to -20 mV) was remarkably similar to that of voltage-dependent Ca^{2+} channels. There is an e-fold increase in STOCs for a 5 mV depolarization. This is inseparable from the voltage-dependent Ca^{2+} channel. This suggests that Ca^{2+} influx is providing some trigger. Whether this is direct (in the sense that it is activating the ryanodine receptor), or indirect (in that it is changing bulk average Ca^{2+} or altering luminal Ca^{2+}), is difficult to tell.

Sanders: Tom Bolton, when you collaborated with Imaizumi, I seem to recall that you tried nicardipene, and it didn't block this voltage-dependent effect. I asked you this question in Japan a couple of years ago and I thought you said that you had tried nicardipene and you didn't change this voltage-dependent spark activity.

Bolton: If I didn't know the answer then, I don't know it now.

Nelson: We have seen Ca^{2+} channel blockers block the voltage dependence of the frequency of the STOCs.

Sanders: Ca^{2+} entry is critical to set the sparks off.

Nelson: The amplitude is another issue. It goes up with depolarization.

Brading: Fabiato showed years ago in the heart that quite a lot of Ca^{2+} is needed in the SR before there were these spontaneous releases of Ca^{2+}.

Somlyo: He also showed that the rate of the increase was important. A fast rate gives good Ca^{2+}-induced Ca^{2+} release (CICR), whereas a slow rate does not induce CICR.

Bolton: When you put something on that depleted the Ca^{2+} store and your STOCs disappeared, afterwards you had an increased size of STOC. Why?

McCarron: It may be that there are other second messengers generated in addition to $InsP_3$ that sensitize the K_{Ca} or the ryanodine receptor to release Ca^{2+}.

Bolton: Does it always happen, or is there some difference between agents that deplete? For example, is caffeine different from activating with a receptor stimulant?

McCarron: I'm not sure.

Brading: Does tetracaine have an effect on K^+ channels?

Burdyga: It blocks them.

McCarron: We looked at the outward current activated by $InsP_3$ in the presence and absence of tetracaine, and they were similar (McCarron et al 2002).

Blaustein: 2-aminoethoxydiphenyl borate (2-APB) is used as a selective blocker of $InsP_3$ receptors. However, I think I have seen recent papers showing that 2-APB also blocks Ca^{2+} entry in some way. Do you think it blocks Ca^{2+} entry and reduces the tonic component in this way?

McCarron: No, although we only have indirect evidence for this. Our reasoning is that the tonic component is blocked by inhibitors of the voltage-dependent Ca^{2+} channel, and it is not blocked by the SKF96365 which blocks store-operated channels.

van Breemen: I think there is some direct effect of 2-APB on some voltage-gated channels. I showed a small effect in the open position, so we did the same experiment in the basilar artery, where we had a much greater sensitivity of the voltage-gated Ca^{2+} channels to 2-APB. At $50\,\mu M$ it caused a 50% block of the high K^+-induced Ca^{2+}, which is not related to the $InsP_3$ receptor.

McCarron: We also did a similar experiment with the high K^+ contraction.

Wier: We don't find an effect of APB on high K$^+$ contraction if phentolamine is present. With respect to your mechanism of tonic contraction, which you indicated was dependent on Ca^{2+} influx, in the mesenteric small artery that we study, which develops a tonic contraction, this isn't accompanied by tonic elevation of Ca^{2+}, but rather by these asynchronous Ca^{2+} waves that we have already seen. Nevertheless, when we add up all these Ca^{2+} indicator signals we get something that looks like a steady elevated level of Ca^{2+}. We would say that in that tissue the dependence on Ca^{2+} influx is to keep the SR going, to keep generating these waves. Is this relevant to your tissue?

McCarron: If the SR is refilled again during the course of our carbachol additions, STOCs would reappear. In effect we could make a similar argument.

Wier: What you need is direct evidence by examining what is going on in the individual cells.

Somlyo: What happens physiologically? It might be worth taking other processes into account as well. For example, the fact that both the Rho inhibitor DC3B and the Rho kinase inhibitor Y-27639 inhibit the tonic phase of the contraction, suggests that two things are stimulated by agonists during the tonic phase: one causes subminimal elevation of Ca^{2+} and the other inhibits myosin phosphatase through Rho kinase or another mechanism. The absence of either could result in phasic decay.

McCarron: I agree that there will be many second messenger and channel systems that will be active.

Somlyo: I know we live in a Ca^{2+}-centric universe, but we are learning about other mechanisms.

Fry: The voltage dependence of these STOCs that you have shown is interesting, because there is now more evidence for T-type Ca^{2+} currents in smooth muscle cells. Kenton Sanders, you said that it was in the window current for the L-type Ca^{2+} channels, but it is probably about 10 mV too negative for that. Is there any evidence that T-type channel activity will be involved?

Sanders: -30 to -40 mV is right in the window for L-type channels.

Fry: John McCarron was saying that the voltage required for activation of STOCs was a bit more negative than this.

McCarron: It is -20 to -30 mV.

Nelson: This window is just a way to describe the current–voltage relationship. The open probability is finite; it doesn't drop to zero. The question is, is enough Ca^{2+} coming in? The open probability of the Ca^{2+} channel is increasing as you go from -100 to -50 mV. The question is whether enough Ca^{2+} is entering.

Isenberg: I have a more general problem. I don't think it is very fair to compare the space underneath the cell-attached membranes with the cytosol, when you are measuring Ca^{2+}.

McCarron: The question is that when we have a little bleb of membrane inside a patch pipette is the microdomain there comparable to the microdomain underneath the rest of the sarcolemma? This is a difficult question to answer. The only indirect evidence we have that the microdomain in the patch electrode is not changing during the course of the experiment is the amplitude of the unitary K^+ current. Presumably, if this bleb was almost sealed off and so the contents not in free diffusional equilibrium with the rest of the cell, the local K^+ concentration might change during depolarization and K^+ channel activity. If so, the unitary current might be expected to change with it. The unitary K^+ current remains completely unchanged throughout the course of the experiment.

Isenberg: In cell-free patches the SR is adjacent to the surface membrane. Does K^+ channel activity vary quite a lot from patch to patch?

McCarron: There is a variation. I know that channels are reported to be completely randomly distributed, but we don't always see them when we put patches on the cell.

Iino: You lost the entire response to caged $InsP_3$ after caffeine and ryanodine treatment. One reservation I have is that the $InsP_3$-induced Ca^{2+} release is a regenerative process. $InsP_3$ has a Ca^{2+} dependence, so Ca^{2+} release will have some positive feedback effect on the $InsP_3$ receptor. This is somewhat analogous to an action potential Na^+ channel corresponding to $InsP_3$ receptor and K^+ channel to SERCA. If you have a very small amount of Ca^{2+} in the store, it will probably have a very small positive feedback effect. It may not really trigger the $InsP_3$-induced Ca^{2+} release. Therefore, you need a very high $InsP_3$ concentration to show that the $InsP_3$-sensitive store is completely depleted. Indeed, we have done some experiments using single isolated intestinal smooth muscle cells. When we treated the cells with caffeine and ryanodine, the cells lost the response to $30\,\mu M$ carbachol. However, when the same cells were subsequently whole-cell-clamped to introduce a high concentration of $InsP_3$ from the pipette, we observed a definite $InsP_3$-induced Ca^{2+} response (Yamazawa et al 1992).

McCarron: In this particular experiment it was completely gone, but you are right; mostly it is pretty well wiped out, but there is 5–7% remaining. With respect to the regulation of the $InsP_3$ receptor by Ca^{2+}, the only thing we can say is that the experimental conditions were such that the bulk averaged Ca^{2+} concentration remained unchanged, whether the store was depleted via the ryanodine receptor or not. Of course, this doesn't rule out local Ca^{2+} events near the $InsP_3$ receptor which may regulate Ca^{2+} release.

Kotlikoff: Did you interpret your experiment as suggesting that there is a separate store?

Iino: Yes. We think there are two stores. One has both $InsP_3$ and ryanodine receptors; the other has only $InsP_3$ receptors. In the case of the uterus, the ryanodine receptor-expressing store constitutes about 5%. This is the reason we don't see a very big caffeine response.

Blaustein: Looking at the different stores, we have just heard the other view here that it is the ryanodine store that is separate and the other one that is combined. When we look at different cells, I think we'll see a whole spectrum of different stores.

Bolton: There's an implicit assumption of physical separateness of stores in the cell. The assumption here is that it is possible to deplete one and leave another filled, and therefore they are not connected. But if they are connected — as has been shown in other cells — and Ca^{2+} can move freely within the lumen of the SR, you have to come up with some other explanation. I alluded to this earlier in our discussions. It depends on the efficacy with which a system is connected to the opening of channels. If you apply something that is not very efficacious at producing InsP$_3$, it will only open a few channels and the SR Ca^{2+} pump will work so hard that you will only get partial depletion. On the other hand, if you activate a system that is well connected to InsP$_3$, or apply something like caffeine that opens up lots of ryanodine receptors, you will overwhelm the SR Ca^{2+} pump, and so there will be no Ca^{2+} in the store. I think the whole argument of separate Ca^{2+} stores is ignoring the interactions between the SR Ca^{2+} pump and how many pores are being opened in the membrane. It's a naïve idea that may be completely wrong.

McCarron: We do our best to have maximal concentrations of both InsP$_3$ and caffeine. We apply 10 mM caffeine from a puffer pipette. The cell receives a very high caffeine concentration, which is presumably saturating almost instantaneously. For the InsP$_3$ response we have set up our conditions to mimic the size of the transients that we get from caffeine. It is a maximal response that we are getting, at least as far as we can see from discharge from the store.

Bolton: Your conclusion is that it is the same store.

McCarron: No. We believe there are two stores. The store that is responsible for STOCs contains both receptors. There is also a separate store that contains ryanodine receptors alone that doesn't contribute to the generation of STOCs. When we deplete the ryanodine-sensitive stores using a combination of caffeine and ryanodine the response to InsP$_3$ is almost completely abolished. Then the counter protocol is to deplete the InsP$_3$-sensitive store by repetitively applying high concentrations of InsP$_3$ in the absence of external Ca^{2+}. Very quickly we lose the response to InsP$_3$ under these conditions. Under the conditions in which the InsP$_3$-sensitive store has been depleted the response to caffeine remains at 60–70% of its control value. It is difficult to explain these data in any other way than there being two separate stores.

Blaustein: And you require Ca^{2+} from the outside to refill the store once it empties, so this is different from the ryanodine system.

McCarron: It is essential to have external Ca^{2+} to refill the $InsP_3$ store, whereas the ryanodine-only store seems to be able to be replenished from the cytoplasmic Ca^{2+} available.

Bolton: If you hit it with carbachol, which generates a lot of $InsP_3$, can you completely deplete the store so there is no caffeine response?

McCarron: We haven't tried this.

Bolton: I would suggest you probably would. We have done this experiment and this is what we find.

Blaustein: Again, it may well depend on the cell type. We have looked at a totally different cell, an astrocyte, in culture and have seen exactly the same thing. We can empty the $InsP_3$ store and it requires extracellular Ca^{2+} to get it back, but the caffeine-sensitive store is still responsive. We can deplete it by using an agonist such as ATP.

McCarron: The difficulty with using agonists to generate $InsP_3$ is the additional second messengers that they produce. It is plausible that the carbachol could suppress ryanodine receptor function by activating protein kinase C. You would block the response to caffeine without gaining any information about the interaction of the two receptors on stores.

Iino: The experiment I was talking about earlier, regarding the compartments of the Ca^{2+} store, was carried out in skinned fibres in the absence of ATP (Iino 1990). Therefore, there is no interference from Ca^{2+} uptake. We just looked at the unidirectional efflux of Ca^{2+} from the Ca^{2+} store. We saw no caffeine response after high $InsP_3$ application, and we did see $InsP_3$ responses after 50 nM caffeine. It can be complicated in intact cells.

References

Bonev AD, Jaggar JH, Rubart M, Nelson MT 1997 Activators of protein kinase C decrease Ca^{2+} spark frequency in smooth muscle cells from cerebral arteries. Am J Physiol Cell Physiol 273:C2090–2095

Iino M 1990 Biphasic Ca^{2+} dependence of inositol 1,4,5-trisphosphate-induced Ca^{2+} release in smooth muscle cells of the guinea pig taenia caeci. J Gen Physiol 95:1103–1122

Jaggar JH, Nelson MT 2000 Differential regulation of Ca^{2+} sparks and Ca^{2+} waves by UTP in rat cerebral artery smooth muscle cells. Am J Physiol Cell Physiol 279:C1528–C1539

McCarron JG, Craig JW, Bradley KN, Muir TC 2002 Agonist-induced phasic and tonic responses in smooth muscle are mediated by $InsP_3$. J Cell Sci 115:2207–2218

Perez GJ, Bonev AD, Patlak JB, Nelson MT 1999 Functional coupling of ryanodine receptors to K_{Ca} channels in smooth muscle cells from rat cerebral arteries. J Gen Physiol 113:229–238

Perez GJ, Bonev AD, Nelson MT 2001 Micromolar Ca^{2+} from sparks activates Ca^{2+}-sensitive K^{+} channels in rat cerebral artery smooth muscle. Am J Physiol Cell Physiol 281:C1769–1775

Yamazawa T, Iino M, Endo M 1992 Presence of functionally different compartments of the Ca^{2+} store in single intestinal smooth muscle cells. FEBS Lett 301:181–184

Additional fluxes of activator Ca^{2+} accompanying Ca^{2+} release from the sarcoplasmic reticulum triggered by $InsP_3$-mobilizing agonists

Luc Raeymaekers, Bernd Nilius, Thomas Voets, Ludwig Missiaen, Kurt Van Baelen, Jo Vanoevelen and Frank Wuytack

Laboratorium voor Fysiologie, Katholieke Universiteit Leuven, Campus Gasthuisberg O/N, Herestraat 49, B3000 Leuven, Belgium

Abstract. Activation of phospholipase C (PLC)-linked receptors leads not only to Ca^{2+} release from the sarcoplasmic reticulum (SR) by inositol-1,4,5-trisphosphate ($InsP_3$), but also to Ca^{2+} entry via opening of receptor-activated Ca^{2+} channels (RACCs) and store-operated Ca^{2+} channels (SOCs), in addition to possible contributions of Ca^{2+} release from non-SR stores. We review recent results on these non-SR Ca^{2+} fluxes. In A7r5 smooth-muscle cells (SMCs), high $InsP_3$ concentrations release Ca^{2+} from a thapsigargin-insensitive store. Presumably this store corresponds to the Golgi and is filled by a Pmr1-type Ca^{2+} pump. Molecular candidates for RACCs and SOCs are found among the members of the TRPC channel family. Inoue and colleagues have recently demonstrated that in vascular SMCs TRPC6 is an essential part of a RACC that is activated by α-adrenergic stimulation via the diacylglycerol branch of phosphatidylinositol-4,5-bisphosphate hydrolysis. In TRPC4 knockout mice, contractility of SMCs appears unaffected. However, endothelium-dependent relaxation is impaired mainly due to lack of a SOC activity in endothelial cells. The best-characterized SOC current, mainly observed in blood cells, is I_{crac}. Recently, it has been proposed that CaT1 (TRPV5) forms at least part of the pore of CRAC. This view is challenged by data from our laboratory.

2002 Role of the sarcoplasmic reticulum in smooth muscle. Wiley, Chichester (Novartis Foundation Symposium 246) p 71–80

Agonist binding to many receptors leads to the activation of phospholipase C (PLC) enzymes in the plasma membrane, and the production of inositol-1,4,5-trisphosphate ($InsP_3$) and diacylglycerol (DAG) from phosphatidylinositol-4,5-bisphosphate (PIP_2), resulting in profound effects on the cytoplasmic Ca^{2+} concentration ($[Ca^{2+}]_i$). The best known of these effects is the release of Ca^{2+} from the endoplasmic reticulum (ER) via the opening of $InsP_3$ receptors and ryanodine receptors in the ER membrane. We will focus on additional less-well characterized effects initiated by the hydrolysis of PIP_2 that may contribute together with Ca^{2+} release from the ER to changes of $[Ca^{2+}]_i$. Additional mechanisms may include the release of Ca^{2+} from the Golgi membranes and the influx of Ca^{2+} via two different types of channels: receptor-activated Ca^{2+} channels (RACCs) opened by DAG, and store-operated channels (SOCs), opened by a signal originating in the ER when the luminal Ca^{2+} content is decreased following $InsP_3$-induced Ca^{2+} release.

Ca^{2+} in the Golgi apparatus of A7r5 smooth-muscle cells

Cultured A7r5 cells with the plasma membrane selectively permeabilized by using saponin accumulated $^{45}Ca^{2+}$ in an ATP-dependent way in the presence of mitochondrial inhibitors. Most of this $^{45}Ca^{2+}$ uptake presumably occurred in the ER via SERCA-type Ca^{2+} pumps because it was inhibited for about 90% by a low dose of the SERCA-specific inhibitor thapsigargin (TG). The identity of the TG-insensitive Ca^{2+} uptake mechanism is not clear. A possible candidate is a Ca^{2+} pump belonging to the Pmr1 family of P-type Ca^{2+}-transport ATPases that has been characterized in yeast, where it is localized in the membranes of the Golgi apparatus (Dürr et al 1998). In order to compare the TG-insensitive Ca^{2+} uptake in A7r5 cells to that of Pmr1-mediated transport, we have overexpressed the PMR-1 Ca^{2+} pump of *Caenorhabditis elegans* in COS-1 cells and studied the $^{45}Ca^{2+}$ uptake in the presence of TG following permeabilization of the plasma membrane. Overexpression of Pmr1 resulted in the appearance of an additional component of Ca^{2+} uptake that was insensitive to inhibition by TG (Van Baelen et al 2001, Missiaen et al 2001). Concentrations of the SERCA inhibitor cyclopiazonic acid (CPA, 10–100 μM) that inhibited ER Ca^{2+} uptake to the same extent as TG, had no effect on the Ca^{2+} uptake that remains in the presence of TG, demonstrating that these concentrations did not affect the Pmr1 pump. However, increasing the concentration of CPA above 100 μM inhibited the Pmr1-mediated component, with full inhibition at 1 mM. A very similar inhibition profile by CPA was observed for the TG-resistant $^{45}Ca^{2+}$ uptake in the A7r5 cells, indicating that this compartment could similarly be filled by a Ca^{2+} pump of the Pmr1 family. The expression of this Ca^{2+} pump in A7r5 cells was verified at the mRNA level by RT-PCR amplification.

The Ca^{2+} taken up in the TG-insensitive compartment could be released by the Ca^{2+} ionophore A23187. Only a partial release was observed with $InsP_3$. The concentration of $InsP_3$ required for half-maximal release was about two times higher than that required for half-maximal activation of Ca^{2+} release from the compartment filled by the SERCA pump. A very similar pattern was seen for the $InsP_3$ effect on the additional Ca^{2+} compartment created in COS-1 cells by overexpression of the worm Pmr1. Although still indirect, these data strongly suggest the presence in A7r5 cells of a Ca^{2+} store, presumably the Golgi, that can partially be released by InsP$_3$.

Agonist-induced Ca^{2+} entry into smooth-muscle cells

In most cell types, including smooth-muscle cells, activating agonists induce an entry of Ca^{2+}. An important component of the Ca^{2+} entry is the basis for the sustained elevation of [Ca^{2+}]$_i$ that follows the initial rapid rise of [Ca^{2+}]$_i$ caused by Ca^{2+} release from the ER. Recently it has become clear that members of the large family of proteins that are structurally related to the *Drosophila* TRP channels function in mammalian cells as Ca^{2+}-permeable cation channels and may contribute to the agonist-induced Ca^{2+} influx. The TRP family can be divided in three subfamilies, indicated as TRPC, TRPV and TRPM (for review see Clapham et al 2001). The mammalian members of the TRPC subfamily consists of seven paralogues related to the *Drosophila* TRP and TRPL. Of these, mainly TRPC3, TRPC4 and TRPC6 are expressed in smooth muscle cells (Walker et al 2001). Recent functional data on these three types of channels will be presented. In addition, evidence that CaT1 (TRPV5) may represent the Ca^{2+}-release-activated Ca^{2+} (CRAC) channel (Yue et al 2001) will be reconsidered.

Mammalian TRPC3 and TRPC6 contribute to Ca^{2+} entry via RACCs

Cells of the bovine endothelial CPAE cell line do not express TRPC3. These cells respond to purinergic activation with a transient rise of [Ca^{2+}]$_i$. This transient signal was converted into a more sustained increase of [Ca^{2+}]$_i$ after transfection with a TRPC3 construct. Concomitantly, an inward current appeared that presented a slightly positive reversal potential, whose amplitude was augmented in divalent cation-free solution (DVF) and which was blocked by further removal of Na$^+$ ions. The SERCA inhibitors TG and BHQ failed to activate the current, whereas a phospholipase C inhibitor inhibited the current that had been activated by ATP. These data strongly suggest that TRPC3 forms non-selective cation channels that are opened by receptor activation independent of store depletion (Kamouchi et al 1999). In line with these data, it was shown by Hofmann et al (1999) that transfection of CHO-K1 cells with TRPC3 or the closely related TRPC6 induces a non-selective cation channel that is activated by DAG, one of the products of phospholipase C activation. TRPC6 was further characterized both after transfection and in the native state in vascular smooth-muscle cells (Inoue et al 2001). Confirming previous data, heterologously expressed TRPC6 induced an inward current and Ca^{2+} entry that was activated by DAG. Activation of α-adrenergic receptors in cultured rat portal vein smooth-muscle cells induced a current with similar properties, concomitantly with Ca^{2+} entry.

These effects were inhibited specifically by antisense nucleotides directed against TRPC6, whereas no effect was observed after antisense treatment directed against TRPC1 and TRPC7, which were found to be coexpressed with TRPC6 in these cells. These results have elucidated a complete pathway of activation of one of the RACCs, starting with α-adrenergic receptor activation and opening of a molecularly characterized channel via activation of phospholipase C and the production of DAG.

TRPC4 may form part of a store-operated channel (SOC) complex

TRPC4 knockout mice have been generated by Freichel et al (2001). Isolated aortas from these animals show a marked defect in endothelium-dependent relaxation. This defect could be attributed to a diminished store-operated Ca^{2+} entry in the endothelial cells. The defect is accompanied by a decreased Ca^{2+} entry and by the lack of a current which is mainly carried by Ca^{2+}. Ca^{2+} release was not affected in control endothelial cells by passive store depletion using SERCA inhibitors in combination with active store depletion by dialysing the cells with $30\,\mu M$ $InsP_3$. SOC currents and Ca^{2+} entry in Ca^{2+} reapplication experiments were blocked by $1\,\mu M$ La^{3+}. In line with the proposed role of this Ca^{2+} entry in the later phase of the Ca^{2+} signal, the agonist-induced rise of $[Ca^{2+}]_i$ was reduced in endothelial cells of TRPC4-deficient mice. This effect, as well as the endothelium-dependent relaxation of the aorta, were blocked by $1\,\mu M$ La^{3+}, as observed for the store-operated Ca^{2+} entry and the Ca^{2+} current.

Is CaT1 (ECaC2) the I_{CRAC} channel?

One of the best characterized SOC currents is I_{CRAC}, a Ca^{2+} selective current, that has been observed mainly in blood cells in conditions of depletion of $InsP_3$-sensitive Ca^{2+} stores combined with very low $[Ca^{2+}]_i$ (Parekh & Penner 1997). It has been proposed by Yue et al (2001) that TRPV5 (CaT1, ECaC2) comprises all or part of the I_{CRAC} pore. A study of the related TRPV6 (CaT2, ECaC1) did not provide any evidence for a store-operated activation mechanism for this heterologously expressed protein (Vennekens et al 2000, Nilius et al 2000). Therefore, a detailed study was undertaken of the activation mechanism and of the pore properties of CaT1, expressed in HEK-293 cells, and compared to those of I_{CRAC} in RBL-2H3 cells under identical experimental conditions (Voets et al 2001). Confirming data of Yue et al (2001), many similarities between CRAC and CaT1 were observed. However, many of these common characteristics, such as the anomalous mole fraction behaviour and the single-channel conductance to Na^+ ions, are also shared by voltage-operated Ca^{2+} channels. CRAC but not CaT1 expressed in HEK-293 cells is sensitive to store depletion: CRAC in RBL cells is activated by application of ionomycin, whereas currents through CaT1 are inhibited due to blocking elevation of intracellular Ca^{2+} (Nilius et al 2001). More importantly, both channels can be distinguished by several other features concerning pore properties: the current in DVF is maintained over more than ten minutes in CaT1-transfected HEK cells but rapidly decays when CRAC channels are permeated by monovalent cations. CaT1 but not CRAC shows a voltage-dependent gating at hyperpolarizing voltage steps which is mediated by unblocking from intracellular Mg^{2+}. CRAC gating in DVF solution is completely voltage independent in a wide range of intracellular Mg^{2+} concentrations. CRAC and CaT1 have completely different permeation properties for Cs$^+$ and can be discriminated by their striking differences in rectification of monovalent currents at positive potentials. Furthermore, CRAC and CaT1 show a variety of different pharmacological features, among them a completely different response to 2-APB. The most likely interpretation of these data is that CRAC and CaT1 are two distinct channels with a number of common properties but encoded by different genes.

References

Clapham DE, Runnels LW, Strübing C 2001 The TRP ion channel family. Nat Rev Neurosci 2:387–396

Dürr G, Strayle J, Plemper R et al 1998 The medial-Golgi ion pump Pmr1 supplies the yeast secretory pathway with Ca^{2+} and Mn^{2+} required for glycosylation, sorting, and endoplasmic reticulum-associated protein degradation. Mol Biol Cell 9:1149–1162

Freichel M, Suh SH, Pfeiffer A et al 2001 Lack of an endothelial store-operated Ca^{2+} current impairs agonist-dependent vasorelaxation in TRP4$^{-/-}$ mice. Nat Cell Biol 3:121–127

Hofmann T, Obukhov AG, Schaefer M, Harteneck C, Guderman T, Schultz, G 1999 Direct activation of human TRPC6 and TRPC3 channels by diacylglycerol. Nature 397:259–263

Inoue R, Okada T, Onoue H et al 2001 The transient receptor potential protein homologue TRP6 is the essential component of vascular α_1-adrenoreceptor-activated Ca^{2+}-permeable cation channel. Circ Res 88:325–332

Kamouchi M, Philipp S, Flockerzi V et al 1999 Properties of heterologously expressed hTRP3 channels in bovine pulmonary artery endothelial cells. J Physiol 518:345–358

Missiaen L, Van Acker K, Parys JB et al 2001 Baseline cytosolic Ca^{2+} oscillations derived from a non-endoplasmic reticulum Ca^{2+} store. J Biol Chem 276:39161–39170

Nilius B, Vennekens R, Prenen J, Hoenderop JGJ, Bindels RJ, Droogmans G 2000 Whole-cell and single cell monovalent cation currents through the novel rabbit epithelial Ca^{2+} channel ECaC. J Physiol 527:239–248

Nilius B, Vennekens R, Prenen J, Hoenderop JGJ, Droogmans G, Bindels 2001 The single pore residue Asp752 determines Ca^{2+} permeation and Mg^{2+} block of the epithelial Ca^{2+} channel. J Biol Chem 276:1020–1025

Parekh AB, Penner R 1997 Store depletion and calcium influx. Physiol Rev 77:901–930

Van Baelen K, Vanoevelen J, Missiaen L, Raeymaekers L, Wuytack F 2001 The Golgi PMR1 P-type ATPase of *Caenorhabditis elegans*. Identification of the gene and demonstration of calcium and manganese transport. J Biol Chem 276:10683–10691

Vennekens R, Hoenderop JGJ, Prenen J et al 2000 Permeation and gating properties of the novel epithelial Ca^{2+} channel. J Biol Chem 275:3963–3969

Voets T, Prenen J, Fleig A et al 2001 CaT1 and the calcium-release activated calcium channel manifest distinct pore properties. J Biol Chem 276:47767–47770
Walker RL, Hume JR, Horowitz B 2001 Differential expression and alternative splicing of TRP channel genes in smooth muscles. Am J Physiol 280:C1184–C1192
Yue L, Peng J-B, Hediger MA, Clapham DE 2001 CaT1 manifests the pore properties of the calcium-release-activated calcium channel. Nature 410:705–709

DISCUSSION

Lompré: I have a question concerning the knockout mice that show defective relaxation. How can this be related to the work of Shull's group (Liu et al 1997) on the SERCA3-defective mice? They also see that the endothelial-dependent relaxation is depressed in these mice.

Raeymaekers: We don't have any data on this. We didn't look at SERCA isoforms in endothelial cells. It is an interesting question though.

Lompré: Normally endothelial cells have SERCA2B and SERCA3. Liu et al (1997) showed that SERCA3-defective mice have the same defect in relaxation that you describe. If they add NO donors, they also restore relaxation as for TRP4. Could it be that there is a relationship between this pool of Ca^{2+} SERCA3 and TRP4?

Raeymaekers: The data indicate that it is SERCA3 that fills the stores that are involved in the endothelium-dependent relaxation. At face value, the separate data indicate that this store could be connected to TRPC4.

Taylor: My understanding of the David Clapham paper was that the evidence suggesting that CaT1 might be I_{CRAC} was based on the comparison of electrophysiology and how similar they look. If this falls apart, then so does the argument. It seems to me that when people are expressing TRPs, they often get a different answer according to which cell they express them in. They end up explaining it away with heterologerization or whatever. Is any of this likely to contribute to the differences that you see?

Raeymaekers: It might be important in this respect to mention that there is a recent paper indicating that many previous measurements of I_{CRAC} have been compromised by contribution of another current, which is due to TRPL7 (Nadler et al 2001). This is a non-specific Ca^{2+} channel that is activated by MgATP. In the data presented here, possible contributions of this current have been excluded.

Sanders: All those expression systems have their own TRPs. There aren't any pure expression systems for TRP channels except for baculovirus.

Isenberg: I have a question that relates to the first part of your paper. Could you help us to understand better the role of the Golgi network in storing Ca^{2+}? This information about Ca^{2+}-binding proteins in the lumen of the Golgi is interesting. Does Golgi have the same Ca^{2+}-binding proteins as the SR?

Raeymaekers: No, it has a specific set of Ca^{2+} binding proteins. I don't know of any calreticulin in the Golgi. This protein also has a KDEL retrieval signal.

Somlyo: Proteins packaged in the Golgi and stored in vesicles can have a high Ca^{2+} content. For example the prohormone of atrial natriuretic factor contains a bunch of vicinal glutamates that result in fairly high Ca^{2+} binding activity. I wonder whether these are the proteins that may be cell specific and are binding Ca^{2+} in the Golgi. I have a related question: did you see whether the uptake into this compartment was oxalate facilitated? Oxalate-treated permeabilized smooth muscle has huge calcium oxalate deposits in the perinuclear region, presumably within the Golgi.

Raeymaekers: The Ca^{2+} uptake is stimulated by oxalate, but we can't fully exclude the possibility that by overexpression there is some overflow of PMR1 into the ER.

Somlyo: The ER would also be oxalate-facilitated. You would have to be able to separate the two.

Sanders: Another structural question concerns the idea from Lutz Birnbaumer that the InsP$_3$ receptor is regulating TRP channels directly, and that there is a binding site in TRP channels for part of the InsP$_3$ receptor. It seems to me that spatially the SR is too far from the plasma membrane for this to occur: 20 nm at best is too far for a molecular interaction.

Somlyo: There is another problem. We only have relatively low-resolution information about where these channels are. We can say that they are on the SR. But if you ask me about a 15 nm wide SR section, whether the InsP$_3$ receptor is on the plasma-membrane-related side or not, I couldn't say. Or more to the point, where the SR is surrounding a mitochondrion, separated from it by about the same distance, I couldn't tell you whether the InsP$_3$ release channels are adjacent and close to the mitochondrion or whether they are on the other surface, away from the mitochondrion. This will make a lot of difference whether the SR releases Ca^{2+} close to the mitochondrion or acts as a barrier protecting the mitochondrion from Ca^{2+}. High-resolution information is not available.

Sanders: But if the closest membrane is 20 nm away from the SR, that's too far for a molecular interaction. What about this possibility that InsP$_3$ receptors are at the plasma membrane?

Somlyo: I'll stay away from that issue until the paper is published.

Brading: How close would the caveolae membrane get to the SR where it butts through?

Somlyo: They can get quite close.

Brading: So if the TRP channels were in the caveolae they would be close to the SR.

Raeymaekers: There is strong evidence that they are in

Somlyo: Are they caveolae or rafts?

Taylor: This whole argument about conformational coupling has to live with the fact that there is the DT40 cell line in which all InsP$_3$ receptors have been knocked out, and capacitative Ca^{2+} entry is unaffected.

Somlyo: But Ogawa recently published a paper about skeletal muscle, showing that there is some thing similar happening there: SOCs mediate Ca^{2+} entry (Kurebayashi & Ogawa 2001). InsP$_3$ does not release Ca from the SR of striated muscle.

Walsh: I have a question related to the DAG activation of TRP6. What is known about this? Is it a direct activation? I thought you said that it wasn't affected by PKC inhibitors. Just by sequence gazing, can you see a DAG binding site?

Raeymaekers: I'm not aware of any information on this. It could be an indirect effect.

Walsh: Is the α1 adrenoceptor activating the channel via DAG specifically?

Raeymaekers: Yes.

Hirst: I would like this all put into context of how transmitters work. You talked about cation-selective channels with TRP receptors; we have heard earlier about InsP$_3$ levels in intestinal muscle. I know no evidence that either of these things is activated by neuronally released transmitters in a wide range of tissues. So I don't know of a junction potential that uses a cation-selective channel, nor do I know of an intestinal muscle where InsP$_3$ levels go up. As far as I know neuronally released transmitters go through interstitial cells of Cajal (ICCs). No transmitter gets to the smooth muscle.

McCarron: In a knockout mouse lacking the type 1 InsP$_3$ receptor, the excitatory junction potentials are reduced in gastrointestinal smooth muscle.

Hirst: That tissue transmission goes through ICCs, and if you pull the ICCs out there is no transmission.

McCarron: The response to acetylcholine is also reduced. The membrane depolarization is knocked down by about 50% when there is no type 1 InsP$_3$ receptor.

Hirst: What relevance does that have? No one is denying that there is an InsP$_3$-dependent pathway there. What I'm asking is when is it used? When are these cation selective channels used?

Kotlikoff: Didn't William Large show the two components of the postsynaptic response, one being the cation-selective component, the other being a Cl$^-$ selective component?

Hirst: He did when he added noradrenaline. He got a Cl$^-$ response and an increase in membrane resistance. You don't see a cation-selective channel activated by a neuronally released transmitter.

Kotlikoff: So your concern is not that the neurotransmitter can't release it; it is whether it does it when the nerve is being stimulated.

Nelson: There is a difference between stimulating a nerve and drowning a piece of tissue in 1 mM noradrenaline.

Hirst: Or acetylcholine. As far as I know, acetylcholine doesn't get to intestinal muscles when it comes out of the nerves. I want the observations put in context.

Bolton: What am I supposed to say? We have not done the experiments.

Sanders: If you get rid of the interstitial cells there is no further response to nerve stimulation. You can get a nice response to acetylcholine. Then you go back to the wild-type animal and you can block the breakdown of acetylcholine and then get a smooth muscle response. The point is that while there are synaptic connections in GI smooth muscle, synaptic connections between the motor neurons and smooth muscles are not so common. The major synaptic connections are between the motor neurons and the interstitial cells.

Somlyo: Moving away from the intestine, I seem to remember that many years ago Geoff Burnstock convinced a poor electron microscopist to serially section a smooth muscle cell in the vas deferens, which he thought would be about 25 μm long. It turned out to be several times longer than this, and they showed that in the vas deferens smooth muscle there were neural synapses.

Brading: It's when you get these densely innervated smooth muscles where the body wants to activate them by nerves and nothing else, rather than regulate them, that you get close junctions and excitatory junction potentials. But in the gut, where the nerves are not actually doing things and it isn't densely innervated, they are modulating ongoing activity that is being generated presumably through these interstitial cells. Then you have a completely different pattern.

Sanders: Where in the gut are the nerves not doing anything?

Brading: They are not doing it directly onto the smooth muscles. You don't use your nerves to contract a bit of your longitudinal gut muscle in the same way as you switch on your vas deferens when you want to ejaculate, do you?

Sanders: That's a personal question.

Bolton: When you stimulate a nerve, you say the acetylcholine acts on receptors that are on ICCs, but if you separate all the cells, presumably we have always been working on ICCs.

Sanders: No, the smooth muscle cells have the receptors. It's the old concept of spare receptors. Whether or not they are coupled to the same mechanisms, I think David Hirst's point is that neurally released transmitters may do something quite different to bath-applied transmitters.

Brading: We are going back 10 years to where the cardiac muscle field was. It's the same argument; we are just running a bit late.

Hirst: I was hoping that someone could put this modern work into context.

Bolton: What you are saying is that all these receptors are there but they have no function, or at least not the function that we are attributing to them.

Kotlikoff: One should be careful about extrapolating experiments from the GI tract to all tissues. Those experiments haven't been done in many tissues. One of the things that Professor Ryugi Inoue has done very nicely is show that at least the single cell responses to neurotransmitter are very well mimicked biophysically by the TRP analogue. The next question is how important is this physiologically? This isn't known. But we do now have a molecular identity for what we see in single cells. One other phenomenon that we have discussed widely is the store-operated Ca^{2+} release in smooth muscle. However, we lack a biophysical analogue for this and in most single cell experiments we don't have a reliable response that can be identified as a store-release response in itself. If we release Ca^{2+} in many single smooth muscle cells we don't see a current that many people have reported for stimulation of neurotransmitter receptors. This is one distinction that we should make in terms of recognizing what we do know.

Brading: Professor Iino is one of the few people actually trying to look at what happens when nerves are stimulated, as opposed to just applying excess transmitter. He does do some very elegant work stimulating nerves and actually seeing what is happening, although I might argue with the frequencies with which he stimulates them (they're probably a little high for real life).

Nelson: The pressure-induced depolarization in arteries is blocked by treatment with antisense to TRP6 (Welsh et al 2002).

Hirst: I suppose that is actually what I'm asking. I hear about all these channels, and I have no idea what they are doing. That's the first time I've heard someone describe a TRP channel doing something that I regard as a physiological event.

Brading: It is a reflection of the fact that the molecular biology is often a lot easier to do than the electrophysiology.

References

Kurebayashi N, Ogawa Y 2001 Depletion of Ca^{2+} in the sarcoplasmic reticulum stimulates Ca^{2+} entry into mouse skeletal muscle fibres. J Physiol 533:185–199

Liu LH, Paul R J, Sutliff RL et al 1997 Defective endothelium-dependent relaxation of vascular smooth muscle and endothelial cell Ca^{2+} signalling in mice lacking sarco(endo)plasmic reticulum Ca^{2+}-ATPase isoforms. J Biol Chem 272:30538–30545

Nadler M J, Hermosura MC, Inabe K et al 2001 LTRPC7 is a Mg.ATP-regulated divalent cation channel required for cell viability. Nature 411:590–595

Welsh DG, Morielli AD, Nelson MT, Brayden JE 2002 Transient receptor potential channels regulate myogenic tone of resistance arteries. Circ Res 90:248–250

Molecular candidates for capacitative and non-capacitative Ca^{2+} entry in smooth muscle

Ryuji Inoue and Yasuo Mori*

*Department of Pharmacology, Graduate School of Medical Sciences, Kyushu University, Fukuoka 812-8582 and *School of Life Science, The Graduate University for Advanced Studies, Okazaki 444-8585, Japan*

Abstract. Recent investigations have revealed that mammalian homologues of transient receptor potential (TRP) protein (TRP1–7) are promising candidates for Ca^{2+} entry mechanisms (or channels) associated with various metabotropic G protein-coupled receptors (GPCRs) in smooth muscle, stimulation of which generates lipid second messengers and depletes internal stores. RT-PCR and immunocytochemical experiments have demonstrated that although the level of expression varies depending on tissues, the major TRP isoforms expressed in smooth muscle are TRP4, 6 and 7. In some vascular preparations, the significant expression of TRP1 mRNA and protein is also detected. Consistent with these findings, recent functional studies using TRP6- and TRP1-specific antisense oligonucleotides and antibodies have suggested that TRP6 is the essential component of α_1-adrenoceptor activated, store depletion-independent Ca^{2+} entry channels, while TRP1 is partly involved in Ca^{2+} entry associated with store depletion or capacitative Ca^{2+} entry. In addition, coexpression of different TRP isoforms results in the appearance of cation channels showing novel properties reminiscent of some native GPCR-activated Ca^{2+}-permeable non-selective cation channels. Thus, at present, TRP proteins may be the most important clues for elucidating the molecular entities of receptor- and store-operated Ca^{2+} entry mechanisms in smooth muscle and their roles in smooth muscle functions.

2002 Role of the sarcoplasmic reticulum in smooth muscle. Wiley, Chichester (Novartis Foundation Symposim 246) p 81–90

In most types of smooth muscle cells (SMCs), stimulation of G protein-coupled receptors (GPCRs) results in a large Ca^{2+} transient and a sustained increase in intracellular Ca^{2+} concentration, which are thought to be mediated by inositol-1,4,5-trisphosphate (InsP$_3$)-mediated Ca^{2+} release and accompanying trans-membrane Ca^{2+} entry, respectively. Various experimental techniques and protocols have revealed that although there appear to be some differences in the extent of their contribution, at least three distinct Ca^{2+}-mobilizing mechanisms can account for this Ca^{2+} entry (Carl et al 1996, Kuriyama et al 1998, Kotlikoff et al 1999, Sanders 2001). These include receptor-operated Ca^{2+}-permeable cation channels (ROCCs), voltage-operated Ca^{2+} channels (VOCCs) that are indirectly activated by membrane depolarization resulting from cation entry through ROCCs, and a Ca^{2+} entry mechanism associated with store depletion (i.e. capacitative Ca^{2+} entry; CCE). The importance of VOCCs in inducing smooth muscle contractions during GPCR stimulation, has been widely accepted by the finding that dihydropyridines such as nifedipine, highly selective blockers for L-type VOCCs, significantly reduce the sustained elevation in $[Ca^{2+}]_i$ induced by GPCR agonists and attenuate concomitant contractions (Karaki et al 1997, but see Curtis & Schofield 2001). In contrast, the absence of selective blockers for ROCCs and CCE has strongly hampered their distinction from other Ca^{2+}-transporting mechanisms and thus a clear understanding about their roles in regulating smooth muscle functions. Previous electrophysiological experiments have shown that Ca^{2+}-permeable channels recorded from smooth muscle are almost exclusively of ROCC type that are activated by lipid second messengers such as diacylglycerol or direct interaction with G proteins in a store depletion-independent manner (in other words, non-capacitative Ca^{2+} entry pathway). These channels exhibit relatively large unitary conductances (tens of picosiemens) and a weak permeation preference for divalent cations (e.g. Ca^{2+}) to monovalent cations (Na^+). They are thus thought to serve as membrane depolarizers indirectly triggering voltage-dependent Ca^{2+} entry and presumably as direct Ca^{2+} entry routes (Carl et al 1996, Kuriyama et al 1998, Kotlikoff et al 1999, Sanders 2001). In contrast, in most smooth muscle tissues in which CCE has been demonstrated by the Ca^{2+} fluorimetric method, it has proved difficult to record Ca^{2+} currents of the sort of amplitude that could account for a large sustained increase in $[Ca^{2+}]_i$, although there are some exceptions where inward currents are induced by pharmacological agents capable of depleting intracellular stores (Gibson et al 1998, Table 1). However, in most of these cases, the observed inward currents are non-selective to cations and their magnitude is very small even at maximum activation. In other cases, recording conditions used seem inappropriate for eliminating the secondary induction of Ca^{2+}-dependent conductances. It is therefore uncertain whether CCE in smooth muscle would really reflect Ca^{2+} entry through concurrently activated, small store depletion-associated cation channels (see below).

The molecular entities of VOCCs have been well characterized. Ten distinct genes have so far been cloned, six of which have been identified in smooth muscle (Hofmann et al 1999, Ertel et al 2000). In contrast, the molecular counterparts of ROCCs and CCE have long been elusive, due probably to the lack of specific ligands/blockers and the presence of complex entangled regulatory mechanisms. The transient receptor potential (TRP) protein gene

TABLE 1 Properties of store depletion-activated currents (channels) in smooth muscle

Cell type	Current amplitude	Reversal potential (Er)	Conductance	$I\text{–}V$	Activators	Blockers	Recording conditions	References
Mouse anococcygeus	-3 pA at -40 mV	$+31$ mV [$+18$ mV]	—	sl. I/R	CPA caffeine CCh ryanodine	Cd^{2+}(IC_{50}:68 μM) Ni^{2+}(IC_{50}:98 μM) SKF(IC_{50}:6.4 μM) La^{3+}:no effect	W/C (10BAPTA or EGTA)	Waymann et al 1996 Waymann et al 1998
Mouse and rabbit aortae	50 pA (Ca-free)	$\sim$0 mV	3 pS	sl. O/R (linear)	TG CIF BAPTA	La^{3+}(2 mM) Ni^{2+}(2 mM)	C/A I/O	Trepakova et al 2000 Trepakova et al 2000
Pulmonary artery	$\sim$400 pA at -60 mV	0 mV ($\sim$20 mV)	5.3 pS	sl. O/R	Phe CPA	La^{3+}(50 μM) Ni^{2+}(1 mM)	W/C (1.15 mM EGTA +1 mM Ca)	McDaniel et al 2001
Choroidal artery	-10 pA at -80 mV	$+29$ mV [$+7$ mV]		I/R	CPA Caffeine	nifedipine (1 μM)	P-W/C	Curtis & Scholfield 2001
Rabbit portal vein		20 mV	2.5 pS	—	Spont. act. CPA CaM-ant.	—	C/A	Albert & Large 2001

I/R, inward rectification; O/R, outward rectification; CPA, cyclopiazonic acid; TG, thapsigargin; CCh, carbamylcholine; CIF, Ca^{2+} influx factor; CAM-ant., calmodulin antagonist; W/C, whole-cell recording; C/A, cell-attached recording; P-W/C, perforated whole-cell recording.

(*trp*) and its homologue *trp*-like (*trpl*) were originally identified in the investigation of abnormal visual transduction in *Drosophila* eye, and encode Ca^{2+}-permeable cation channels activated in a store depletion-dependent and independent fashion, respectively (Hardie & Minke 1993). Since *Drosophila* phototransduction involves a cascade-like activation of rhodopsin, G_q G protein and phospholipase $C\beta$, and Ca^{2+} mobilization from both intracellular stores and the extracellular space (Hardie & Minke 1993), *trp* genes were soon recognized to be promising candidate genes for Ca^{2+} entry mechanisms associated with GPCR-mediated phosphoinositide hydrolysis in vertebrates. There are now seven mammalian homologues of TRP (TRP1–7) cloned on the basis of homology searches of expressed sequence tag databases and PCR using primers annealing to conserved regions of TRP (Birnbaumer et al 1996, Okada et al 1999). Amongst them, TRP1, 4 and 5 have been found to be activated by agents depleting internal stores, whereas TRP3, 6 and 7, closely related members of TRP subfamily, can be activated by stimuli other than store depletion signals such as diacylglycerol and Ca^{2+} (Hofmann et al 2000, Clapham et al 2001, Montell 2001). The distribution pattern of TRP isoforms depends on the tissue. For example, TRP4 and 5 seem to be preferentially expressed in the brain, while TRP3, 6 and 7 are found in extra-brain tissues such as the heart, lung and eye (Clapham et al 2001, Montell 2001). In smooth muscle tissues, quantitative RT-PCR technique has shown that TRP4 (and its short splice variant), TRP6 and TRP7 (and its two short forms of splice variants) are expressed in both gut and vascular smooth muscles such as stomach, intestine, colon and pulmonary and renal arteries (Walker et al 2001). Abundant TRP6 mRNA expression relative to TRP1, 3 and 4 has also been demonstrated in portal vein myocytes (Inoue et al 2001), and TRP1 transcripts have been detected in several different vascular tissues including the aorta, pulmonary artery, pial artery, human left internal mammary artery, and portal vein (Xu & Beech 2001,McDaniel et al 2001). It is unlikely that these TRP mRNA transcripts are the artefacts of overamplification by PCR, since immunoreactivity to TRP isoform-specific antibodies can also be detected in some gastrointestinal and vascular smooth muscle tissues by laser scanning confocal microscopy (Inoue et al 2001, Xu & Beech 2001, Fig. 1).

The possible function of TRP proteins in smooth muscle has been investigated by a number of recent studies. Inoue et al (2001) performed a detailed comparison of currents due to recombinantly expressed TRP6 in HEK293 cells and one of native ROCCs, α_1-adrenoceptor activated cation currents previously recorded from rabbit portal vein myocytes (Helliwell & Large 1997). Both currents are activated by diacylglycerol and show almost identical current–voltage relationships, cation permeabilities and pharmacological sensitivities. Furthermore, TRP6 immunoreactivity, which is localized near the sarcolemma of portal vein myocytes, was markedly reduced after 3–5 day treatment with

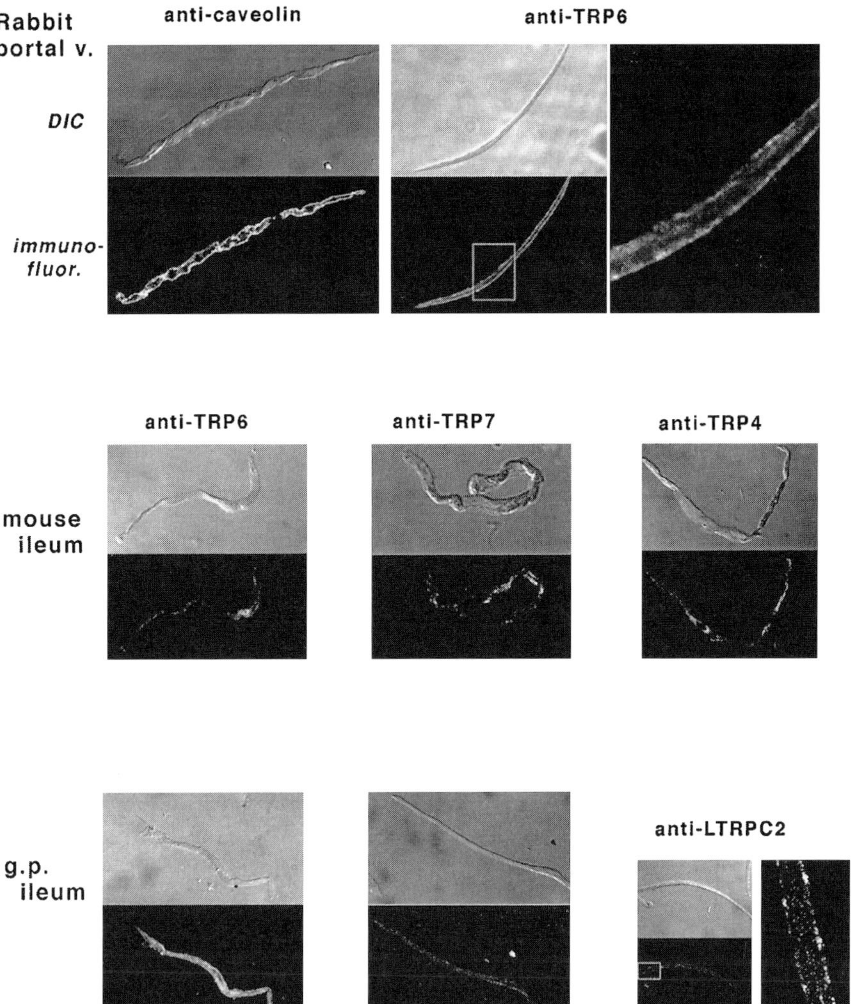

FIG. 1. Expression of TRP proteins in various smooth muscles. Single myocytes dissociated from rabbit portal vein, guinea-pig and mouse ileum were immunostained with antibodies specific to TRP4, TRP6, TRP7 and LTRPC2. Differential interference contrast and immunofluorescence images with a laser scanning confocal microscope. For further details see Inoue et al (2001).

TRP6-specific antisense oligonucleotides, and consequently, cationic current and Ba^{2+} influx evoked in response to α_1-adrenoceptor activation were both greatly attenuated. Importantly, Ba^{2+} influx induced by thapsigargin, which is an indication of store operated divalent cation entry, was not significantly affected by this procedure, thus suggesting that TRP6 is the essential component of native ROCC which may serve as a store depletion-independent Ca^{2+} entry pathway during α_1-adrenoceptor stimulation.

There is, however, little supportive evidence that TRP6 may also contribute to other types of ROCCs in smooth muscle. For example, muscarinic cation channels ubiquitously found in the whole gut as well as in tracheal smooth muscle show considerably different properties as compared with recombinant TRP6. These channels are activated by M_2 muscarinic receptor stimulation through pertussis toxin-sensitive G proteins without mediation of lipid messengers, inhibited by flufenamate, and exhibit a characteristic hyperpolarization-induced current relaxation (Carl et al 1996, Kuriyama et al 1998, Kotlikoff et al 1999). This overall profile is very different from that observed for TRP6. However, an important suggestion to reconcile these apparent discrepancies has been provided by a recent study: TRP isoforms would form heteromultimeric complexes and could generate cationic channels having entirely new properties. It has been reported that coexpression of TRP1 with TRP4 or TRP5 results in a hyperpolarization-inhibitable cation current (Strübing et al 2001) which resembles muscarinic cation currents described above and also those identified in some regions of the CNS (Haj-Dahmane & Andarde 1996). We therefore investigated how heterologous coexpression of TRP isoforms identified from gut smooth muscle affect the properties of cationic currents arising due to respective TRP proteins. Surprisingly, coexpression of TRP6 with LTRPC2, a TRP-related protein expressed in guinea-pig ileal myocytes (Fig.1, Perraud et al 2001), reproduced a number of hallmarks of muscarinic cation currents such as a rapid activation, pertussis toxin sensitivity and inhibition by flufenamate. These results give us an important intuition that native ROCCs in smooth muscle may be more-or-less in a heterooligomeric configuration comprised of different TRP isoforms where TRP6 may play a central role. This certainly requires further scrutiny and more detailed studies.

Another interesting finding about the roles of TRP proteins in smooth muscle has come from a recent study using the TRP1-specific antibody targeted to the putative outer vestibule of TRP1 channels residing between the fifth and sixth putative membrane-spanning domains. Treatment of rabbit pial arterioles with this antibody resulted in a significant reduction of Ca^{2+} and Ba^{2+} influxes evoked by thapsigargin, thus suggesting that functional blockade of ion permeation through the TRP1 channel pore attenuated Ca^{2+} entry activated by store depletion (Xu & Beech 2001). A similar contribution of TRP1 to store-operated Ca^{2+} entry has also been implicated in proliferation of cultured pulmonary artery myocytes, where both the magnitude of CCE and the level of TRP1 mRNA expression are enhanced in a parallel fashion (Golovina et al 2001). These results seem consistent with the idea that TRP1 may serve as a pore-forming subunit of the

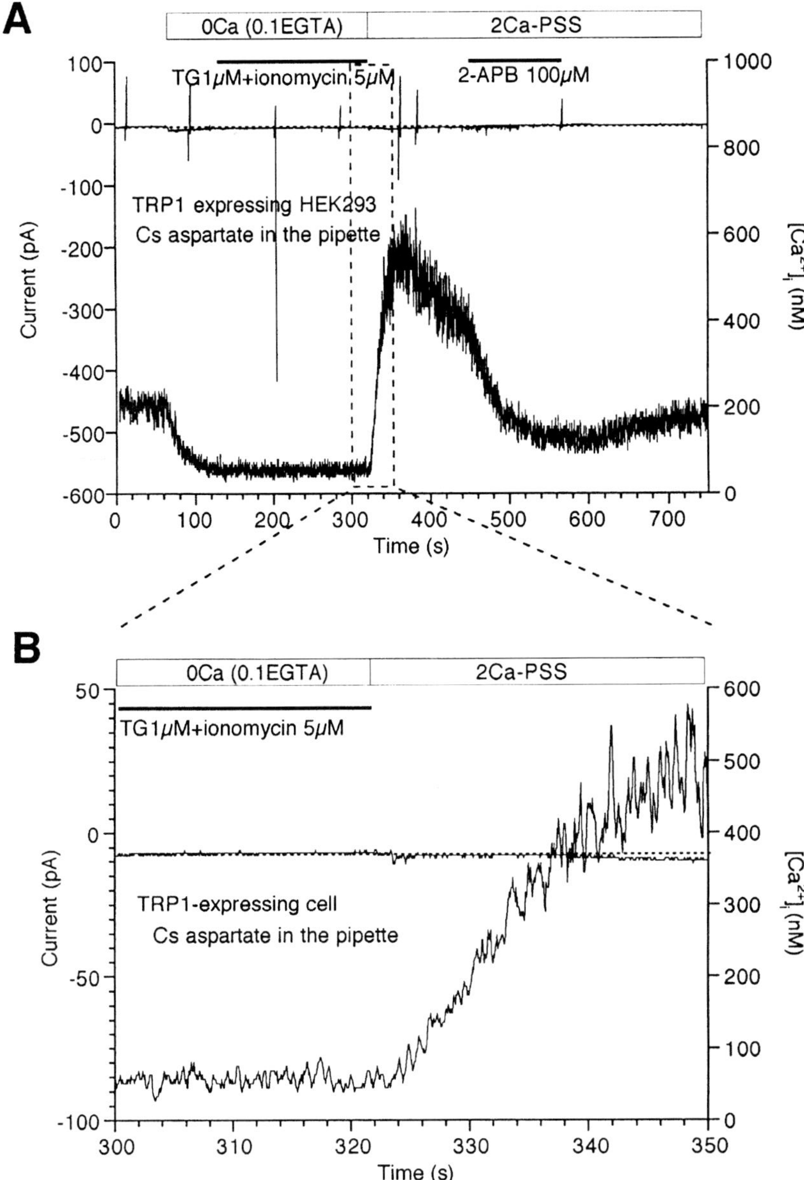

FIG. 2. Simultaneous recording of membrane currents and Ca^{2+} fluorescence. (A) Upper and lower traces indicate the time courses of membrane current and [Ca^{2+}]$_i$ respectively. Cells were voltage-clamped at −60 mV. Pipette contained Cs aspartate internal solution supplemented with 50 M fura-2. (B) Expanded time-courses of membrane current and [Ca^{2+}]$_i$ form the dotted box in A. TG, thapsigargin; 2-APB, 2-aminoethoxydiphenyl borate.

CCE pathway (Xu & Beech, 2001). However, caution is warranted with regard to this interpretation. No electrophysiological data have yet been obtained to validate directly the contribution of TRP1 to CCE as an electrogenic pathway. As described above, in many smooth muscle tissues exhibiting CCE, it is difficult to demonstrate a Ca^{2+}-selective current large enough to cause a significant $[Ca^{2+}]_i$ increase. This may imply that CCE would not necessarily reflect the opening of store-depletion activated Ca^{2+}-permeable channels, i.e. it might occur through an electroneutral pathway. In fact, when TRP1 was overexpressed in HEK293 cells and membrane currents and $[Ca^{2+}]_i$ were measured simultaneously under the conditions in which Ca^{2+}-dependent Cl^- currents were minimized, readmission of Ca^{2+} into the bath after store depleting procedures produced only a marginal inward current while a large rapid increase occurred in $[Ca^{2+}]_i$ (Fig. 2A). Expanding the rising phase of $[Ca^{2+}]_i$ increase due to CCE clearly shows that there is almost no correlation between the rate of $[Ca^{2+}]_i$ increase and the magnitude of membrane current (Fig. 2B). A similar approach has previously been taken in mouse anococcygeus muscle to prove the causal relationship between store depletion activated current and $[Ca^{2+}]_i$ increase. However, the observed membrane current was just mirrored or even preceded by an increase in $[Ca^{2+}]_i$ upon store depletion (Wayman et al 1999), rather suggesting that the current may merely reflect a Ca^{2+}-dependent conductance secondarily induced by $[Ca^{2+}]_i$ increase. It would therefore be fair not to conclude now that TRP1 is the channel-forming subunit of CCE in smooth muscle. Alternatively, it would be more accurate to say that TRP1 serves as an important up-regulator of CCE in some smooth muscle cells.

In summary, mammalian TRP proteins exhibit many essential features characteristic of native ROCCs and CCE in smooth muscle. In particular, TRP6 and TRP1 seem to play a pivotal role as a molecular component of some ROCCs and up-regulators of CCE. Further efforts to investigate various heteromeric combinations of TRP isoforms and to associate them with native ROCCs/CCE will bring us more complete understanding about the mechanisms of Ca^{2+} mobilization during GPCR stimulation in smooth muscle.

References

Albert AP, Large WA 2001 Store-operated inward ion channels in rabbit portal vein myocytes. J Physiol 531:80P–81P

Birnbaumer L, Zhu X, Jiang M et al 1996 On the molecular basis and regulation of cellular capacitative calcium entry: roles for Trp proteins. Proc Natl Acad Sci USA 93:15195–15202

Carl A, Lee HK, Sanders KM 1996 Regulation of ion channels in smooth muscle by calcium. Am J Physiol 271:C9–C34

Clapham DE, Runnels LW, Strübing C 2001 The TRP ion channel family. Nature Rev Neuroci 2:387–396

Curtis TM, Scholfield CN 2001 Nifedipine blocks Ca^{2+} store refilling through a pathway not involving L-type Ca^{2+} channels in rabit arteriolar smooth muscle. J Physiol 532:609–623

Ertel EA, Campbell KP, Harpold MM et al 2000 Nomenclature of voltage-gated calcium channels. Neuron 25:533–535

Gibson A, McFadzean I, Wallance P, Wayman CP 1998 Capacitative Ca^{2+} entry and the regulation of smooth muscle tone. Trends Pharmacol 19:266–269

Golovina VA, Platoshyn O, Bailey CL et al 2001 Upregulated *TRP* and enhanced capacitative Ca^{2+} entry in human pulmonary artery myocytes during proliferation. Am J Physiol 280:H746–H755

Haj-Dahmane S, Andrade R 1996 Muscarinic activation of a voltage-dependent cation nonselective current in rat association cortex. J Neurosci 16:3848–3861

Hardie RC, Minke B 1993 Novel Ca^{2+} channels underlying transduction in *Drosophila* photoreceptors: implications for phosphoinositide-mediated Ca^{2+} mobilization. Trends Neurosci 16:371–376

Helliwell RM, Large WA 1997 Alpha1-adrenoceptor activation of a non-selective cation current in rabbit portal vein by 1,2-diacyl-sn-glycerol. J Physiol 499:417–428

Hofmann F, Lacinova L, Klugbauer N 1999 Voltage-dependent calcium channels: from structure to function. Rev Physiol Biochem Pharmacol 139:33–87

Hofmann T, Schaefer M, Schultz G, Gundermann T 2000 Transient receptor potential channels as molecular substrates of receptor-mediated cation entry. J Mol Med 78:14–25

Inoue R, Okada T, Onoue H et al 2001 The transient receptor potential protein homologue TRP6 is the essential component of vascular α_1-adrenoceptor-activated Ca^{2+}-permeable cation channel. Circ Res 88:325–332

Karaki H, Ozaki H, Hori M et al 1997 Calcium movements, distribution, and functions in smooth muscle. Pharmacol Rev 49:157–230

Kotlikoff MI, Herrera G, Nelson MT 1999 Calcium permeant ion channels in smooth muscle. Rev Physiol Biochem Pharmacol 134:147–199

Kuriyama H, Kitamura K, Itoh T, Inoue R 1998 Physiological features of visceral smooth muscle cells, with special reference to receptors and ion channels. Physiol Rev 78:811–920

McDaniel SS, Platoshyn O, Wang J et al 2001 Capacitative Ca^{2+} entry in agonist-induced pulmonary vasoconstriction. Am J Physiol 280:L870–L880

Montell C 2001 Physiology, phylogeny, and functions of the TRP superfamily of cation channels. In: Science's STKE, July, p1–17

Okada T, Inoue R, Yamazaki K et al 1999 Molecular and functional characterization of a novel mouse transient receptor potential protein homologue TRP7. Ca^{2+}-permeable cation channel that is constitutively activated and enhanced by stimulation of G protein-coupled receptor. J Biol Chem 274:27359–27370

Perraud A-L, Fleig A, Dunn CA 2001 ADP-ribose gating of the calcium-permeable LTRPC2 channel revealed by Nudix motif homology. Nature 411:595–599

Sanders KM 2001 Mechanisms of calcium handling in smooth muscles. J Appl Physiol 91:1438–1449

Strübing C, Krapininsky G, Krapivinsky L, Clapham DE 2001 TRPC1 and TRPC5 form a novel cation channel in mammalian brain. Neuron 29:645–655

Trepakova ES, Csutora P, Hunton DL, Marchase RB, Cohen RA, Bolotina VM 2000 Calcium influx factor directly activates store-operated cation channels in vascular smooth muscle cells. J Biol Chem 275:26158–26163

Trepakova ES, Gericke M, Hirakawa Y, Weisbrod RM, Cohen RA, Bolotina VM 2001 Properties of a native cation channel activated by Ca^{2+} store depletion in vascular smooth muscle cells. J Biol Chem 276:7782–7790

Walker RL, Hume JR, Horowitz B 2001 Differential expression and alternative splicing of TRP channel genes in smooth muscles. Am J Physiol 280:C1184–C1192

Wayman CP, McFadzean I, Gibson A, Tucker JF 1996 Two distinct membrane currents activated by cyclopiazonic acid-induced calcium store depletion in single smooth muscle cells of the mouse anococcygeus. Br J Pharmacol 117:566–572

Wayman CP, Gibson A, McFadzean I 1998 Depletion of either ryanodine- or IP_3-sensitive calcium stores activates capacitative calcium entry in mouse anococcygeus smooth muscle cells. Pflüger's Arch 435:231–239

Wayman CP, Wallace P, Gibson A, McFadzean I 1999 Correlation between store-operated cation current and capacitative Ca^{2+} influx in smooth muscle cells from mouse annococcygeus. Eur J Pharmacol 376:325–329

Xu S-Z, Beech DJ 2001 TrpC1 is a membrane-spanning subunit of store-operated Ca^{2+} channels in native vascular smooth muscle cells. Circ Res 88:84–87

Regulation of Ca^{2+} entry pathways by both limbs of the phosphoinositide pathway

Colin W. Taylor

Department of Pharmacology, Tennis Court Road, Cambridge, CB2 1TP, UK

Abstract. All inositol 1,4,5-trisphosphate (InsP$_3$) receptors are biphasically regulated by cytosolic Ca^{2+}. For type 2 InsP$_3$ receptors, InsP$_3$ binding controls whether a stimulatory Ca^{2+}-binding site (exposed after InsP$_3$ binding) or an inhibitory Ca^{2+}-binding site (exposed only in the absence of InsP$_3$) is accessible. Ca^{2+} therefore inhibits these InsP$_3$ receptors only after InsP$_3$ has dissociated. The capacitative Ca^{2+} entry (CCE) pathway is activated by depletion of Ca^{2+} stores, but the local increase in cytosolic [Ca^{2+}] as Ca^{2+} flows through these channels could cause long-lasting inhibition of InsP$_3$ receptors and so termination of the signal that activates CCE. However, the duration of the openings of CCE channels is matched to the behaviour of InsP$_3$ receptors such that during the brief openings of CCE channels, active InsP$_3$ receptors are unlikely to lose enough InsP$_3$ for them to become inhibited. In A7r5 vascular smooth muscle cells, CCE and a non-capacitative Ca^{2+} entry (NCCE) pathway, which is activated by arachidonic acid released from diacylglycerol by diacylglycerol lipase, can be distinguished by their different permeation properties and sensitivity to selective blockers. Arachidonic acid also inhibits CCE and so ensures that during receptor activation only the NCCE pathway mediates Ca^{2+} entry, while CCE contributes only after removal of the agonist.

2002 Role of the sarcoplasmic reticulum in smooth muscle. Wiley, Chichester (Novartis Foundation Symposium 246) p 91–107

Regulation of Ca^{2+} entry and Ca^{2+} mobilization

In smooth muscle, just as in other cells, extracellular stimuli typically evoke an increase in cytosolic Ca^{2+} concentration ([Ca^{2+}]$_c$) by stimulating release of Ca^{2+} from intracellular stores or Ca^{2+} entry across the plasma membrane. Members of two closely related families of intracellular Ca^{2+} channels, ryanodine receptors and the receptors for inositol-1,4,5-trisphosphate (InsP$_3$) appear to be expressed in all smooth muscle cells and they provide the most important route for release of Ca^{2+} from the sarcoplasmic reticulum (Sutko & Airey 1996). The roles of additional Ca^{2+}-mobilizing messengers, such as nicotinic acid adenine dinucleotide phosphate (NAADP) and sphingosine-1-phosphate, are less clear; both have been shown to release Ca^{2+} from the intracellular stores of several cell types via mechanisms that appear not to require ryanodine or $InsP_3$ receptors (Lee 2000, Pyne & Pyne 2000). The physiological significance of these novel messengers in smooth muscle (Iizuka et al 1998), as in other tissues, remains unclear.

Stimulation of Ca^{2+} entry across the plasma membrane is the second means whereby extracellular stimuli can evoke an increase in cytosolic $[Ca^{2+}]$. The channels that mediate this Ca^{2+} entry include voltage-gated Ca^{2+} channels; ligand-gated ion channels; store-operated Ca^{2+} channels that open in response to a signal generated by depletion of intracellular Ca^{2+} stores (Putney 1999) (these are also known as capacitative Ca^{2+} entry [CCE] channels); and a diverse collection of channels with varying degrees of selectivity for Ca^{2+} that open in response to signals generated within the cell, the so-called 'second messenger-operated Ca^{2+} channels'. Here, I consider only the latter two categories of Ca^{2+} channel and address two specific issues:

(1)   The extent to which regulation of $InsP_3$ receptors by cytosolic Ca^{2+} influences the ability of empty Ca^{2+} stores to activate CCE.
(2)   The relative contributions of Ca^{2+} entry via CCE and second messenger-operated pathways to the Ca^{2+} signals evoked by physiological stimuli in the A7r5 vascular smooth muscle cell line. For simplicity, I refer to the second of these pathways as a non-capacitative Ca^{2+} entry (NCCE) pathway.

Regulation of InsP$_3$ receptors by cytosolic Ca^{2+}: lessons from superfusion experiments

The first evidence that increases in cytosolic $[Ca^{2+}]$ biphasically regulate the activity of $InsP_3$ receptors came from studies of smooth muscle (Iino 1990). Despite some intervening controversies, it now seems reasonably clear that all three subtypes (types 1–3) of the mammalian $InsP_3$ receptor are stimulated by modest increases in cytosolic $[Ca^{2+}]$ and inhibited by more substantial increases (Mak et al 2001, Swatton et al 1999). The details of the mechanisms underlying the regulation of $InsP_3$ receptors by cytosolic Ca^{2+} may, however, differ between subtypes. This biphasic pattern of Ca^{2+} regulation is widely supposed to contribute to the complex patterns of Ca^{2+} signals recorded in intact cells (Berridge 1997), but it may also be important in determining the ability of intracellular Ca^{2+} stores to regulate the CCE pathway.

Rapid superfusion of permeabilized hepatocytes loaded with $^{45}Ca^{2+}$ has allowed us to examine the kinetics of $InsP_3$-evoked Ca^{2+} mobilization with high temporal resolution under conditions where the concentrations of $InsP_3$ and cytosolic Ca^{2+} can be rapidly changed (Marchant & Taylor 1997). The results of these studies have

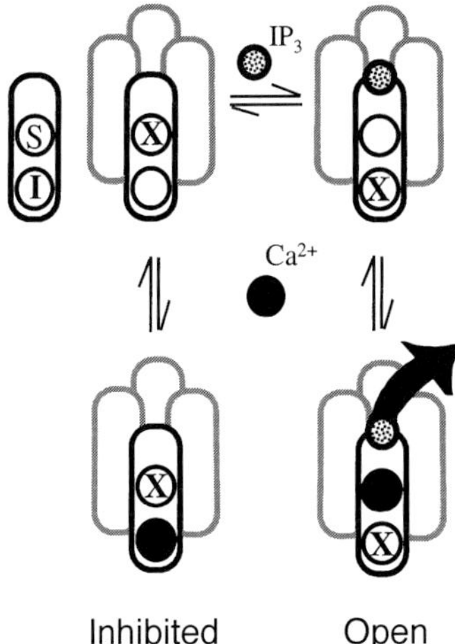

FIG. 1.   Regulation of type 2 InsP$_3$ receptors by Ca^{2+} and InsP$_3$. Two Ca^{2+}-binding sites, a stimulatory (S) and inhibitory (I) site, are associated with each of the four subunits of the type 2 InsP$_3$ receptor; it remains unclear whether the sites reside on the receptor itself or on accessory proteins. Binding of InsP$_3$ is proposed to cause exposure of the stimulatory Ca^{2+}-binding site and occlusion of the inhibitory site. When all four subunits have completed the sequential binding of InsP$_3$ and then Ca^{2+} (the figure shows the sequence for only one subunit), the Ca^{2+} channel opens.

led to the model shown in Fig. 1. We suggest, at least for the type 2 InsP$_3$ receptors expressed in hepatocytes, that binding of InsP$_3$ to its receptor controls which of two Ca^{2+}-binding sites is exposed. With no InsP$_3$ bound to the receptor, only an inhibitory Ca^{2+}-binding site is accessible; an increase in [Ca^{2+}]$_c$ in the absence of InsP$_3$ therefore inhibits the channel. When InsP$_3$ binds to each of the four subunits of its receptor, the inhibitory Ca^{2+}-binding sites are concealed, and stimulatory Ca^{2+}-binding sites are exposed. The channel of the InsP$_3$ receptor opens when Ca^{2+} binds to those stimulatory Ca^{2+}-binding sites (Adkins & Taylor 1999, Marchant & Taylor 1997). When the receptor has InsP$_3$ bound, therefore, an increase in [Ca^{2+}]$_c$ can stimulate channel opening, but not inhibit it. I return to these properties later in the context of CCE, but one further feature is relevant to the subsequent discussion, and that is that inhibition by cytosolic Ca^{2+} in the absence of InsP$_3$ occurs very rapidly (half-time about 50 ms), but it reverses only slowly (half-time about 400 ms) (Adkins & Taylor 1999).

Capacitative Ca^{2+} entry: another role for InsP$_3$ receptors

The evidence that empty intracellular Ca^{2+} stores can stimulate Ca^{2+} entry into cells via the CCE pathway is unassailable (Putney 1997). However, neither the signal responsible for communicating the state of the stores to the Ca^{2+} entry channel,

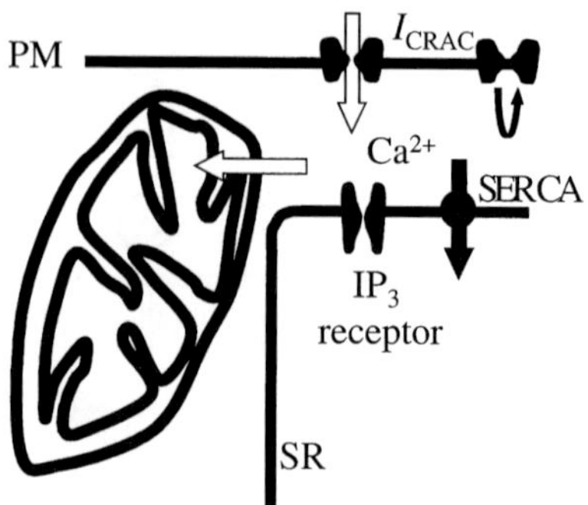

FIG. 2.   The capacitative Ca^{2+} entry complex. The SR, with its resident Ca^{2+} pump (SERCA) and InsP$_3$ receptors; the plasma membrane, with the channels (I_{CRAC}) regulated by empty stores; and mitochondria are all proposed to be intimately associated in this capacitative Ca^{2+} entry complex. Inevitably, as Ca^{2+} enters via I_{CRAC}, InsP$_3$ receptors and SERCA will be transiently exposed to very high $[Ca^{2+}]_c$; the implications are discussed in the text.

nor the molecular identities of the CCE channels have been unequivocally identified. Conformational coupling between the InsP$_3$ receptor and the CCE channel (Kiselyov et al 1999), the involvement of diffusible messengers (Randriamampita & Tsien 1993), or the insertion of channels into the plasma membrane by a secretory process (Yao et al 1999), have each been suggested as possible links between empty stores and activation of CCE. The CCE channels themselves are likely to be members of the large family of trp-related ion channels (Clapham et al 2001).

Irrespective of whether the InsP$_3$ receptor directly communicates with the CCE channel, the InsP$_3$ receptor clearly plays a central role in allowing those receptors that stimulate InsP$_3$ formation to empty intracellular Ca^{2+} stores and to thereby activate CCE. Figure 2 illustrates the intimate relationship proposed to exist between the intracellular Ca^{2+} stores, the Ca^{2+} pumps (the sarcoplasmic/endoplasmic reticulum ATPase, SERCA), and InsP$_3$ receptors resident in the membrane of the stores, mitochondria and the CCE channels (Gilabert & Parekh 2000). Because the intracellular stores activate CCE only when they are almost completely empty (Fierro & Parekh 2000), it is immediately apparent that if CCE is to be sustained, the InsP$_3$ receptor must be capable of keeping the intracellular store empty in the face of a substantial local increase in $[Ca^{2+}]_c$. This increase in $[Ca^{2+}]_c$ will not only fuel Ca^{2+} uptake into the stores by the SERCA, but more importantly it might be expected to lead to rapid, complete and only slowly reversible inhibition of the InsP$_3$ receptors. Within milliseconds of Ca^{2+} entering the cell via the CCE pathway, therefore, we might expect to initiate a long-lasting inhibition of InsP$_3$ receptors that would allow refilling of the stores and so termination of CCE. The outcome then might be that CCE channels would be active for only a tiny fraction of the period during which the extracellular stimulus was present.

I_{CRAC} channels, the most thoroughly investigated of CCE pathways, are rapidly inhibited by the Ca^{2+} that passes though them and then rapidly recover once the channel has closed (Zweifach & Lewis 1995). Even during sustained stimulation by empty stores, therefore, each of these channels is likely to open for only about 100 ms before closing and re-opening (Fig. 3). I return now to the effects of Ca^{2+} on InsP$_3$ receptors. Because Ca^{2+} inhibits InsP$_3$ receptors only if they have no InsP$_3$ bound, it is clear that if an open InsP$_3$ receptor (where all four subunits have InsP$_3$ bound) is to be inhibited by Ca^{2+} entering via CCE, then InsP$_3$ must dissociate from its receptor within the 100 ms openings of the CCE channels— the period when the local [Ca^{2+}]$_c$ is high. By measuring the rate at which InsP$_3$ receptors close after removal of InsP$_3$ in superfusion experiments (Marchant & Taylor 1998), and recognizing that only one of the four occupied subunits needs to lose InsP$_3$ for the channel to close, we estimate that InsP$_3$ dissociates from the active state of its receptor with a half-time of about 1 second. It is not yet clear how many of the inhibitory Ca^{2+}-binding sites must be occupied by Ca^{2+} for the InsP$_3$ receptor to enter its long-lasting Ca^{2+}-inhibited state: it is likely to be at least two (Mak et al 1998). If Ca^{2+} entering via CCE is to cause long-lasting inhibition of the InsP$_3$ receptor, at least two of the four bound InsP$_3$ molecules must therefore dissociate from the receptor during a typical 100 ms opening of the CCE channel. Figure 3 suggests that during such a brief interval very few InsP$_3$ receptors would lose sufficient InsP$_3$ to allow them to become susceptible to inhibition by Ca^{2+}. I suggest, therefore, that auto-inhibition of CCE channels by Ca^{2+}, which serves to limit their bouts of opening to about 100 ms, is an adaptation that minimizes the risk that Ca^{2+} entry will inhibit InsP$_3$ receptors and so terminate the signals that activate CCE. Paradoxically, therefore, local Ca^{2+} feedback inhibition of CCE channels may be the mechanism that allows sustained activation of CCE.

Reciprocal regulation of CCE and NCCE in A7r5 cells

In A7r5 vascular smooth muscle cells, vasopressin, via the V$_{1A}$ receptor, stimulates phospholipase C leading to formation of both InsP$_3$ and diacylglycerol (DAG), each of which leads to activation of a distinct Ca^{2+} entry pathway. InsP$_3$ stimulates release of Ca^{2+} from intracellular stores and thereby generates the signal required for activation of CCE. In A7r5 cells, as in many other cells (Putney 1997), CCE can also be activated by emptying the intracellular stores using a combination of thapsigargin (to inhibit the SERCA) and ionomycin (a Ca^{2+} ionophore) (Byron & Taylor 1995). The CCE pathway in A7r5 cells is permeable to Ca^{2+}, Ba^{2+} and Mn^{2+}, but not to Sr^{2+}, and is selectively blocked by 1 μM Gd^{3+} or 2-aminoethoxydiphenylborate (2-APB, 100 μM). Arachidonic acid,

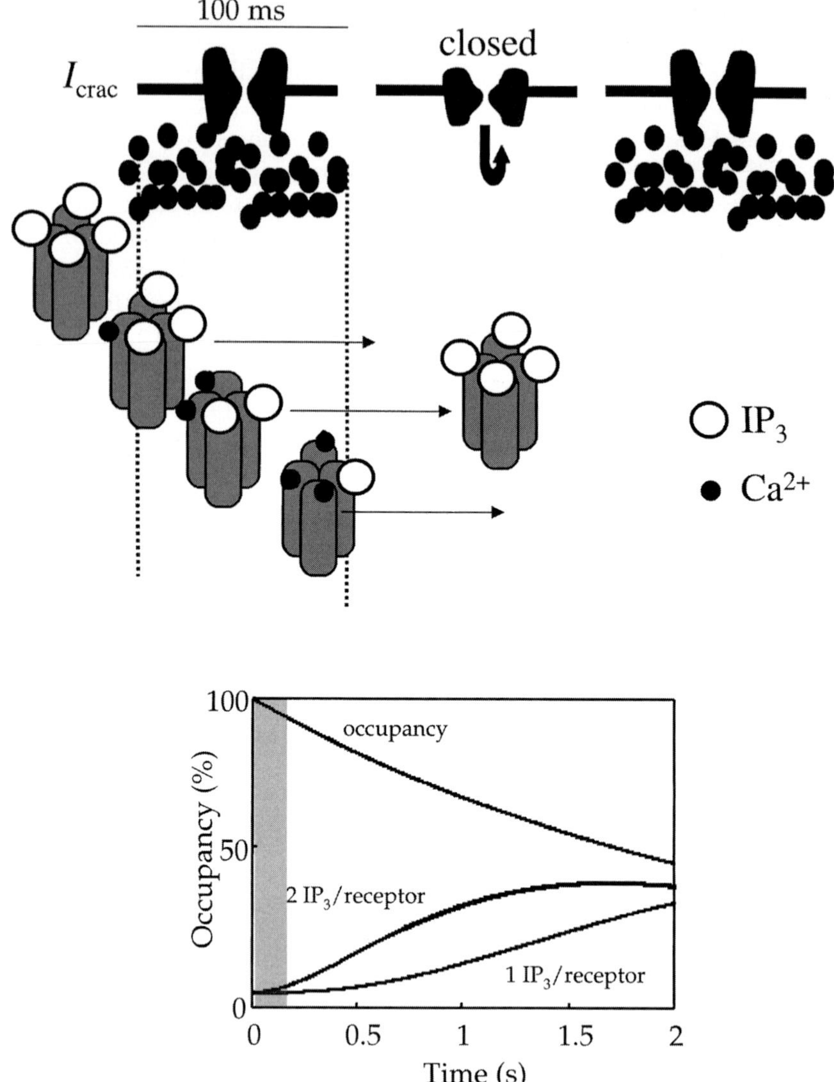

FIG. 3.   Protecting InsP$_3$ receptors from Ca^{2+} entry: a matter of timing. Ca^{2+} entry via I_{CRAC} can inhibit active InsP$_3$ receptors only if they are able to lose two or more of their four bound InsP$_3$ molecules during the period (about 100 ms) when I_{CRAC} channels are open (auto-inhibition by Ca^{2+} then closes the I_{CRAC} channel). The lower panel shows the measured rate of InsP$_3$ dissociation from active InsP$_3$ receptors (half-time = 1 s), and the times over which the occupancy of the InsP$_3$ receptors fall to two or one InsP$_3$ molecule bound to each tetrameric receptor. The results demonstrate that very few InsP$_3$ receptors would lose enough InsP$_3$ during the brief opening of an I_{CRAC} channel for them to become susceptible to Ca^{2+} inhibition. See text for further details.

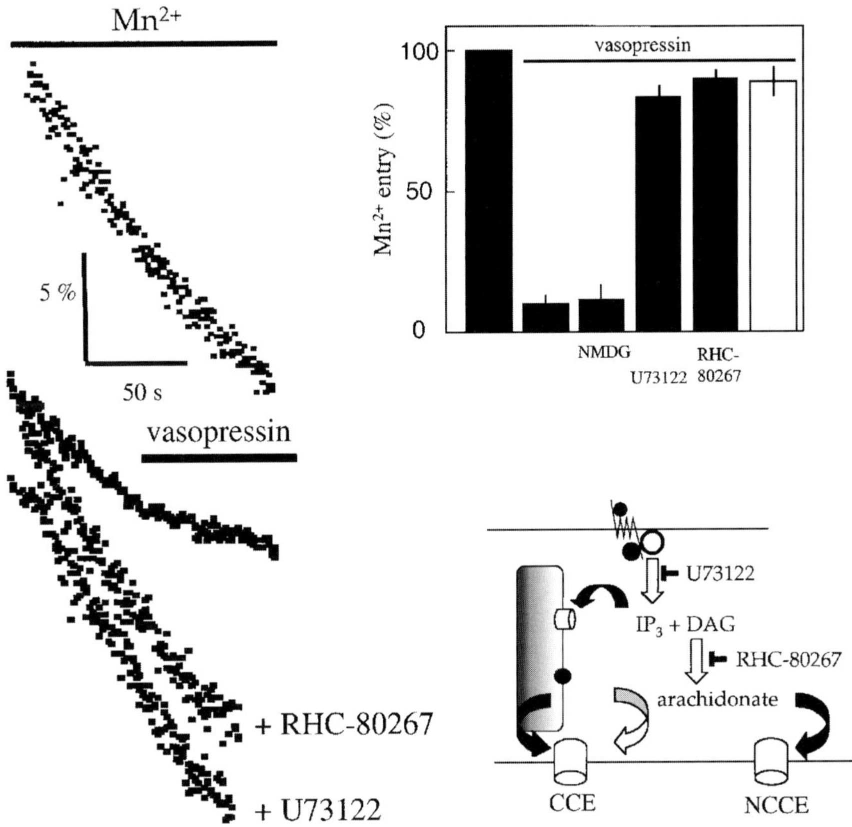

FIG. 4.   Inhibition of capacitative Mn^{2+} entry by vasopressin in A7r5 cells. Mn^{2+} entry into thapsigargin-treated A7r5 cells was monitored by recording quench of Fura-2 fluorescence. Vasopressin abolished Mn^{2+} entry via the CCE pathway, and that inhibition was reversed by inhibition of phospholipase C (U73122) or DAG lipase (RHC-80267), but it persisted when extracellular Na$^+$ is replaced by NMDG. Vasopressin failed to inhibit Mn^{2+} entry in cells lacking DAG lipase (open bar in histogram). Arachidonic acid, released from DAG by DAG lipase, is proposed to simultaneously activate NCCE and inhibit (white curved arrow) CCE. Data from Moneer & Taylor (2002).

produced from DAG by DAG lipase (Migas & Seversen 1996), activates a second Ca^{2+} entry pathway in A7r5 cells. This NCCE pathway is permeable to Sr^{2+}, Ca^{2+} and Ba^{2+}, though not to Mn^{2+}, and it is reversibly blocked by LOE-908 (30 μM) or by 100 μM Gd^{3+}. Activation of NCCE by vasopressin is prevented by RHC-80267, a selective inhibitor of DAG lipase (Broad et al 1999). In A7r5 cells, therefore, both limbs of the phosphoinositide pathway regulate distinct Ca^{2+} entry pathways: InsP$_3$ activates CCE by emptying intracellular Ca^{2+} stores, and DAG provides the arachidonic acid that causes activation of NCCE. Recent results suggest that

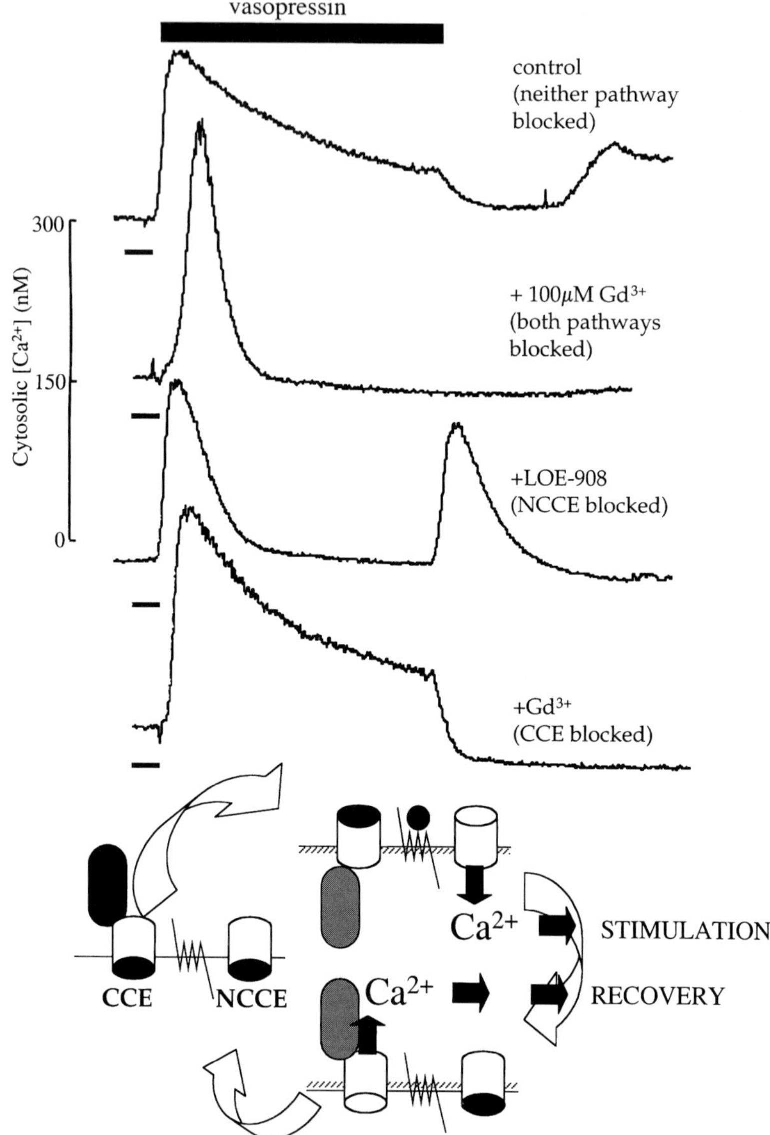

FIG. 5. Only non-capacitative Ca^{2+} entry occurs during receptor activation. Cells were stimulated with vasopressin (30 nM for 300 s) under conditions where both, neither or only one of the Ca^{2+} entry pathways were blocked. The same scale applies to all four traces, with the origin shown on each by a thick line. (Results taken from Moneer & Taylor [2002].) The cartoon indicates that reciprocal regulation of the two pathways may allow rapid switching between stimulation (contraction evoked by Ca^{2+} entering via NCCE) and recovery (relaxation evoked by Ca^{2+} entering via CCE)

in both A7r5 (Moneer & Taylor 2002) and other cells (Luo et al 2001, Mignen et al 2001), CCE and NCCE pathways do not operate independently.

In A7r5 cells, concentrations of vasopressin that stimulate NCCE, simultaneously inhibit CCE (Fig. 4). Both effects of vasopressin depend upon its ability to stimulate phospholipase C and on subsequent metabolism of the DAG to arachidonic acid by DAG lipase. Evidence in support of this conclusion includes the use of selective inhibitors of phospholipase C and DAG lipase, the inability of vasopressin to either stimulate NCCE or inhibit CCE in A7r5 cells that lack DAG lipase, and the ability of exogenous arachidonic acid to both activate NCCE and inhibit CCE (Moneer & Taylor 2002). Arachidonic acid therefore co-ordinates the activities of the two vasopressin-regulated Ca^{2+} entry pathways by stimulating NCCE and inhibiting CCE (Fig. 4).

In Fig. 5, responses to stimulation with a submaximal concentration of vasopressin were recorded under conditions where both, neither or only one of the two Ca^{2+} entry pathways was selectively blocked. The results demonstrate that when vasopressin is present, all Ca^{2+} entry occurs via the NCCE pathway. When vasopressin is removed, NCCE rapidly terminates, and only then is there a large transient Ca^{2+} entry via the CCE pathway. I suggest, therefore, that both addition *and* removal of vasopressin trigger abrupt switches between Ca^{2+} entry pathways: NCCE rapidly switches on when vasopressin is added, and when vasopressin is removed NCCE rapidly switches off as CCE switches on (Fig. 5). Because Ca^{2+} entering cells via different Ca^{2+} entry pathways can selectively couple to different cellular responses (Fagan et al 1998, Lin et al 2000), I suggest that reciprocal regulation of two Ca^{2+} entry pathways may allow an extracellular stimulus to abruptly initiate a response (via NCCE) when it binds to its receptor and then actively reverse it by activating another response (via CCE) when the stimulus dissociates.

Acknowledgements

Supported by grants from the Wellcome Trust, British Heart Foundation, and the Biotechnology and Biological Sciences Research Council.

References

Adkins CE, Taylor CW 1999 Lateral inhibition of inositol 1,4,5-trisphosphate receptors by cytosolic Ca^{2+}. Curr Biol 9:1115–1118

Berridge MJ 1997 Elementary and global aspects of calcium signalling. J Physiol 499:291–306

Broad LM, Cannon TR, Taylor CW 1999 A non-capacitative pathway activated by arachidonic acid is the major Ca^{2+} entry mechanism in rat A7r5 smooth muscle cells stimulated with low concentrations of vasopressin. J Physiol 517:121–134

Byron KL, Taylor CW 1995 Vasopressin stimulation of Ca^{2+} mobilization, two bivalent cation entry pathways and Ca^{2+} efflux in A7r5 rat smooth muscle cells. J Physiol 485:455–468

Clapham DE, Runnels LW Strübing C 2001 The trp ion channel family. Nature Reviews Neuroscience 2:387–396

Fagan KA, Mons N Cooper DMF 1998 Dependence of the Ca^{2+}-inhibitable adenylyl cyclase of C6-2B glioma cells on capacitative Ca^{2+} entry. J Biol Chem 273:9297–9305

Fierro L, Parekh AB 2000 Substantial depletion of the intracellular Ca^{2+} stores is required for macroscopic activation of the Ca^{2+} release-activated Ca^{2+} current in rat basophilic leukaemia cells. J Physiol 522:247–257

Gilabert JA, Parekh AB 2000 Respiring mitochondria determine the pattern of activation and inactivation of the store-operated Ca^{2+} current I_{CRAC}. EMBO J 19:6401–6407

Iino M 1990 Biphasic Ca^{2+} dependence of inositol 1,4,5-trisphosphate-induced Ca^{2+} release in smooth muscle cells of the guinea pig taenia caeci. J Gen Physiol 95:1103–1122

Iizuka K, Yoshii A, Dobashi K, Horie T, More M, Nakazawa T 1998 $InsP_3$, but not novel Ca^{2+} releasers, contributes to agonist-initiated contraction in rabbit airway smooth muscle. J Physiol 511:915–933

Kiselyov K, Mignery GA, Zhu MX, Muallem S 1999 The N-terminal domain of the IP_3 receptor gates store-operated hTrp3 channels. Mol Cell 4:423–429

Lee HC 2000 NAADP: An emerging calcium signaling molecule. J Membr Biol 173:1–8

Lin S, Fagan KA, Li K-X, Shaul PW, Cooper DMF, Rodman DM 2000 Sustained endothelial nitric-oxide synthase activation requires capacitative Ca^{2+} entry. J Biol Chem 275:17979–17985

Luo D, Broad LM, Bird GSJ, Putney JW Jr 2001 Mutual antagonism of calcium entry by capacitative and arachidonic acid-mediated calcium entry pathways. J Biol Chem 276:20186–20189

Mak D-O, McBride S, Foskett JK 1998 Inositol 1,4,5-trisphosphate activation of inositol trisphosphate receptor Ca^{2+} channel by ligand tuning of Ca^{2+} inhibition. Proc Natl Acad Sci USA 95:15821–15825

Mak D-O, McBride S Foskett JK 2001 Regulation by Ca^{2+} and inositol 1,4,5-trisphosphate ($InsP_3$) of single recombinant type 3 $InsP_3$ receptor channels. Ca^{2+} activation uniquely distinguishes types 1 and 3 $InsP_3$ receptors. J Gen Physiol 117:435–446

Marchant JS, Taylor CW 1997 Cooperative activation of IP_3 receptors by sequential binding of IP_3 and Ca^{2+} safeguards against spontaneous activity. Curr Biol 7:510–518

Marchant JS, Taylor CW 1998 Rapid activation and partial inactivation of inositol trisphosphate receptors by inositol trisphosphate. Biochemistry 37:11524–11533

Migas I, Seversen DL 1996 Diacylglycerols derived from membrane phospholipids are metabolized by lipases in A10 smooth muscle cells. Am J Physiol 271:C1194–1202

Mignen O, Thompson JL Shuttleworth TJ 2001 Reciprocal regulation of capacitative and arachidonate-regulated noncapacitative Ca^{2+} entry pathways. J Biol Chem 276: 35676–35683

Moneer Z, Taylor C 2002 Reciprocal regulation of capacitative and non-capacitative Ca^{2+} entry in A7r5 vascular smooth muscle cells: only the latter operates during receptor activation. Biochem J 362:13–21

Putney JW Jr 1997 Capacitative calcium entry. Springer-Verlag, Heidelberg

Putney JW Jr 1999 TRP, inositol 1,4,5-trisphosphate receptors, and capacitative calcium entry. Proc Natl Acad Sci USA 96:14669–14671

Pyne S, Pyne NJ 2000 Sphingosine 1-phosphate signalling in mammalian cells. Biochem J 349:385–402

Randriamampita C, Tsien RY 1993 Emptying of intracellular Ca^{2+} stores releases a novel small messenger that stimulates Ca^{2+} influx. Nature 364:809–814

Sutko JL, Airey JA 1996 Ryanodine receptor Ca^{2+} release channels: does diversity in form equal diversity in function? Physiol Rev 76:1027–1071

Swatton JE, Morris SA, Cardy TJA, Taylor CW 1999 Type 3 inositol trisphosphate receptors in RINm5F cells are biphasically regulated by cytosolic Ca^{2+} and mediate quantal Ca^{2+} mobilization. Biochem J 344:55–60
Yao Y, Ferrer-Montiel AV, Montal M, Tsien RY 1999 Activation of store-operated Ca^{2+} current in *Xenopus* oocytes requires SNAP-25 and but not a diffusible messenger. Cell 98:475–485
Zweifach A, Lewis RS 1995 Rapid inactivation of depletion-activated calcium current (I_{CRAC}) due to local calcium feedback. J Gen Physiol 105:209–226

DISCUSSION

Fry: Is it arachidonic acid that stimulates the NCCE pathway or something downstream from this, such as prostaglandins?

Taylor: It is not prostaglandins because we can block the metabolism of arachidonate, and so prevent formation of prostaglandins and leukotrienes, but we still get activation of NCCE by vasopressin, indeed there is a modest potentiation of the response consistent with lesser degradation of the arachidonate.

McCarron: Does the NCCE pathway fail to refill the store because of the continued presence of InsP$_3$?

Taylor: I don't know. One of the issues we are keen to address is the nature of the preferred relationships between each Ca^{2+} entry pathway and a cellular response. I don't know whether NCCE simply isn't as good at refilling stores as other Ca^{2+} entry pathways, or as you say it isn't providing enough Ca^{2+} to counteract the still-present InsP$_3$ during that submaximal stimulation. It could be either.

Brading: A technical question. When you use the Ca^{2+}-free solutions and add Ca^{2+} back, what do you do about other divalent cations?

Taylor: That is all we do. Mg^{2+} is there throughout, of course.

Brading: In the early days, when we took Ca^{2+} off cells, we always found vigorous contraction as soon as we put it back on again. I don't know whether this was due to the fact that we were losing stores. In my day, when one hadn't any idea about these sorts of things, we were worried about surface charge and what we have to do when replacing Ca^{2+} to neutralize the surface charges. I remember some experiments that Bülbring & Tomita (1969) did when they looked at the effect of removing Ca^{2+} on membrane resistance. They found that they had to raise the Mg^{2+} content of the medium to 12 mM before the membrane resistance went back to what it was before Ca^{2+} was removed. One of the explanations for this could be that there are big changes in the surface charge, which are determining the opening and closing of various voltage-sensitive channels in the membrane.

Taylor: Certainly, if we make transient changes in extracellular Ca^{2+} that are relatively brief we see no cytosolic Ca^{2+} signal. We can deplete the stores without thapsigargin and ionomycin if we leave them in Ca^{2+}-free solution for long enough, but this takes quite a long time in these cells. Merely changing the extracellular Ca^{2+} doesn't evoke a Ca^{2+} signal in A7r5 cells.

Fry: With the Ca^{2+} paradox in the heart, you have to put EGTA on and really reduce the Ca^{2+}; just taking the Ca^{2+} away on its own would give you about 20 μM free Ca^{2+}, and this wouldn't evoke a Ca^{2+} paradox.

Taylor: There's no EGTA here for the simple reason that it would chelate some of the inhibitors we use, such as Gd^{3+}. But with that said, we don't need to chelate extracellular Ca^{2+} with EGTA to prevent Ca^{2+} entry: we get no detectable Ca^{2+}-entry whether we simply omit Ca^{2+} or replace it with EGTA.

Brading: Another issue concerns what your Gd^{3+} might be doing to your surface charges.

Somlyo: What is Gd^{3+} doing to L-type channels, and what if arachidonic acid in smooth muscle inhibits L-type channels?

Taylor: I should have said at the outset that all of these experiments are done with the L-type channels deliberately blocked by nifedipine or verapamil.

Sanders: What induces this long-lasting inhibitory state of InsP$_3$ receptors?

Taylor: The experimental observation from which this comes is as follows. First, we pretreat the permeabilized cells in our superfusion apparatus with 100 μM Ca^{2+} for about a second, to drive the cells to the state where they can no longer respond to InsP$_3$ (even at heroic concentrations). Next, we wash out the Ca^{2+} to a low cytosolic level of 200 nM. Then we look for how long it takes for the normal response to a maximal concentration of InsP$_3$ to recover. That takes a long time.

Sanders: You talk about this like it was something pathological: you said there is a disastrous state that you get to. I'm wondering whether it is such a disaster: would it be the essence of an InsP$_3$-driven clock. If you get to this state and the InsP$_3$ receptors become refractory for a period until that inhibition can be removed, then this might be the basis of an oscillation. There are some important clock mechanisms driven by InsP$_3$ receptors that have quite a long refractory period to them.

Hirst: We have been looking at slow-wave generation in the gut. The only way that we can fit our data for the recovery of InsP$_3$-dependent pathway of release is to use your kinetics (Marchant & Taylor 1998). We need a 15 s time constant for recovery of InsP$_3$-dependent Ca^{2+} responses in interstitial cells of Cajal. Your kinetics fit our observations perfectly. I'd go along with Kenton Sanders and say that this might not be pathology; it might instead be physiology.

Taylor: Let me explain why I thought this was a significant possibility. First I should qualify my suggestion by saying that I think the pattern of Ca^{2+} regulation is not the same for all InsP$_3$ receptor subtypes; they are all biphasically regulated by Ca^{2+}, but there is some fine tuning going on. But there are experiments in cells that express the same (type 2) InsP$_3$ receptor subtype that we have examined in hepatocytes, where another ligand of the InsP$_3$ receptor that may dissociate from it 10 times more slowly (adenophostin A) seemed to be more effective at bringing about CCE. But it was more effective only in cells where the free Ca^{2+} was not buffered. So it looked as if there was something about the affinity of ligands for the InsP$_3$ receptor that interfered somehow with their susceptibility to Ca^{2+} inhibition and this impacted upon CCE. This is what got me thinking that it might be important for InsP$_3$ to protect the InsP$_3$ receptor from inhibition by Ca^{2+} if stores are not to rapidly refill when CCE occurs. Whether you want to couch this in pathological terms or not depends on what side of the fence you are on.

Paul: I was struck by the fact that in order to get the CCE, you have to nearly completely empty the stores. I wonder how often this occurs in real physiology? In the more tonic vessels that I have worked with, there does seem to be an SR effect that is very long lasting. We never really empty the stores. Does this happen a lot?

Taylor: I deliberately drew the whole of my argument on CCE from RBL cells, which express type 2 InsP$_3$ receptors and where some of the best measures of I_{CRAC} have been made. In this situation there is evidence that substantial depletion of a subset of the stores is needed to activate CCE. Our own work in the A7r5 cells shows a much closer coupling between store depletion and Ca^{2+} entry: there seems to be the same dose-dependent mobilization of Ca^{2+} stores as there is activation of CCE, suggesting that there is no threshold phenomenon: as the stores empty they become progressively better at bringing about activation of CCE. So it may differ between cells, but the particular argument that I was making was for RBL cells.

Burdyga: There is a long-standing observation from smooth muscle cells placed in high K^+ solution. A steady state of Ca^{2+} and sustained contraction is reached, and then when agonists are applied there is a massive release of Ca^{2+}, which means that the InsP$_3$ receptors are not disabled. What concentration are you talking about in your case? Is it locally high, in order to disable the InsP$_3$ receptors?

Taylor: In the experiments I showed we deliberately used a high Ca^{2+}. This is partly a technical feature of the way that the superfusion works, but also we wanted to completely inhibit. The EC_{50} for this effect depends on how long the cells are left incubating in it, but it is in the order of $1.5\,\mu M$.

Blaustein: With regard to the stores releasing Ca^{2+}, here you are talking about indirect functional observations when you study the releasable pool. If we actually look at the store itself with furaptra and use high concentrations of an agonist or maximal doses of caffeine, it releases some Ca^{2+}, but the Ca^{2+} doesn't go down much below $50\,\mu M$. You can release more if you use the Ca^{2+}-free solution or ionomycin. The SR is not really empty. Our feeling is that there are multiple stores. There may be some stores that can be depleted relatively easily, whereas other stores are more difficult to deplete. I have some evidence that the 'junctional' stores may be the ones that you are talking about; release of Ca^{2+} is probably not 'complete'.

Taylor: Perhaps I should add that if we empty the stores with vasopressin, which will completely empty them, as opposed to thapsigargin and ionomycin, we can get the same degree of activation of CCE as long as we inhibit DAG lipase. It is just that normally one never delivers the store-depletion stimulus in isolation: cells would always feel the accelerator (store depletion) and the brakes (arachidonic acid from DAG). Thapsigargin is only the accelerator (in terms of emptying stores and so activating CCE), and vasopressin under those conditions (DAG lipase inhibited) is deprived of the brakes as well. My argument would be that physiologically you never get the one without the other: the accelerator (store depletion) is always paired with the brakes (arachidonic acid to inhibit CCE).

Blaustein: Usually when we try to empty the stores and then refill them, we do things that cause massive emptying. But if your suggestion is right that you really have to empty the store in order to get this mechanism to work and refill again, this is an unusual system. Has anyone tried looking at a small amount of release with a low dose of agonist for a short period, and seen evidence for refilling?

van Breemen: I made an observation in endothelial cells where we depleted the ER Ca^{2+} with SERCA inhibition, or with ryanodine, or caffeine. We measured the Ca^{2+} entry with Mn^{2+}. I found if I used the SERCA inhibitor I needed to deplete only by 20% to get a maximum effect on the rate of Mn^{2+} quenching of Fura-2. If I used ryanodine or caffeine there was no increase in the rate of Mn^{2+} entry even if the stores were depleted by 50%.

Lompré: If there are two pools, thapsigargin inhibits both pools, and ryanodine or caffeine only one.

Young: We have data showing that induction of a Ca^{2+} wave in cultured myometrial cells that lack L-type Ca^{2+} channels causes activation of Ca^{2+}-activated K^+ channels that we think are store-operated. This is just with passage of a Ca^{2+} wave without thapsigargin.

Taylor: Anant Parekh has some nice data speaking to Mordy Blaustein's concern. He infused 2,4,5-InsP$_3$ through a patch pipette into a cell and then saw activation of I_{CRAC}. Even under these conditions with supramaximal concentrations of an InsP$_3$ receptor agonist, he could still accelerate the rate of I_{CRAC} activation when he added thapsigargin. His argument from this was that even with maximal activation of the InsP$_3$ receptor, the local Ca^{2+} entry into that complex can fuel the SERCA enough to still have some compensatory uptake into the stores, such that until you take the SERCA right out of action you may not be seeing the full activation of I_{CRAC}.

Eisner: Does this mean that the SR is not fully depleted?

Taylor: That's his argument; even with InsP$_3$ present there is still some compensating SERCA that puts a bit of Ca^{2+} back into the stores again.

Kotlikoff: Would your prediction be that at the end of an agonist response or nerve depolarization, at its termination you would get CCE, and presumably some large spark of Ca^{2+}?

Taylor: There were a couple of papers earlier on in this meeting where I thought that people were showing CCE to be a rather ineffective contractile stimulus.

Indeed, I thought there was relaxation being triggered by a CCE signal in someone's earlier presentation.

Kotlikoff: At least in isolated cells, we don't commonly see a spiking Ca^{2+} signal at the end of an application of agonist where we completely release Ca^{2+}. But this would be the prediction of your model.

Taylor: This is what we do see. Under normal conditions when we stimulate with a submaximal concentration of vasopressin, we do get Ca^{2+} entry via NCCE for as long as vasopressin is present, but that Ca^{2+} entry seems to be insufficient to refill the stores. Of course, while vasopressin is present we don't get any CCE, despite the partially emptied stores, because the CCE pathway remains inhibited by arachidonic acid. Then when we remove vasopressin we see a delay (for which we presently have no explanation, although it's very reproducible) before we get a transient Ca^{2+} entry through CCE because it's no longer inhibited by arachidonic acid

Somlyo: Are these experiments done in the presence of normal extracellular Ca^{2+}? When you are measuring the currents, aren't you buffering intracellular Ca^{2+}?

Taylor: I am not measuring any currents. These are all Fura-2 measurements with normal extracellular Ca^{2+}.

Somlyo: My understanding is that even in unexcitable cells, when people try to measure what they believe to be capacitative currents, they use very high concentrations of Ca^{2+} buffers because these are low conductance channels and are very difficult to detect. This raises the question posed by Alison Brading about whether the Ca^{2+} influx in depolarized smooth muscle has something to do with what many of us saw years ago by removing Ca^{2+} and adding back Ca^{2+} to smooth muscle depolarized with high K$^+$. In other words, are you really measuring currents or are these assumed currents based on these measurements?

Taylor: I am not interested in measuring currents; I am measuring cytosolic Ca^{2+}. The only way you can make current measurements in a setting like this is by so perturbing the system that it would no longer have any bearing on real physiology.

Hellstrand: If you do this experiment in the presence of thapsigargin or after thapsigargin, do you get an enhanced rebound Ca^{2+} response after washing out the agonist?

Taylor: We haven't done that here. With vasopressin, if we block the DAG lipase pathway vasopressin is perfectly capable of giving a CCE that is as large as that evoked by thapsigargin.

Nelson: To carry this further, I think it would be important to measure currents, or at least to do voltage clamp. These cells are in a non-voltage-clamped situation, so as you carry out these manoeuvres the voltage is changing in some unknown way which would contribute to changes in Ca^{2+} fluxes as well. I would be hard pressed to interpret some of this without at least voltage clamping the cells and seeing what happens.

Taylor: The voltage clamp is more useful than measuring currents. Nonetheless, a cell that is responding to vasopressin is not voltage clamped in a physiological situation.

Brading: Under voltage clamp you get huge currents that are not normal under physiological conditions.

Somlyo: Did I understand you correctly that you think that the InsP$_3$ channel opens with a Hill coefficient of 4?

Taylor: The measured Hill coefficient is just under 3. It varies in different laboratories.

Somlyo: We did some work with David Trentham in which we looked for this in terms of modelling the lag phases, and at best we could come up with a Hill coefficient of 2. This was in smooth muscle, not Jurkat cells. My belief is that it is closer to 1. David wanted to be more conservative and said that it could be 2, but never 4 (Somlyo et al 1992).

Iino: I also think that it is close to 1. In our system, which is skinned fibres or permeabilized DT40 cells, we can control the Ca^{2+} concentration around the InsP$_3$ receptor. Under these conditions, the Hill coefficient is 1. For the bilayer measurements it is also 1.

Taylor: Our measured Hill coefficient is not 4; it is about 2.8.

Somlyo: You might want to consider what the adenine nucleotide concentrations are when these measurements are made. As you know, the InsP$_3$ channel is also highly adenine nucleotide sensitive, and is also sensitive to a few other things in the cells that we may not even know about.

Iino: I don't think that adenine nucleotides would have any direct effect on the Hill coefficient.

Sanders: There are discrete sites of innervation of smooth muscles. Therefore, it is unlikely you ever get store depletion with neurotransmission. The SR is all connected: it is a continuous tubular system. Releasing Ca^{2+} at one site is not going to empty the store in any way.

Fry: Unless there were separate stores.

Sanders: There would have to be a separate store at the site of neurotransmisson.

Hirst: And the probability of release at each junction would have to be 1, instead of 0.01.

Sanders: The measures used to induce this phenomenon of CCE are pretty extreme.

Taylor: I'm not sure they are so extreme. The fact that we can't put a whisker between the dose–response curves describing any of the phenomena does suggest that we cannot recruit one without the other, as long as we are globally applying the signal. I live with the concerns that if the signal is locally applied, then a widely distributed ER may never deplete sufficiently.

Nelson: Does depolarization raise Ca^{2+} as well?

Taylor: If we don't block the L-type channels then yes.

Nelson: If you depolarize or hyperpolarize in the presence of vasopressin, does the Ca^{2+} go up or down? Does it follow a Ca^{2+} electrochemical gradient, as you would expect for a non-voltage-dependent ion channel?

Taylor: When we did the experiment in which we went into 135 mM extracellular K$^+$, we had to up the extracellular Ca^{2+} concentration about 10- or 12-fold to bring the CCE signal back up to where it would have been before.

Nelson: That is also a good test, because if you hyperpolarized and your Ca^{2+} went down, then you might examine how good your block is of the Ca^{2+} channels.

Taylor: All we have done is stepped to high K$^+$ in the presence of nimodipine or verapamil and either of these completely blocks the deplorarization-evoked Ca^{2+} signal.

References

Bülbring E, Tomita T 1969 Effect of calcium, barium and manganese on the action of adrenaline in the smooth muscle of the guinea-pig taenia coli. Proc R Soc Lond B Biol Sci 172:121–136

Marchant JS, Taylor CW 1998 Rapid activation and partial inactivation of inositol trisphosphate receptors by inositol trisphosphate. Biochemistry 37:11524–11533

Somlyo AV, Horiuti K, Trentham DR, Kitazawa T, Somlyo AP 1992 Kinetics of Ca^{2+} release and contraction induced by photolysis of caged D-myo-inositol 1,4,5-trisphosphate in smooth muscle: the effects of heparin, procaine, and adenine nucleotides. J Biol Chem 267:22316–22322

Calcium release by ryanodine receptors in smooth muscle

M. I. Kotlikoff, Yong-Xiao Wang, Hong-Bo Xin and Guanju Ji

Department of Biomedical Sciences, College of Veterinary Medicine, Cornell University, Ithaca, NY 14853-6401, USA

Abstract. Recent experiments have revealed an unanticipated complexity in Ca^{2+} release processes in smooth muscle. While Ca^{2+} release via stimulation of phospholipase C activity and the gating of inositol-1,4,5-trisphosphate ($InsP_3$) receptor Ca^{2+} channels has been well characterized, the role of the homologous family of sarcoplasmic reticulum (SR) Ca^{2+} channels, ryanodine receptors (RyRs), in excitation–contraction coupling in smooth muscle is less clear. These Ca^{2+}-gated SR Ca^{2+} channels produce unitary Ca^{2+} sparks, which in turn open Ca^{2+}-activated membrane ion channels such as Ca^{2+}-activated Cl^- channels (Cl_{Ca}) and Ca^{2+}-activated K^+ channels. In this manner local Ca^{2+} sparks trigger spontaneous transient inward and outward currents, thereby driving physiological electrical activity in smooth muscle. Here we summarize some of our recent findings on Ca^{2+}-induced Ca^{2+} release (CICR) and stretch-induced Ca^{2+} release (SICR) in smooth muscle.

2002 Role of the sarcoplasmic reticulum in smooth muscle. Wiley, Chichester (Novartis Foundation Symposium 246) p 108–124

Over the past several years the understanding of the molecular processes underlying Ca^{2+} release in muscle has been substantially advanced. One important finding has been the identification of spontaneous, localized Ca^{2+} release events (Ca^{2+} sparks) in smooth muscle. High-speed confocal microscopy has allowed the visualization of ryanodine receptor (RyR)-mediated spontaneous Ca^{2+} sparks and Ca^{2+} waves in smooth muscle cells by a number of groups (Nelson et al 1995, Kannan et al 1997, ZhuGe et al 1998, Mironneau et al 1996, Wier et al 1997, Imaizumi et al 1998, Collier et al 2000, Gordienko et al 2001). These events have been observed in isolated myocytes and in intact smooth tissues, suggesting that they play a role in the regulation of smooth muscle tone. The fact that 'spontaneous' Ca^{2+} release from the sarcoplasmic reticulum (SR) drives electrical activity represents a unique paradigm for the generation of electrical activity in smooth muscle, which may be uncoupled from neural control. Moreover, the demonstration of Ca^{2+}-induced Ca^{2+} release (CICR) in certain smooth muscles (Ganitkevich & Isenberg 1992, Imaizumi et al 1998, Collier et al 2000) provides

further evidence for a complex interaction between the biochemically coupled inositol-1,4,5-trisphosphate (InsP$_3$) and the Ca^{2+}-coupled RyR Ca^{2+} release systems.

Regulation of [Ca^{2+}]$_i$ in smooth muscle

In smooth muscle, the relationship between tension and Ca^{2+} has long been appreciated, and a rise in cytosolic Ca^{2+} has been shown to be a necessary and sufficient event for contraction. While Ca^{2+}-independent modulatory processes exist in smooth muscle (Kitazawa et al 1989), full activation of contractile proteins does not occur in the absence of a rise in bulk cytosolic free Ca^{2+} concentration ([Ca^{2+}]$_i$) and following force development, decreases in [Ca^{2+}]$_i$ result in muscle relaxation (Somlyo & Himpens 1989). The regulation of cytosolic Ca^{2+} concentration in smooth muscle is characterized by numerous Ca^{2+}-permeant ion channels mediating Ca^{2+} flux across the sarcolemma and SR, and by a substantial diversity between tissues with regard to the extent that individual channels contribute to excitation–contraction (EC) coupling. The relative complexity of Ca^{2+} signalling in smooth muscle is immediately apparent if one compares the processes underlying Ca^{2+} transport during EC coupling between skeletal and smooth muscle. During excitation of skeletal muscle a single neurotransmitter (acetylcholine) binds to a single type of receptor/ligand-gated ion channel (nicotinic receptor), and mediates Ca^{2+} flux from the SR to the cytosol via a single type of Ca^{2+} channel — RyR1. By contrast, smooth muscle EC coupling is marked by redundancy at every level of activation. Multiple neurotransmitters and autocoids bind to cognate receptors that include ligand-gated (ionotropic) cation channels with variable Ca^{2+} permeability, and G protein-coupled receptors. The former receptor channels are analogous to the nicotinic receptor in skeletal muscle in that they generate a postsynaptic potential that alters the membrane potential, thereby regulating the activity of voltage-dependent channels, including voltage-dependent Ca^{2+} channels. Ca^{2+} release from the SR in smooth muscle is similarly redundant, mediated by two tetrameric intracellular Ca^{2+} channels, InsP$_3$ receptor (InsP$_3$R) and RyR channels. Both of these channels are Ca^{2+} sensitive and can support spatially transmitted Ca^{2+} signalling (Ca^{2+} waves), although the extent to which InsP$_3$ channels are activated by a rise in [Ca^{2+}]$_i$ in the absence of stimulation of phospholipase C (PLC) is uncertain. Release of SR Ca^{2+} plays two prominent, and possibly discrete, roles. Release of Ca^{2+} can produce a global rise in [Ca^{2+}]$_i$ and contraction of the syncytial tissue, but also serves to gate Ca^{2+}-activated sarcolemmal channels that are highly expressed in smooth muscle, thereby indirectly modulating the activity of voltage-dependent ion channels such as voltage-dependent Ca^{2+} channels. These complex responses to extracellular signals are imposed upon a pre-existing

Ca^{2+} homeostasis that results from graded Ca^{2+} influx through voltage-dependent Ca^{2+} channels (Fleischmann et al 1994) and spontaneous intracellular Ca^{2+} release through ryanodine receptors. Thus the myocyte integrates numerous Ca^{2+} inputs and the concentration of cytosolic free Ca^{2+} ($[Ca^{2+}]_i$) at any given time is the net result of the activity of these Ca^{2+} channels, as well as that of Ca^{2+} pumps and exchangers that remove Ca^{2+} ions from the cytosol.

Ryanodine receptors and CICR in smooth muscle

Ca^{2+} efflux from the SR in muscle is mediated by two important Ca^{2+} permeant channels: $InsP_3Rs$ and RyRs. These channels share substantial sequence homology, protein topology and likely fourfold symmetry (Marks et al 1990, Chadwick et al 1990). While neurotransmitter activation of PLC-linked receptors and attendant activation of $InsP_3R$ is a well characterized mechanism by which activation of many smooth muscles occurs, the role of RyR-mediated Ca^{2+} release in EC coupling in smooth muscle is less well understood. RyRs are Ca^{2+}-permeant SR channels that are gated by an increase in $[Ca^{2+}]_i$; RyRs show a biphasic Ca^{2+} dependence, with increases in $[Ca^{2+}]_i$ below 1 μM resulting in channel gating and higher levels causing channel inactivation (Bezprozvanny et al 1991). In cardiac muscle, Ca^{2+} influx through voltage-dependent Ca^{2+} channels activates closely associated RyR2 located in the junctional SR to release Ca^{2+} into the cytoplasm, a process termed CICR (Fabiato 1983). The coupling process underlying CICR has been shown to involve a local increase in $[Ca^{2+}]_i$ in the microdomain of the L-type Ca^{2+} channel, which is sensed by RyR, resulting in RyR gating and a localized Ca^{2+} release, termed a Ca^{2+} spark (Cheng et al 1993, Lopez-Lopez et al 1995, Cannell et al 1995).

RyR are expressed in smooth muscle and functionally coupled to the SR, as evidenced by numerous studies demonstrating RyR-mediated Ca^{2+} release following exposure to caffeine, and specific experiments demonstrating RyR-mediated Ca^{2+} release (Nelson et al 1995, ZhuGe et al 1998, Collier et al 2000). Evidence indicates that specific RyR isoforms are differentially expressed in smooth muscle, a fact that could underlie observed differences in EC coupling between these tissues. Three RyR isoforms are encoded in the mammalian genome; RyR1 is principally expressed in skeletal muscle, where Ca^{2+} entry is not required for Ca^{2+} release (Armstrong et al 1972), but gating of the skeletal muscle L-type Ca^{2+} channel is physically coupled to RyR opening (Tanabe et al 1990, Nakai et al 1998). RyR2 is the major isoform expressed in heart cells and mediates CICR (Nabauer et al 1989) and the underlying localized Ca^{2+} release events, termed Ca^{2+} sparks (Cheng et al 1993, Cannell et al 1995). Finally, RyR3 is a ubiquitously expressed isoform whose function in many cells is unknown, although it appears to play a key role in Ca^{2+} signalling in some non-excitable cells (Giannini et al 1995).

RyR3 has been reported to be the major isoform in some smooth muscle tissues (Neylon et al 1995), but RyR2 has been reported to be the primary isoform expressed in urinary bladder smooth muscle (Chambers et al 1999), in which CICR has been definitively established (Ganitkevich & Isenberg 1992, Imaizumi et al 1998, Collier et al 2000). RyR1 has also been reported in smooth muscle at the expression and functional levels (Neylon et al 1995). It should be noted that determination of isoform expression has largely been reported using non-quantitative RT-PCR methods, with associated interpretive difficulties.

Our studies in the urinary bladder have identified a novel form of CICR. As described above, CICR in heart cells involves an extremely tight coupling between Ca^{2+} ions permeating L-type Ca^{2+} channels and RyR, and evidence indicates that the opening of a single L-type channel is sufficient to elicit a Ca^{2+} spark (Cannell et al 1995, Collier et al 1999). By contrast, CICR is characterized by what we have termed 'loose coupling'. As shown in Fig. 1, depolarization of a single rabbit urinary bladder myocyte activates an L-type Ca^{2+} current (middle traces) and Ca^2 release. This release was shown to be due to the gating of RyR receptors, as it was inhibited in the presence of ryanodine or ruthenium red, but was unaffected by dialysis with heparin. However, there are several unique features about this process in smooth muscle, relative to RyR-mediated CICR in heart. First, unlike in cardiac myoyctes, the release is focal, arising from individual areas of the cell, which are also the site of spontaneous Ca^{2+} sparks and correspond to frequent discharge sites (FDSs), or sites of repeated Ca^{2+} spark initiation (Bolton & Gordienko 1998). Second, there is a clear delay between the onset of the Ca^{2+} current and the beginning of the Ca^{2+} spark. Third, as shown in Fig. 2, L-type Ca^{2+} currents can be evoked without triggering CICR if the time of opening of the channel is kept short or the percentage of channels that open is small, resulting in a net flux of Ca^{2+} ions that is insufficient to trigger release. Fourth, the probability of CICR is a function of net Ca^{2+} flux, rather than L-type channel open probability or single-channel amplitude. These differences suggested to us that CICR in smooth muscle does not result from a close molecular coupling between L-type channels and RyR gating as occurs in striated muscle, but takes a non-obligate form in which the event triggering Ca^{2+} release from RyRs is not a local rise in Ca^{2+} at the mouth of the L-type channel, but an increase in [Ca^{2+}]$_i$ sufficient to activate clusters of RyRs at the specialized release sites.

To test this hypothesis, we performed several experiments. In one set of experiments we attempted to maximize the single-channel amplitude of L-type channel openings by stepping to very positive potentials near E$_{Ca}$ to maximally increase channel open probability in the absence of Ca^{2+} flux, and then stepping back to very negative potentials where the single-channel current amplitude was very high before the channels rapidly closed. Using this strategy we were unable to elicit individual sparks, indicating that even at single-channel current

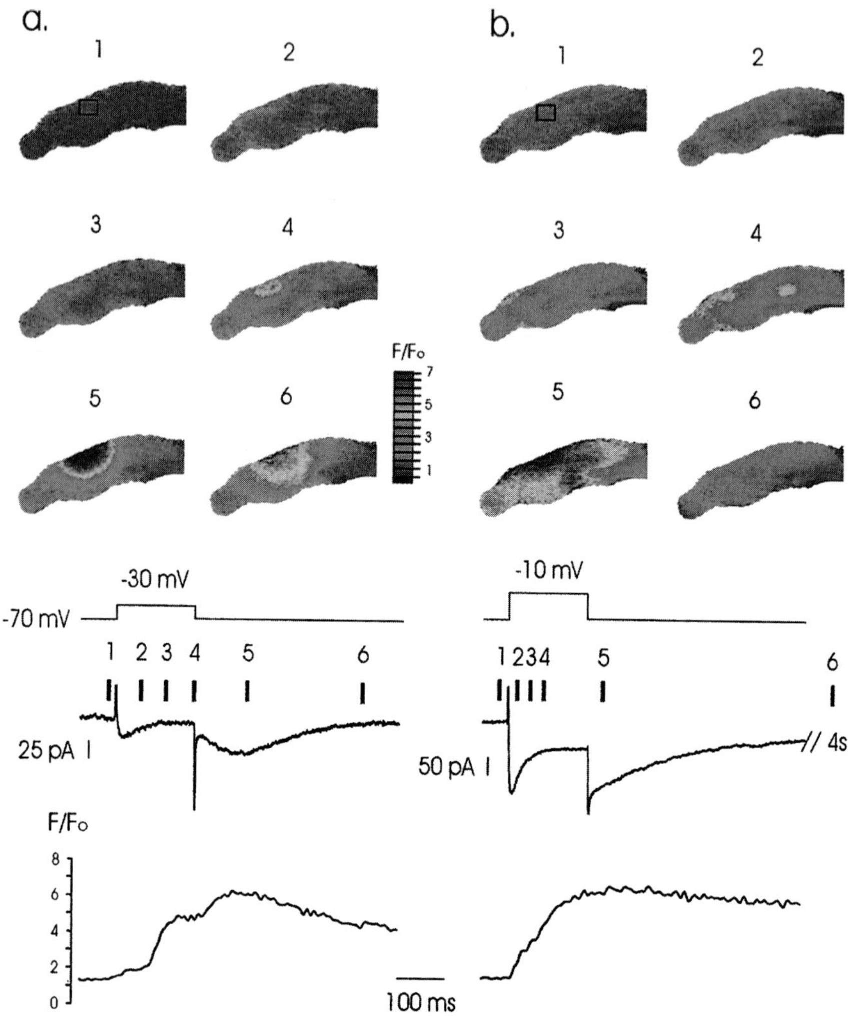

FIG. 1.   CICR in smooth muscle. CICR in a myocyte depolarized to $-30\,mV$ (left) or $-10\,mV$ (right). Sequential images are shown (above) at the time points indicated on the current traces (middle). Mean fluorescence for the pixels from the box above show the time course of the spark (below). Note the delay in onset of the spark relative to the current. (From Collier et al 2000.)

amplitudes much higher than encountered physiologically, Ca^{2+} entering through the channel was not sufficient to reach RyRs and activate Ca^{2+} sparks. In a second group of experiments, we used high concentrations of the mobile Ca^{2+} buffer EGTA to attempt to uncouple I_{Ca} and RyR gating. In these experiments both ventricular cardiomyocytes and bladder smooth muscle cells were loaded with

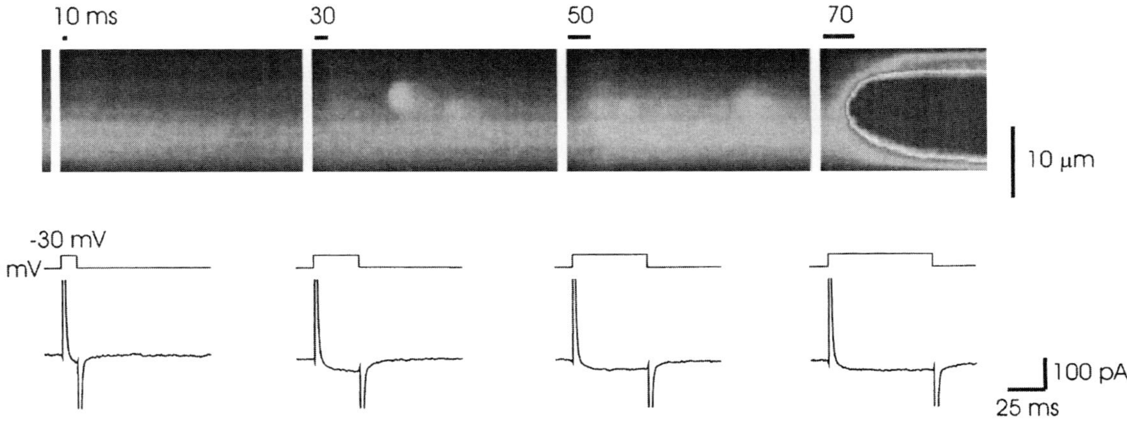

FIG. 2. Ca^{2+} channels can open without activating Ca^{2+} sparks. Individual sparks are not evoked with very brief depolarizations to activate I_{Ca}. With longer depolarizations, delayed Ca^{2+} sparks (middle panels) or a propagated wave (far right panel) are evoked. X–T linescan image is shown above, voltage and current, below. (From Collier et al 2000.)

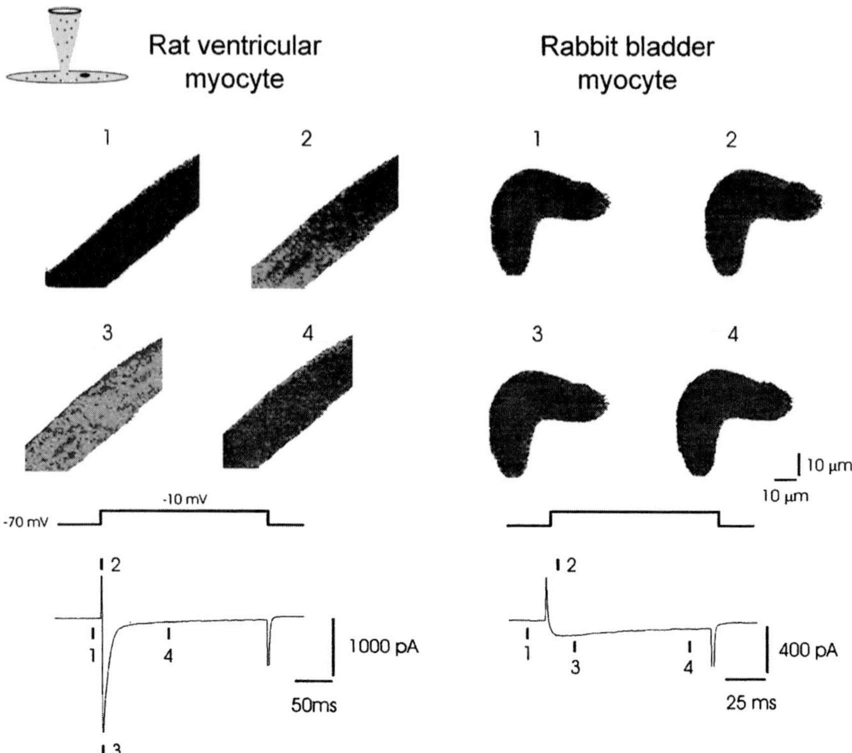

FIG. 3.   Tight coupling in cardiac but not smooth muscle myocytes. Cells dialysed with 17 nM mobile Ca^{2+} buffer (EGTA). When depolarized, brief Ca^{2+} release events are seen in ventricular myocytes, indicating RyR gating occurs before the mobile buffer can scavenge the gating Ca^{2+} ions. Conversely, in smooth muscle cells CICR is completely blocked. The simplest interpretation of these data is that Ca^{2+} ions must traverse a distance of at least 100 nm, the distance beyond which the mobile buffer can prevent a rise in Ca^{2+}. (From Collier et al 2000.)

EGTA and Fluo-4 and simultaneous recordings of I_{Ca} and Ca^{2+} fluorescence were made. As expected for tight coupling in cardiac myocytes, brief Ca^{2+} sparks were recorded following activation of I_{Ca} (Fig. 3). Conversely, however, coupling between L-type Ca^{2+} channels and RyRs was completely abrogated in bladder myocytes recorded under identical conditions.

Taken together, our studies indicate a 'loose coupling' between I_{Ca} and RyR gating. The physiological significance of this coupling is yet to be determined. However, we suggest that this system provides for a graded Ca^{2+} release that depends on integrated neural activity. That is, single or low frequency synaptic depolarizations will not be amplified by intracellular Ca^{2+} release, whereas repeated spike activity will be integrated, in that the attendant increase in global

E-C Coupling in Muscle

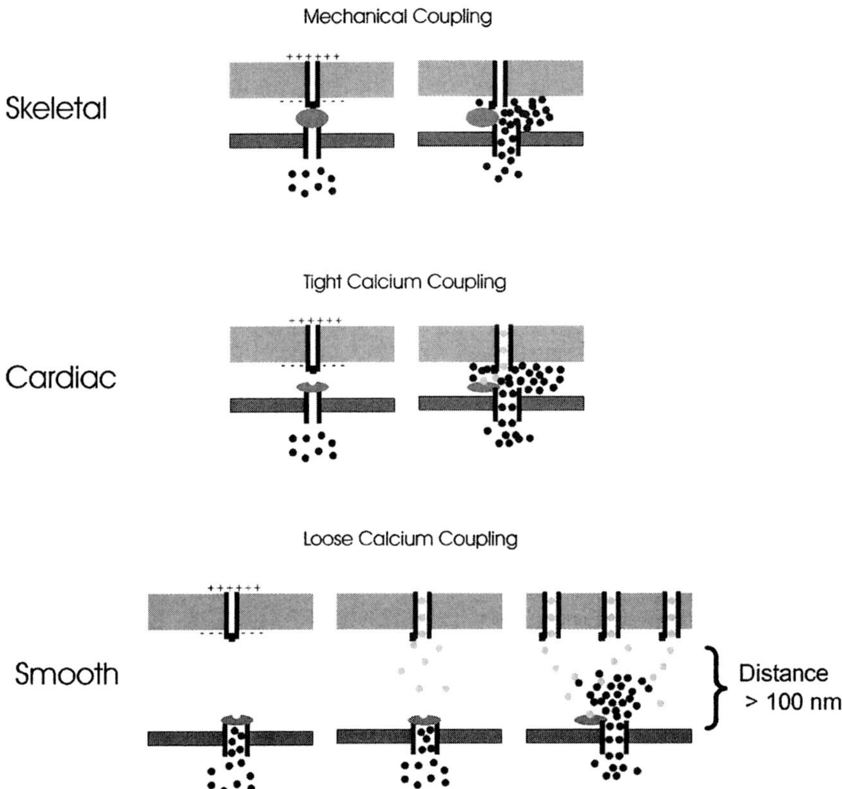

FIG. 4. Model of RyR-mediated Ca^{2+} release in muscle. Evidence indicates allosteric coupling between the L-type Ca^{2+} channel and RyR1 in skeletal muscle, tight coupling between the channels in cardiac (RyR2), and loose coupling to localized RyR (probably RyR2) in smooth muscle.

[Ca^{2+}]$_i$ will activate CICR. This represents a substantial departure from coupling processes in striated muscle, as summarized in Fig. 4.

Stretch-induced Ca^{2+} release in smooth muscle

Recent experiments have revealed a second process leading to RyR-mediated Ca^{2+} release. The application of linear stress to single rabbit urinary bladder myocytes by stretching single cells at each end resulted in the consistent activation of repeated

Ca^{2+} sparks (Ji et al 2002). Single Ca^{2+} sparks with kinetics quite similar to spontaneous events are evoked from FDSs following moderate stretch of cells to about 20% of its resting length, by means of two pipettes attached at each end of the cell. In these experiments, stretch-induced Ca^{2+} sparks occur from the same sites as those at which spontaneous Ca^{2+} sparks are observed before the application of stretch. Equivalent results have been obtained from mouse urinary bladder myocytes, indicating that stretch-induced Ca^{2+} release (SICR) may be a widespread phenomenon.

Our initial hypothesis was that Ca^{2+}-permeant, stretch-activated cation channels, which had been reported in smooth muscle (Kirber et al 1988, Wellner & Isenberg 1993), were activated by linear stress, resulting in a form of CICR similar to that observed by activating L-type Ca^{2+} channels. Surprisingly, however, we found that SICR occurred in external solutions in which Ca^{2+} was omitted or buffered to nanomolar levels and in the presence of $100\,\mu M$ Gd^{3+} ions, which block stretch-activated cation channels (Wellner & Isenberg 1994). Experiments also demonstrated that SICR is not blocked by heparin dialysis, but is completely blocked by exposure of cells to ryanodine, thus confirming the role of RyR in this process. However, in experiments similar to those shown in Fig. 3, dialysis of cells with $17\,mM$ EGTA was not sufficient to eliminate brief stretch-induced Ca^{2+} sparks. Thus SICR appears to be a process separate from CICR, in that an increase in $[Ca^{2+}]_i$ is not required for RyR gating.

The demonstration of Ca^{2+} release events of equivalent size and duration arising from FDSs suggests that linear stretch activates RyR gating in an undetermined manner, but one independent from extracellular Ca^{2+} influx; this process is somewhat reminiscent of the mechanical or allosteric gating of RyR1 in skeletal muscle (Fig. 4). Future studies will be aimed at the elucidation of the coupling mechanisms underlying this process, the RyR isoforms involved, and, most importantly, the physiological relevance of SICR for processes such as the generation of active tone following the expansion of smooth muscle containing organs and blood vessels.

Conclusion

Recent revelations concerning the complexity of Ca^{2+} release in smooth muscle, particularly the finding of spontaneous and length-associated Ca^{2+} release, suggest a model of excitability that differs substantially from conventional assumptions about the control of electrical activity in airway smooth muscle and point to the importance of factors that regulate intracellular Ca^{2+} release through RyR channels, in addition to neurotransmitter regulation of Ca^{2+} release through PLC-linked receptors. As shown in Fig. 5, rather than electrical activity resulting solely from postsynaptic responses to descending neural control, activity may be

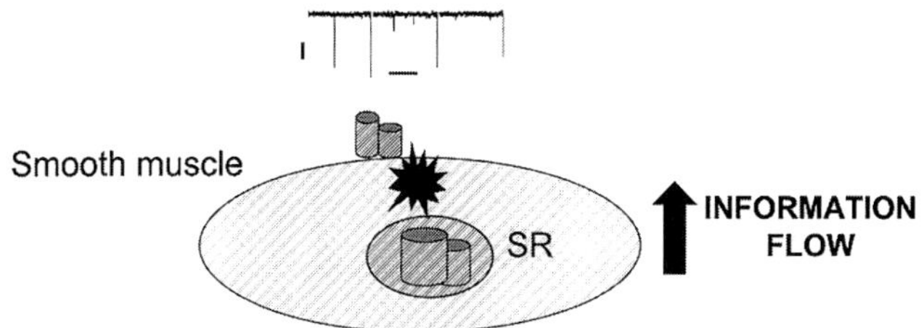

FIG. 5.   Ca^{2+} sparks drive electrical activity in myocytes. In contrast to the traditional concept of electrical activity deriving from descending neural control via postsynaptic responses, 'spontaneous' Ca^{2+} release results in electrical activity in smooth muscle. The figure shows a Ca^{2+} spark activating sarcolemmal Ca^{2+}-activated Cl$^-$ channels and spontaneous transient inward currents (STICs) (current trace above). Whether Ca^{2+} sparks activate outward STOCs (Ca^{2+}-activated K$^+$ currents) or STICs will depend on the proportion of channels expressed and the resting potential of the myocyte.

dominated by intracellular Ca^{2+} release events which drive sarcolemmal Ca^{2+}-activated ion channels, and thereby dictate electrical activity. An elegant example of the importance of this regulatory system was recently provided by Brenner and colleagues, who demonstrated the consequences of altering the sensitivity of the target ion channel on systemic blood pressure (Brenner et al 2000).

RyRs do not exist as isolated SR ion channels, but as protein complexes subject to modulation by cellular metabolites, [Ca^{2+}]$_i$, kinases and other factors (e.g. Marx et al 2000). Currently, very little is known with respect to the expression and localization of RyR isoforms, the regulatory factors underlying sensitivity of the complex to gating, and the effect of luminal (SR) Ca^{2+} on the probability of spontaneous or triggered release. A renewed focus on the role of RyRs in smooth muscle should help move the field from the initial discovery of these exciting phenomena to a clearer understanding of the function of this system in diverse smooth muscle tissues.

References

Armstrong CM, Bezanilla FM, Horowicz P 1972 Twitches in the presence of ethylene glycol bis(-aminoethyl ether)-N,N'-tetraacetic acid. Biochim Biophys Acta 267:605–608

Bezprozvanny I, Watras J, Ehrlich BE 1991 Bell-shaped calcium-response curves of Ins(1,4,5)P$_3$- and calcium-gated channels from endoplasmic reticulum of cerebellum. Nature 351:751–754

Bolton TB, Gordienko DV 1998 Confocal imaging of calcium release events in single smooth muscle cells. Acta Physiol Scand 164:567–575

Brenner R, Perez GJ, Bonev AD et al 2000 Vasoregulation by the β1 subunit of the calcium-activated potassium channel. Nature 407:870–876

Cannell MB, Cheng H, Lederer WJ 1995 The control of calcium release in heart muscle. Science 268:1045–1049

Chadwick CC, Saito A, Fleischer S 1990 Isolation and characterization of the inositol trisphosphate receptor from smooth muscle. Proc Natl Acad Sci USA 87:2132–2136

Chambers P, Neal DE, Gillespie JI 1999 Ryanodine receptors in human bladder smooth muscle. Exp Physiol 84:41–46

Cheng H, Lederer WJ, Cannell MB 1993 Calcium sparks: elementary events underlying excitation–contraction coupling in heart muscle. Science 262:740–744

Collier ML, Thomas AP, Berlin JR 1999 Relationship between L-type Ca^{2+} current and unitary sarcoplasmic reticulum Ca^{2+} release events in rat ventricular myocytes. J Physiol (Lond) 516:117–128

Collier ML, Ji G, Wang Y, Kotlikoff MI 2000 Calcium-induced calcium release in smooth muscle: loose coupling between the action potential and calcium release. J Gen Physiol 115:653–662

Fabiato A 1983 Calcium-induced release of calcium from the cardiac sarcoplasmic reticulum. Am J Physiol (Cell Physiol) 245:C1–C14

Fleischmann BK, Murray RK, Kotlikoff MI 1994 Voltage window for sustained elevation of cytosolic calcium in smooth muscle cells. Proc Natl Acad Sci USA 91:11914–11918

Ganitkevich VY, Isenberg G 1992 Contribution of Ca^{2+}-induced Ca^{2+} release to the $[Ca^{2+}]_i$ transients in myocytes from guinea-pig urinary bladder. J Physiol 458:119–137

Giannini G, Conti A, Mammarella S, Scrobogna M, Sorrentino V 1995 The ryanodine receptor/ calcium channel genes are widely and differentially expressed in murine brain and peripheral tissues. J Cell Biol 128:893–904

Gordienko DV, Greenwood IA, Bolton TB 2001 Direct visualization of sarcoplasmic reticulum regions discharging Ca^{2+} sparks in vascular myocytes. Cell Calcium 29:13–28

Imaizumi Y, Torii Y, Ohi Y et al 1998 Ca^{2+} images and K^+ current during depolarization in smooth muscle cells of the guinea-pig vas deferens and urinary bladder. J Physiol (Lond) 510:705–719

Ji G, Feldman M, Barsotti RJ, Kotlikoff MI 2002 Stretch-induced calcium release in smooth muscle. J Gen Physiol, in press

Kannan MS, Prakash YS, Brenner T, Mickelson JR, Sieck GC 1997 Role of ryanodine receptor channels in Ca^{2+} oscillations of porcine tracheal smooth muscle. Am J Physiol 272:L659–L664

Kirber MT, Walsh JVJ, Singer JJ 1988 Stretch-activated ion channels in smooth muscle: a mechanism for the initiation of stretch-induced contraction. Pflüg Arch Eur J Physiol 412:339–345

Kitazawa T, Kobayashi S, Horiuti K, Somlyo AV, Somlyo AP 1989 Receptor coupled, permeabilized smooth muscle: role of the phosphatidylinositol cascade, G proteins and modulation of the contractile response to Ca^{2+}. J Biol Chem 264:5339–5342

Lopez-Lopez JR, Shacklock PS, Balke CW, Wier WG 1995 Local calcium transients triggered by single L-type calcium channel currents in cardiac cells. Science 268:1042–1045

Marks AR, Fleischer S, Tempst P 1990 Surface topography analysis of the ryanodine receptor/ junctional channel complex based on proteolysis sensitivity mapping. J Biol Chem 265:13143–13149

Marx SO, Reiken S, Hisamatsu Y et al 2000 PKA phosphorylation dissociates FKBP12.6 from the calcium release channel (ryanodine receptor): defective regulation in failing hearts. Cell 101:365–376

Mironneau J, Arnaudeau S, Macrez-Lepretre N, Boittin FX 1996 Ca^{2+} sparks and Ca^{2+} waves activate different Ca^{2+}-dependent ion channels in single myocytes from rat portal vein. Cell Calcium 20:153–160

Nabauer M, Callewaert G, Cleemann L, Morad M 1989 Regulation of calcium release is gated by calcium current, not gating charge, in cardiac myocytes. Science 244:800–803

Nakai J, Tanabe T, Konno T, Adams B, Beam KG 1998 Localization in the II–III loop of the dihydropyridine receptor of a sequence critical for excitation–contraction coupling. J Biol Chem 273:24983–24986.

Nelson MT, Cheng H, Rubart M et al 1995 Relaxation of arterial smooth muscle by calcium sparks. Science 270:633–637

Neylon CB, Richards SM, Larsen MA, Agrotis A, Bobik A 1995 Multiple types of ryanodine receptor/Ca^{2+} release channels are expressed in vascular smooth muscle. Biochem Biophys Res Commun 215:814–821

Somlyo AP, Himpens B 1989 Cell calcium and its regulation in smooth muscle. FASEB J 3:2266–2276

Tanabe T, Beam KG, Adams BA, Niidome T, Numa S 1990 Regions of the skeletal muscle dihydropyridine receptor critical for excitation–contraction coupling. Nature 346:567–569

Wellner MC, Isenberg G 1993 Stretch-activated nonselective cation channels in urinary bladder myocytes: importance for pacemaker potentials and myogenic response. Exper Suppl (Basel) 66:93–99

Wellner MC, Isenberg G 1994 Stretch effects on whole-cell currents of guinea-pig urinary bladder myocytes. J Physiol (London) 480:439–448

Wier WG, ter Keurs HE, Marban E, Gao WD, Balke CW 1997 Ca^{2+} 'sparks' and waves in intact ventricular muscle resolved by confocal imaging. Circ Res 81:462–469

ZhuGe R, Sims SM, Tuft RA, Fogarty KE, Walsh JV Jr 1998 Ca^{2+} sparks activate K$^+$ and Cl$^-$ channels, resulting in spontaneous transient currents in guinea-pig tracheal myocytes. J Physiol (London) 513:711–718

DISCUSSION

McHale: I was fascinated by the stretch-induced firing sparks. Does it matter how the cell is stretched? For example, do you see this with osmotic swelling?

Kotlikoff: We haven't done this. We first glued pipettes onto the end of cells. We then also found that just producing a seal with patch pipettes allows us to stretch cells quite nicely. This allows us to voltage clamp and record the currents at the same time, which we do in one of those experiments, demonstrating activation of a Ca^{2+}-activated Cl$^-$ current.

Paul: Is it length or is it stretch? Does rate of stretch matter?

Kotlikoff: We are just now setting up one of these actuators that will allow us to move the pipette in a way in which we can control the velocity, so I don't know the answer to that yet. Obviously by stretching the cell we are both increasing the length, and also increasing tension on the cell.

Hellstrand: Does this mechanism explain the differences in the relative importance of spark activity in controlling membrane potential in different experiments from different labs? Some people use pressurized arteries and see more of an effect than people who use other preparations.

Kotlikoff: I would hope so. I don't think we have evidence of this, but it is clearly a factor.

Nelson: The pressurized arteries depolarize when Ca^{2+} goes up, too.

Somlyo: Can we relate the kinetics of this process to what happens in whole muscle? As you know, when a smooth muscle is stretched and there is a myogenic contraction, there is a lag phase of about a second. This is surprisingly long for channel activation. How can you relate these kinetics to the kinetics of a lag of 500 ms? Stretch activation has also been suggested to involve PLC activation.

Kotlikoff: Sometimes we see a substantial lag; other times it is shorter than 500 ms. I don't know the answer. Because of the number of differences between a single-cell stretch experiment and real physiology, I don't know whether this explains things such as pressure-induced vascular tone. One thing that will be informative here is if the RyR2 knockouts are viable and we can look at them.

Somlyo: What is the physiology of the FKB12.6 knockouts?

Kotlikoff: These animals have an altered contractile response to agonists *in vitro*. This is about as much as we can say at the moment. We do see a phenotype in terms of their contractile properties. In the RyR2 knockouts we would expect to have a loss of stretch-induced Ca^{2+} release and CICR in smooth muscle. If you have normal pressurized vasomotor responses in those animals, I think this would suggest that the phenomenon that we have described is not essential for myogenic tone.

Nelson: We have examined sparks in pressurized arteries. The effects of pressure on Ca^{2+} sparks appear to be the result of changes in intracellular Ca^{2+}. pH changes RyR open probability; alkalinization increases it. One thing I would look at is whether any of the manoeuvres cause alkalinization.

Kotlikoff: Are you thinking that stretch is somehow alkalinizing?

Nelson: That would be a simple way to explain it.

Bolton: Loose coupling would be another explanation. You are working on rabbit bladder. Imaizumi's group have worked on guinea-pig, and in the vas deferens and urinary bladder they see a spark or hotspot within 10 ms of a depolarization. You can either retreat into a species difference argument, or perhaps they selected their cells very heavily.

Kotlikoff: I have seen these data and would make the following comment. The important thing in my mind is not that you can get sparks to occur very close to the action potential, but that you can get coupled sparks to occur after a substantial decay. To decrease the decay you simply need to increase the current and activate it rapidly, thus choose a voltage at which the current activates maximally. The Imaizumi study depolarized cells to the peak of the Ca^{2+} current, and they are going to minimize this delay. We took the other strategy: I wanted to see how I could maximize it. You can't do this in the heart. I would suggest that if they did those experiments at -40 mV and slowed the Ca^{2+} response, that they would see a different picture. So the decay is one piece of evidence. The other is the EGTA dialysis. I think this speaks very strongly that there is a different physical relationship between the two channels than occurs in heart. A third piece of evidence is that you cannot get coupling at negative potentials in which the amplitude of the current is maximal, but the net flux is minimized.

Bolton: On the question of EGTA, you used 17 mM in your pipette, and you still saw increases in fluorescence when you stretched. How can you get a change in free Ca^{2+} activity with 17 mM EGTA?

Kotlikoff: In the 17 mM EGTA experiment we are looking at release. We dialyse in, and very rapidly after the dialysis we stretch. We are looking over the SR and seeing sparks released from the SR.

Bolton: How can you see a spark with 17 mM EGTA?

Kotlikoff: Because the mobile buffer is not quick enough to block this.

Nelson: BAPTA will abolish this.

Kotlikoff: I don't know whether it would completely abolish it.

Brading: There is a difference between guinea pig and rabbit bladder, because the former doesn't have Ca^{2+}-activated Cl$^-$ channels and the latter does.

Nelson: We have looked at guinea-pig bladder and we have the same results as Mike Kotlikoff in terms of the loose coupling.

McCarron: Is it really CICR or filling up of the stores?

Kotlikoff: The suggestion here is that rather than CICR where Ca^{2+} is binding presumably to the cytosolic component and gating opening, that somehow it is filling luminal SR, which is the gating trigger for RyR-mediated Ca^{2+} release. We don't have experiments that tell us one thing or another there. The only experiments there that we have are the EGTA experiments, whereas if you prevent the rise in cytosolic Ca^{2+} you prevent this response. You could argue that we have somehow prevented it from going somewhere and triggering. We don't have real evidence for this other than we know that it is not communicating directly. I would point out, though, that as has been mentioned, the delay can be quite short, which makes me think that there is not time for extensive ATPase-mediated transport of Ca^{2+} into the SR to markedly alter SR content.

McHale: We can record a Ca^{2+}-activated K$^+$ current within 10 ms of the upstroke of the L-type Ca^{2+} current in the presence of 20 mM EGTA. I don't know whether this is direct activation of the channel.

Kotlikoff: This is a different result, and would suggest coupling between the L-type channel and the BK channel. I think that Mark Nelson will discuss coupling between these channels independent of Ca^{2+} release. We see similar results with respect to the activation of spontaneous inward currents.

Paul: At 20 mM EGTA, do you think there might be other effects of EGTA? In the early years in cardiac muscle people would go to great lengths to avoid using EGTA. They used other kinds of Ca^{2+} chelation because there was some hint that EGTA per se was doing this. That is a high concentration.

Blaustein: Here you have really good evidence for some kind of Ca^{2+}-induced release and the role of RyRs in activation of the muscle. This is very different from the kinds of things that Mark Nelson has shown in vascular smooth muscle, where release of Ca^{2+} from RyRs is involved in relaxation. Isn't this a good example of the enormous diversity in different tissues. Smooth muscle is so diverse when compared with cardiac or skeletal muscle.

Kotlikoff: I would agree, although we don't know what the functional effect of the release is, whether it is relaxation or contraction. One difference here is the prominent presence of inward currents that are depolarizing.

Blaustein: But you get a Ca^{2+} wave. In the vascular smooth muscle no Ca^{2+} waves occur from this. There is a clear hyperpolarization that results from actions at the local domain, where Ca^{2+} release triggers the activation of the BK channel.

Kotlikoff: There are two separate issues here. There does not appear to be CICR in the vascular muscle that I know of. Someone can correct me if I'm wrong, but this process isn't found in many smooth muscle cells. The RyR2-type cells presumably have something that approximates CICR; on the other hand there are tonic tissues that are presumably operated not by something that senses peak Ca^{2+} currents, because these currents are never fully activated. Presumably here the issue is the longer-term, steady-state effects of the voltage-dependent Ca^{2+} channel. We have previously shown that in tonic tissues substantial increase in cytosolic Ca^{2+} occurs within the 'Ca^{2+} window', or the voltage over which steady-state L-type Ca^{2+} channel activity raises intracellular Ca^{2+}. This in turn would be likely to affect RyR gating and Ca^{2+} sparks. The other issue is whether a Ca^{2+} wave necessarily results in contraction. I don't know that we know the answer to that yet.

Wray: Do we know anything about how the numbers or density of RyRs could play a role in the functional differences?

Nixon: Several years ago a study showed there were 10 times more $InsP_3$ receptors than RyRs in the guinea-pig intestinal smooth muscle (Wibo & Godfraind 1994). We have found in most smooth muscles that there are fewer RyRs.

Somlyo: This also depends on how well we detect them. I think we can agree that while there is a diversity in the distribution and function of RyRs in different smooth muscles, they all respond to $InsP_3$.

Sanders: Let me try to provoke you a little about the physiology behind the responses that you showed us. Both of them seem a little inconsistent with organ physiology. First, your data would predict that if you had a volley of action potentials in tissue, this would be followed by a long-lasting Ca^{2+} rise that would initiate Ca^{2+}-activated Cl^- current. In other words, there should be a period after depolarization that should increase the excitability of the tissue. Every time there was an excitable group of events, there would be an increased period of excitability. Is this consistent with what happens from electrical recording in the bladder? If there is a volley of action potentials, do you see an afterdepolarization?

Kotlikoff: We are blocking K$^+$ currents there. These are recorded in Cs.

Sanders: At the resting potential you are going to favour STICs from this increase in sparking.

Kotlikoff: But as you depolarize this will be lost. The relationship to Ca^{2+} is somewhat different as well. I don't know the answer; I think you could postulate that in the absence of Cs, this big wave is not just a depolarizing wave but that you have hyperpolarizing events.

Sanders: The other issue is that when the bladder is stretched, we don't want a lot of inward current generated. This would cause a mess! There must be counteracting phenomena to this primary stretch response.

Eisner: Isn't another complication that this doesn't take account of how the store refills? The first time you stimulate you may release a lot of Ca^{2+} from the store. I guess some of this will get pumped out of the cell, so you are in a different position next time: you have a store with less Ca^{2+} in it. Unless you take account of this, it is hard to predict what will happen.

Sanders: That's an interesting question. From a spark, what proportion of Ca^{2+} is recovered versus lost?

Eisner: For a spark, most of it isn't lost, but from a wave which will see the surface membrane, you will lose a lot. This then comes back to the question of cyclic ADP ribose, FK506 and so on that may potentiate Ca^{2+} release from the SR, but in the presence of these things you will end up with less Ca^{2+} in the SR.

Brading: Bladder has spontaneous action potentials that are Ca^{2+} based, so the stores will fill.

Nelson: We have done experiments on urinary bladder from guinea pig and mouse, and ryanodine does not decrease the phasic contractions, it increases them. It looks like the net effect of activating the RyRs to brake contractility. Presumably it does this by decreasing excitability.

Sanders: So what you are saying is that the resting potential is sitting in a situation where the Ca^{2+} sparks are preferentially activating outward currents, not inward currents.

Nelson: Yes; when ryanodine is added it depolarizes and also regulates the repolarization of the action potential. The net effect of ryanodine is an increase in excitability.

Kotlikoff: I think that's interesting and potentially true; my reservation is that when ryanodine is used in the tissues it is not the best evidence for the role of that process. It will open channels and decrease SR stores, and cause cytosolic Ca^{2+} to rise.

Sanders: In these volume organs, such as the bladder and the gut, things are different. When the gut is stretched, you don't necessarily get a myogenic contraction. It maintains its resting potential. It has to.

Somlyo: If you do a quick stretch you get a contraction. This has been shown in taeniae gut.

Sanders: Taeniae gut may be the only tissue like that. But the circular muscle of the gut has to expand; it has to do this to be a volume organ.

Iino: What do we know about the signal transduction upstream of cADP ribose in smooth muscle cells?

Kotlikoff: That's a good question. It's a complicated situation. One knockout affects both the synthesis and the removal of cADP ribose. To my knowledge we don't have good probes that would allow us to selectively alter this in a way that we could make predictions and test them.

Bradley: Lukyanenko et al (2001) have published a paper in which they suggest that cADP ribose causes its effect by increasing SERCA activity, and the reason that cADP ribose releases Ca^{2+} from the stores is because the store Ca^{2+} content significantly increases. When you knock out FKBP12.6, do you get a significant reduction in your store content? Could it be that cADP ribose isn't actually working on FKBP12.6, but instead on SERCA?

Kotlikoff: The best experiments are in the cardiac myocytes. We do not see an effect on stores in the cardiac myocytes. From everything we see they appear to be loaded quite nicely.

References

Lukyanenko V, Gyorke I, Wiesner TF, Gyorke S 2001 Potentiation of Ca^{2+} release by cADP ribose in the heart is mediated by enhanced SR Ca^{2+} uptake into the sarcoplasmic reticulum. Circ Res 89:614–622

Wibo M, Godfraind T 1994 Comparative localization of inositol 1,4,5-trisphosphate and ryanodine receptors in intestinal smooth muscle: an analytical subfractionation study. Biochem J 297:415–423

Organization of Ca²⁺ stores in vascular smooth muscle: functional implications

Mordecai P. Blaustein, Vera A. Golovina, Hong Song, Jacqueline Choate, Lubomira Lencesova, Shawn W. Robinson* and W. Gil Wier

*Department of Physiology and *Department of Medicine (Cardiology Division), University of Maryland School of Medicine, Baltimore, MD 21201, USA*

Abstract. Much evidence suggests that caffeine/ryanodine (Caf/Ry)-releasable and inositol-1,4,5-trisphosphate (InsP$_3$)-releasable Ca^{2+} stores in the sarcoplasmic reticulum (SR) of smooth muscles are at least partially distinct. We directly visualized SR stores in primary-cultured rat mesenteric artery myocytes with high-resolution digital imaging and the low-affinity Ca^{2+} indicator, Furaptra ($K_d = 75.6\,\mu$M). The SR appears to be a continuous tubular network. Nevertheless, SR Ca^{2+} stores are organized into small, separate, functionally independent compartments. Cyclopiazonic acid (CPA; inhibits SR Ca^{2+} pump) and Caf (or Ry) release Ca^{2+} from different, spatially distinct compartments. Similar heterogeneity is seen with serotonin (acts via InsP$_3$), which unloads only the CPA-sensitive compartments. Some of the SR ('junctional' SR; jSR) lies within 12–15 nm of the plasmalemma (PL). The jSR, the overlying PL microdomains, and the intervening, tiny volume of cytosol form junctional complexes ('PLasmERosomes'). Na$^+$ pumps with high-ouabain-affinity α2 or α3 subunits, Na$^+$/Ca^{2+} exchangers, and store-operated channels are confined to these PL microdomains, whereas Na$^+$ pumps with low-ouabain-affinity α1 subunits and plasma membrane Ca^{2+} pumps are uniformly distributed. As a result of this organization, low-dose ouabain can selectively modulate Na$^+$ and Ca^{2+} concentrations in the PLasmERosomes and jSR Ca^{2+} stores, and can thereby regulate Ca^{2+} signalling.

2002 Role of the sarcoplasmic reticulum in smooth muscle. Wiley, Chichester (Novartis Foundation Symposium 246) p 125–141

There is broad consensus that the sarcoplasmic reticulum (SR) of vascular smooth muscle cells (VSMCs) plays a critical role in concentrating and storing Ca^{2+} during relaxation, and/or releasing this ion during cell activation. The availability of Ca^{2+}-sensitive fluorochromes and digital imaging methods as well as progress in molecular methods during the past decade has greatly advanced our knowledge of Ca^{2+} regulation in cells. Nevertheless, many details of SR Ca^{2+} store organization, and of Ca^{2+} release and refilling, are still poorly understood. It is widely recognized that there are two classes of SR Ca^{2+} release mechanism: those associated with inositol-1,4,5-trisphosphate ($InsP_3$) receptors ($InsP_3Rs$), and those with ryanodine (Ry) receptors (RyRs) (Somlyo & Somlyo 1994). Whether the $InsP_3Rs$ and RyRs are associated with a single, intercommunicating Ca^{2+} store or with functionally independent stores, is, however, controversial. Several groups have provided indirect evidence that, at least in some smooth muscles, the two classes of receptors are associated with functionally distinct Ca^{2+} stores; these data have come from freshly-isolated cells and cultured cells (Yamazawa et al 1992, Tribe et al 1994, Flynn et al 2001, Janiak et al 2001). Other unresolved issues relate to the mechanism of SR Ca^{2+} store refilling following cell activation, and the relationship between the plasmalemma (PL) and the adjacent, sub-PL 'junctional' SR (jSR). The latter topic bears on the observations that VSMCs have specialized PL–SR junctions (Devine et al 1972, Somlyo & Franzini-Armstrong 1985) that resemble, structurally, those of skeletal and cardiac muscle (Franzini-Armstrong et al 1998). This, too, suggests that the SR is not simply a single inter-communicating system of tubules and cisterns with luminal continuity, but rather a more complex organelle with specialized subcompartments. These issues of SR organization and function are addressed below. For additional information, the reader should refer to several recent, comprehensive reviews of SR/endoplasmic reticulum function (Pozzan et al 1994, Meldolesi & Pozzan 1998, Carafoli et al 2001, Blaustein & Golovina 2001).

Do arterial myocytes have functionally independent, spatially distinct SR Ca^{2+} stores?

Low-affinity Ca^{2+} fluorochromes (chlortetracycline, Furaptra and Fura2-FF) were used to visualize, directly, the Ca^{2+} stores within the SR in intact, living, primary cultured arterial myocytes (Tribe et al 1994, Golovina & Blaustein 1997). The application of high spatial resolution imaging to resolve SR subcompartments was enabled by the development of methods to retain the fluorochromes within the SR while promoting extrusion from the cytosol (Golovina & Blaustein 1997, 2000). A key factor is the loading temperature: 37 °C, rather than the more usual 20–25 °C, which fosters dye retention in the cytosol. Confinement of Ca^{2+}-sensitive dye within the SR is revealed by the similarity of the Furaptra and 3,3′-dihexyloxacarbocyanine ($DiOC_6$) images in Figs 1A and B; $DiOC_6$ is a fluorescent lipophylic cation that stains SR and endoplasmic reticulum (ER) (Terasaki 1989).

There have been numerous attempts to measure intra-SR/ER Ca^{2+} concentrations; various methods have been used, and reported values range between $5\,\mu M$ and $5\,mM$ (Meldolesi & Pozzan 1998). Use of ratiometric flurochromes enabled direct measure of the Ca^{2+} concentration within the SR

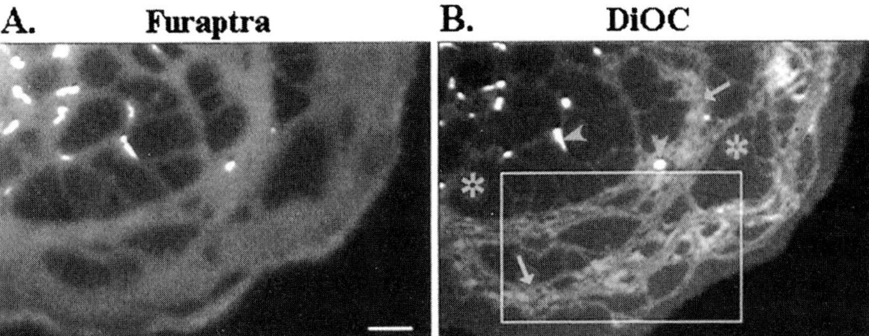

FIG. 1.   Furaptra (A) and DiOC$_6$ (B) images of a primary cultured mesenteric artery myocyte loaded with Furaptra-AM. DiOC$_6$ (3,3′-dihexyloxacarbocyanine) stains SR and mitochondria (very bright spots such as those indicated by arrowheads). This shows that the Furaptra is largely confined to the SR (e.g. regions indicated by arrows in 'B'). Asterisks in 'B' indicate some cytosolic areas that are free of SR. Bar in 'A'=2 μm (from Golovina & Blaustein 1997, with permission).

([Ca^{2+}]$_{SR}$); based on *in situ* calibration of Furaptra (Golovina & Blaustein 2000), [Ca^{2+}]$_{SR}$ in unstimulated myocytes is about 160 μM.

The line-scan data and graph (g) in Fig. 2A show that serotonin (5-HT), which activates the phosphoinositide cascade and promotes InsP$_3$ production, triggers Ca^{2+} release from some portions of the SR, but enhances filling of other portions of the SR. The effects of 5-HT are mimicked by cyclopiazonic acid (CPA), a selective inhibitor of the SR/ER Ca^{2+} pump (SERCA) (Fig. 2Ad–f). In contrast, Fig. 2B shows that the stores that were loaded by 5-HT and CPA were unloaded by caffeine (Caf), which opens Ry-sensitive SR Ca^{2+} channels; Ry had the same effect as Caf (not shown, but see Golovina & Blaustein 1997). The line-scan data in Fig. 2Ab–f reveal that the CPA- (and InsP$_3$-) releasable and Caf-releasable Ca^{2+} stores appear to be contained within contiguous elements of the SR. The two types of stores are less than 0.2 μm apart (indicated by the steep gradients of [Ca^{2+}]$_{SR}$; Fig. 2Ac,e). Such large Ca^{2+} gradients ($>$35 μM) are maintained for many minutes (Fig. 2Ag); this would seem to require some type of physical barrier between the subcompartments, although such a barrier has not yet been observed.

In these cultured VSMCs, about 56% of the SR released Ca^{2+} only in response to CPA (and, presumably, InsP$_3$), and about 22% only in response to Caf; only about 15% of the SR responded to both CPA and Caf. Moreover, the same stores could be refilled and re-emptied during a 30–60 min period (the longest times tested). These results indicate that the SR Ca^{2+} stores are organized into (at least) two classes of small compartments that are relatively stable. It is noteworthy that similar results have been reported in other cell types, including neurons and astrocytes (Golovina

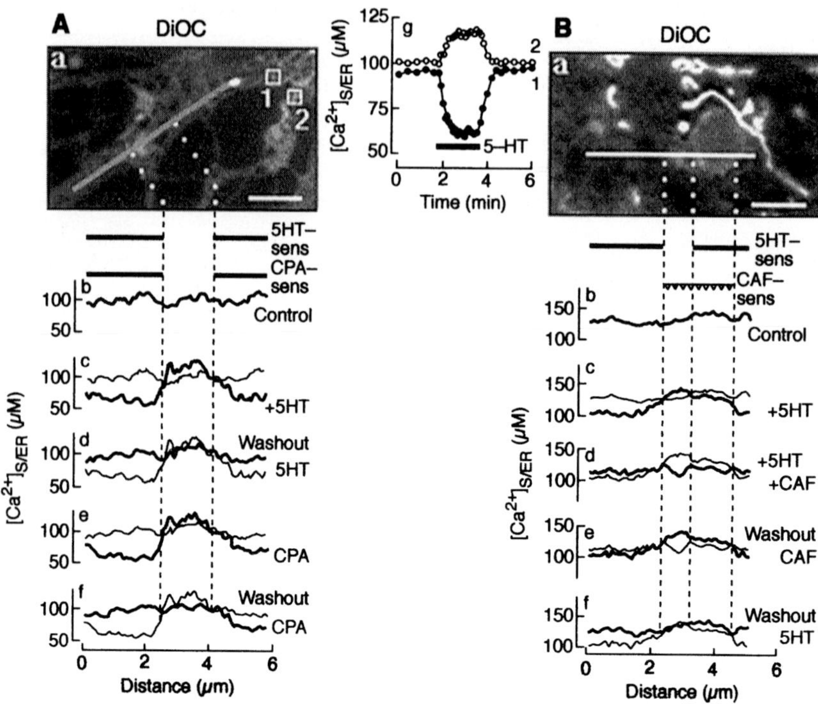

FIG. 2. Effects of serotonin (5-HT), cyclopiazonic acid (CPA), and caffeine (CAF) on Ca^{2+} stores in intact, non-permeabilized, primary cultured mesenteric artery myocytes. The cells were loaded with Furaptra in order to measure $[Ca^{2+}]_{SR}$ (Golovina & Blaustein 1997). (A) Effects of 5-HT and CPA on a myocyte. (Aa) $DiOC_6$ image of a portion of a cell. Long white line shows the 'line scan' position used to determine the Ca^{2+} concentration profile ($[Ca^{2+}]_{SR}$) along an element of the SR for panels b–f ; The analysed area was 5×72 pixels ($0.4\,\mu m \times 5.8\,\mu m$). (b) Control data; in panels c–f, the thin line corresponds to the $[Ca^{2+}]_{SR}$ data from the condition in the preceding panel, in order to show how $[Ca^{2+}]_{SR}$ changed. (c) 5-HT addition; (d) 5-HT washout; (e) CPA addition; and (f) subsequent CPA washout. Bars above graphs indicate SR regions depleted by 5-HT and CPA; note that 5-HT and CPA depleted the same regions. (g) Time-course curves that illustrate the 5-HT-evoked changes in $[Ca^{2+}]_{SR}$ within boxes 1 and 2 in image a. (B) Effects of 5-HT and CAF on a myocyte. (a) $DiOC_6$ image as in (Aa); the long horizontal white line shows the position of the line scan for panels b–f. Analysed area was 5×64 pixels ($0.4\,\mu m \times 5.1\,\mu m$). (b) Control data. (c–f as in Fig. 2A); (c) 5-HT serotonin addition; (d) CAF addition; (e) CAF washout; and (f) subsequent 5-HT washout. Bars above line scan graphs indicate SR regions depleted by 5-HT and by CAF. The CAF-sensitive regions are those in which 5-HT (panel Bc) and CPA (Panel Ac–f) raised $[Ca^{2+}]_{SR}$. $[Ca^{2+}]_{SR}$ decreased by $>6\,\mu M$ in 62% of the SR pixels, and increased by $>6\,\mu M$ in 32% of the SR pixels during 5-HT exposure. The $6\,\mu M$ threshold was chosen because the mean variation in $[Ca^{2+}]_{SR}$ in individual pixels from image to image in unstimulated cells was 5–$7\,\mu M$. Note that, based on a recent, *in situ* calibration of Furaptra (Golovina & Blaustein 2000), all $[Ca^{2+}]_{SR}$ values in this figure should be multiplied by a factor of 1.43 to correct for the use of an *in vitro* calibration factor. Arrowheads in Aa and Ba point to mitochondria; scale bars (lower right corners in Aa and Ba)$=2\,\mu m$ (from Golovina & Blaustein 1997, with permission).

& Blaustein 1997, 2000; reviewed in Blaustein & Golovina 2001). Nevertheless, there must be at least quantitative, and possibly qualitative, variation from cell type to cell type, and perhaps only a single interconnecting Ca^{2+} pool in some cell types (e.g. Janiack et al 2001).

How are SR Ca^{2+} stores organized in small arteries with myogenic tone?

Rat small ($<250\,\mu$m passive diameter) resistance arteries may differ somewhat from larger arteries because the small arteries have little central ('corbular') SR. The SR can be visualized within individual cells in intact, small arteries by loading the cells with Ca^{2+}-sensitive dyes such as Fluo-4 (Miriel et al 1999) or the lower affinity indicators, Fluo-3FF or Fluo-5N. The images reveal that most, if not all of the SR is arranged in small, bacillus-shaped sacs that lie immediately under the PL (Fig. 3); moreover, the SR may form junctions with the PL (see below). The cell diameters in these small arteries rarely exceed 8 μm (Fig. 3) so that no regions of the cytosol lie more than 4 μm from the PL. Thus, Ca^{2+} signals originating at the PL or adjacent jSR may readily and rapidly spread to the cell centre without the need for an extensive SR network.

How are the SR Ca^{2+} stores refilled?

In most types of cells, unloading of the ER or SR opens 'store-operated' Ca^{2+}- (and Na^{+}-) permeable channels (SOCs) that help to refill the Ca^{2+} stores (Putney et al 2001, Clapham et al 2001). The Ca^{2+} release-activated current mediated by these channels is called I$_{CRAC}$ (Parekh & Penner 1997). The refilling phenomenon, known as 'capacitative Ca^{2+} entry' (CCE) was originally believed to function only in non-excitable cells. However, a similar CCE mechanism has now been observed in neurons (reviewed in Blaustein & Golovina 2001) and in smooth muscle (Arnon et al 2000a, Lee et al 2001, McDaniel et al 2001, Potocnik & Hill 2001). A standard method for demonstrating CCE is to incubate the preparation in a Ca^{2+}-free (0Ca^{2+}) medium while blocking SERCA selectively (Inesi & Sagara 1994) with an agent such as thapsigargin (Putney et al 2001), which is relatively irreversible, or CPA, which is readily reversible (Arnon et al 2000a, Golovina & Blaustein 2000). The 0Ca^{2+} medium is used so that the Ca^{2+} that leaks from the SR can be readily extruded from the cells, and does not remain in the cytosol. The SR may then also be depleted by brief application of Caf (often with Ry present). Subsequent restoration of external Ca^{2+} with continued inhibition of SERCA (to prevent Ca^{2+} sequestration) is then associated with a rapid, large rise in the cytosolic free Ca^{2+} concentration ([Ca^{2+}]$_{CYT}$). This rise in [Ca^{2+}]$_{CYT}$ must then be the result of Ca^{2+} entry (i.e. CCE) that, in some cell types, has been measured as a 'Ca^{2+} release-activated current' (I$_{CRAC}$). This entry usually is not prevented by

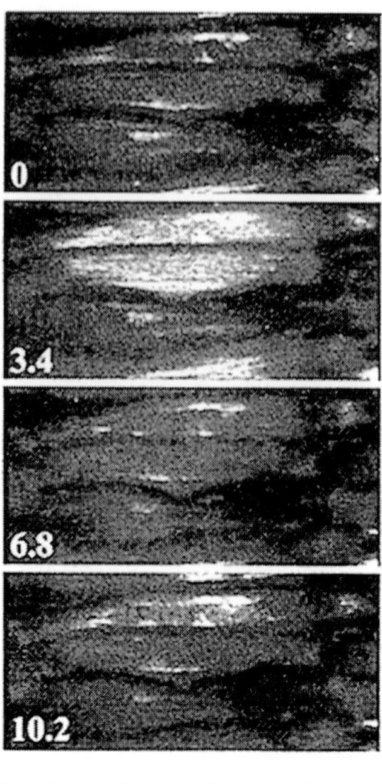

50.0 μm

FIG. 3.   Confocal images showing the location of the SR in live myocytes within an intact, small diameter ($<250\,\mu$m passive diameter), pressurized (70 mmHg) artery from the rat mesenteric artery arcade. The artery was loaded with Fluo-4 as the membrane-permeant acetoxymethyl ester. Some of this high-affinity, Ca^{2+} indicator dye is often sequestered in the SR (cf. Goldman et al 1990). The SR can then be readily visualized, especially when $[Ca^{2+}]_{CYT}$ is low (as in the panels at '0' and 6.8 s), because the intra-SR dye is saturated with Ca^{2+}, and fluoresces brightly. This artery was treated with $1.0\,\mu$m phenylephrine (PE), which caused the $[Ca^{2+}]_{CYT}$ level to oscillate asynchronously in the cells seen in the centre of the panel. The cell outlines are clearly visible when $[Ca^{2+}]_{CYT}$ rises, as in the panels at 3.4 and 10.2 s. Note that nearly all of the SR (the very bright areas, especially in the '0' and 3.4 s panels) lies parallel to, and immediately beneath the PL (from Miriel at al 1999, with permission).

dihydropyridines such as nifedipine (Arnon et al 2000a,b, McDaniel et al 2001, but see Curtis & Scholfield 2001, Stepien & Marche 2000, J. Zhang, W.G. Wier & M.P. Blaustein, unpublished work), and therefore is not mediated by L-type voltage-gated Ca^{2+} channels. Substantial evidence indicates that CCE is mediated by mammalian TRP proteins, homologues of the *Drosophila* transient receptor potential proteins (Birnbaumer et al 2000, but see Clapham et al 2001). There are seven known mammalian TRP isoforms, and several different TRP protein isoforms are expressed in VSMC (McDaniel et al 2001). The SOCs may consist of TRP homo- or heteromultimers. Indeed, knock-down of TRP6 in small cerebral arteries with an antisense oligodeoxynucleotide reduces myogenic tone by 50–70%

(Welsh et al 2002). This is strong evidence that SOCs and CCE play an important role in vascular physiology.

There is evidence that the SOCs, although located in PL, are also associated with jSR/ER: in frog oocytes, SOCs co-sediment with ER elements (Jaconi et al 1997). Recent studies reveal that TRP1 (Xu & Beech 2001, Lee et al 2002a) and several other TRPs (McDaniel et al 2001, Welsh et al 2002) are present in the PL of VSMCs. Moreover, some mammalian TRPs co-immunoprecipitate with InsP$_3$Rs (Birnbaumer et al 2000). These findings all suggest that the TRPs (SOCs) are located in PL microdomains that are closely associated with the jSR (Arnon et al 2000a,b, and see below).

Junctional SR and the overlying plasmalemma: the concept of the PLasmERosome

The latter observations are only some of the more recent evidence that certain PL transport proteins are localized in PL microdomains (or lipid rafts) that overlie, and are functionally associated with, the subjacent jSR. An important clue to the idea of functionally specialized PL microdomains is the evidence that most cells express both plasma membrane Ca^{2+} pumps (PMCA) and Na$^+$/Ca^{2+} exchangers (NCX), as well as two types of Na$^+$ pumps with kinetically different catalytic (α) subunit isoforms (Juhaszova & Blaustein 1997a,b). Immunocytochemistry has revealed that in VSMCs, NCX and Na$^+$ pumps containing high-ouabain-affinity α3 subunits are localized to the PL microdomains that overlie the SR (Juhaszova & Blaustein 1997a,b). In contrast, PMCA and Na$^+$ pumps with low ouabain affinity α1 subunits are uniformly distributed in the PL (Juhaszova & Blaustein 1997b). To confirm the immunocytochemical data on Na$^+$ pump α subunit isoform distribution, primary cultured rat arterial myocytes were transfected with Na$^+$ pump α1 and α3 subunits fused to cyan, green or yellow fluorescent proteins. Such constructs are expressed in the basolateral PL of MDCK cells (S. W. Robinson, personal communication 2001), indicating that the N-terminal green fluorescent protein (GFP) does not interfere with normal targeting. The expressed α1 construct was relatively uniformly distributed over the VSMC surface (Fig. 4A,B) while the α3 construct was confined to PL microdomains that are distributed in a lacy, reticular pattern (Fig. 4C).

Conventional electron microscopy (Devine et al 1972) and freeze-etch (Somlyo & Franzini-Armstrong 1985) of VSMCs reveals that the jSR is separated from overlying PL by a 12–15 nm cytosolic space that is traversed by electron-dense structures. These structures appear similar to the foot processes of cardiac and skeletal muscle (Franzini-Armstrong et al 1998). Indeed, there is striking structural similarity between these PL–jSR regions in VSMC and the diads and triads of cardiac and skeletal muscle (Franzini-Armstrong et al 1998). Moreover,

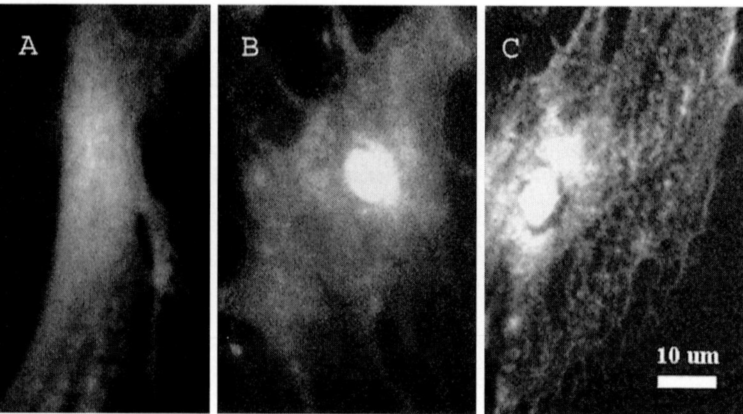

FIG. 4.   Distribution of transfected and expressed Na^+ pump α subunit isoforms $\alpha1$ (A,B) and $\alpha3$ (C) in primary cultured rat mesenteric artery myoctes. Cells in panels A and B were transfected with the green fluorescent protein (GFP) fused to the N-terminal of the rat $\alpha1$ gene and introduced with an adenoviral vector (A) or a gene gun (B). The cyan fluorescent protein (CFP) gene was fused to the N-terminal of the rat $\alpha3$ gene and transfected into myocytes with a gene gun (C). Following 48 h in culture, the cells were fixed and viewed by wide field fluorescence microscopy (488 nm excitation, 519 nm emission for GFP; 458 nm excitation, 480 nm emission for CFP). Note that $\alpha1$-GFP (A,B) is expressed uniformly over the surfaces of the cells. In contrast, $\alpha3$-CFP (C) is expressed in a lacy, reticular pattern (H. Song, S. Robinson, J. Choate, L. Lencesova & M.P. Blaustein, unpublished results) identical to the pattern of SERCA distribution in these cells (Juhaszova & Blaustein 1997a).

comparable PL–ER junctional complexes also have been seen in neurons (Henkart et al 1976, Watanbe & Burnstock 1976). We have named these PL–jSR units 'PLasmERosomes' (Fig. 5). They operate as functional units and play a special role in Ca^{2+} homeostasis and Ca^{2+} signalling in a variety of cell types (Arnon et al 2000b, Blaustein & Golovina 2001). Ca^{2+} can be effectively transferred between the extracellular fluid and the SR through these junctional regions without entering 'bulk' cytosol. The Ca^{2+} is confined within the cytosolic space between the PL and jSR, and is also in the jSR, and does not elevate $[Ca^{2+}]_{CYT}$ when Ca^{2+} is transferred between the jSR and extracellular fluid via NCX (Ashida & Blaustein 1987, Chen & van Breemen 1993). This is the smooth muscle 'Buffer Barrier' model of van Breemen and colleagues (Chen & van Breemen 1993, van Breemen et al 1995, Lee et al 2002b, this volume). It should be noted, however, that this is not really Ca^{2+} 'buffering' but, rather, indicative of the fact that there are two routes for Ca^{2+} to move between the extracellular fluid and the SR: one via the 'bulk' cytosol, and the other through the PL–SR junctional regions. In the latter case, the entering or exiting Ca^{2+} would not be detected by dyes used to measure Ca^{2+} in 'bulk' cytosol. This phenomenon is not limited to smooth muscles, but is very widespread (Golovina et al 2002). Furthermore, as described below, cross-talk

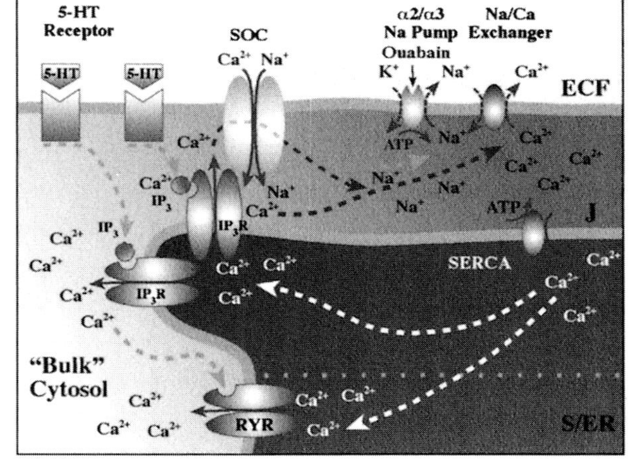
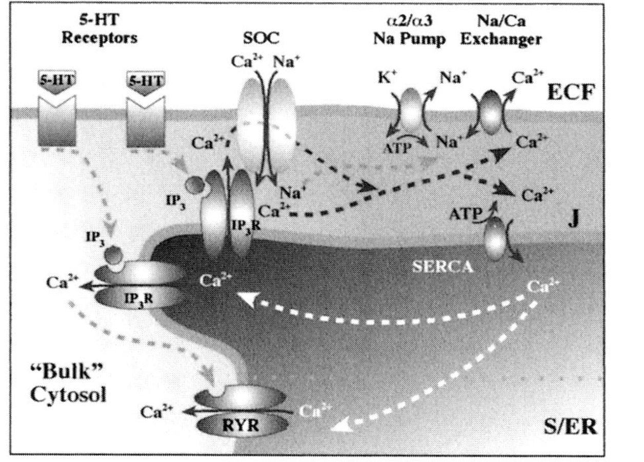

FIG. 5.   Model of the PL–jSR region showing key transport proteins involved in local control of jSR Ca^{2+} stores and modulation of Ca^{2+} signalling. The PL region shows vasoconstrictor (5-HT) receptors, a nearby PL microdomain (containing SOCs, α2/α3 Na^+ pumps, and Na/Ca exchanger), and adjacent jSR (with SERCA and $InsP_3R$ and RyR), and intervening 'diffusion-restricted' junctional cytosolic space ('J'). Left: Normal conditions. Right: After inhibition of α2/α3 Na^+ pumps by low-dose ouabain. Shading indicates relative concentrations of Na^+ and/or Ca^{2+}. ECF = extracellular fluid. α1 Na^+ pumps are widely distributed in the PL, but may be excluded from these microdomains. The model shows a physical association between SOCs and $InsP_3R$ that some authors have suggested (see text) (modified from Arnon et al 2000b, and reproduced with permission).

between the PL and jSR (or ER) apparently plays a key role in modulating Ca^{2+} signals in most types of cells.

Role of the PLasmERosome in the regulation of Ca^{2+} signalling: effects of low-dose ouabain

The vasotonic effects of ouabain illustrate the key functional role of PLasmERosomes. The Na^+ and Ca^{2+} transport systems in the PLasmERosome mediate the vasotonic action of low-dose cardiotonic steroids (Arnon et al 2001b). We postulate that inhibition of the VSMC $\alpha 3$ Na^+ pumps by nanomolar ouabain tends to increase the Na^+ concentration in the junctional cytosol ($[Na^+]_J$) because of uncompensated (net) Na^+ entry via SOCs (Arnon et al 2000a) and NCX. Consequently, the junctional Ca^{2+} concentration ($[Ca^{2+}]_J$) also rises (see Fig. 5), and more Ca^{2+} is stored in the jSR, so that subsequent Ca^{2+} signalling is enhanced. A critical feature of this effect is the assumption that diffusion of cations between the junctional space and 'bulk' cytosol is greatly restricted. This must be the case because Na^+ diffusion in free solution is much too rapid to permit a (local) rise in $[Na^+]_J$. If, however, diffusion is restricted, inhibition of a single Na^+ pump (turnover$=450$ Na^+/s) would be sufficient to double the $[Na^+]_J$ in 6 s if the junctional space has dimensions of 100 nm radius 15 nm thickness (area$=3\times10^4$ nm^2; volume$=5\times10^{-19}$ litres), and the initial $[Na^+]_J$ is the same as that in bulk cytosol (9 mM, or about 2700 Na^+ ions in 5×10^{-19} litres). Indeed, this type of behaviour, if present in cardiac myocytes, can explain how low-dose cardiotonic steroids can induce a cardiotonic effect without elevating bulk $[Na^+]_{CYT}$ (Boyett et al 1986, Levi et al 1994). This also accounts for the simultaneous expression, in the heart, of Na^+ pumps with $\alpha 1$ and those with $\alpha 2$ subunits, which are differently regulated and have different kinetic properties (Blanco & Mercer 1998, James et al 1999).

The model (Fig. 5) illustrates how this structural arrangement contributes to the regulation of Ca^{2+} signalling. Agonists such as serotonin (5-HT) can be expected to elevate $[Ca^{2+}]_{CYT}$ in part as a result of InsP$_3$-mediated SR Ca^{2+} release. This Ca^{2+} release and, thus, the amplitude of the Ca^{2+} signal, is in part dependent upon the amount of Ca^{2+} in the jSR and, in part upon the $[Ca^{2+}]$ in the neighbourhood of the InsP$_3$R (let us assume that the peak InsP$_3$ concentration remains constant). A low (nanomolar) concentration of ouabain can be expected to inhibit only those Na^+ pumps with $\alpha 2$ or $\alpha 3$ subunits, and thus elevate $[Na^+]_J$ and (via NCX) $[Ca^{2+}]_J$, so that more Ca^{2+} will be stored in the jSR. The higher $[Ca^{2+}]_J$ should increase the sensitivity of the InsP$_3$R to InsP$_3$ (Bezprozvanny et al 1991, Iino 2002, this volume) and this, plus the increased jSR Ca^{2+} content, should cause the local Ca^{2+} signal to be amplified in response to low doses of 5-HT (for example). This amplified local signal should then be communicated to the cell interior by diffusion and/or (perhaps) via Ca^{2+}-induced Ca^{2+} release involving RyRs.

Conclusions

The studies reviewed above indicate that the SR in at least some smooth muscles appears to be subdivided into small, functionally independent compartments. By employing imaging methods with low-affinity Ca^{2+} dyes in intact, primary cultured arterial myocytes, we observed that RyR and InsP$_3$R are apparently associated with different SR subcompartments.

Some SR compartments (jSR) lie just beneath special microdomains of PM, and are joined to this PL by electron-dense processes (observed with electron microscopy). Na$^+$ pumps with high-ouabain-affinity α subunits (α2/α3 subunits), NCX and SOCs, appear to be confined to these PL microdomains. These PL microdomains, the subjacent jSR, and the intervening tiny cytosolic volume, form functionally specialized units we call PLasmERosomes. Through the operation of these units (which are apparently present in many types of cells), modulation of Na$^+$ pump activity may have a profound influence on Ca^{2+} signalling in smooth muscles and many other types of cells.

Acknowledgements

This work was supported by NIH grants NS-16106 and HL-45215 (to MPB), HL-60748 (to WGW), and K01-HL-04051 (to SWR), and by a Grant-in-Aid from the AHA Mid-Atlantic Affiliate (to VAG).

References

Arnon A, Hamlyn JM, Blaustein MP 2000a Na$^+$ entry via store-operated channels modulates Ca^{2+} signaling in arterial myocytes. Am J Physiol 278:C163–C173

Arnon A, Hamlyn JM, Blaustein MP 2000b Ouabain augments Ca^{2+} transients in arterial smooth muscle without raising cytosolic Na$^+$. Am J Physiol 279:H679–H691

Ashida T, Blaustein MP 1987 Regulation of cell calcium and contractility in mammalian arterial smooth muscle: the role of sodium–calcium exchange. J Physiol 392:617–635

Bezprozvanny I, Watras J, Ehrlich BJ 1991 Bell-shaped calcium-response curves of Ins(1,4,5)P$_3$- and calcium-gated channels from endoplasmic reticulum of cerebellum. Nature 351:751–754

Birnbaumer L, Boulay G, Brown D et al 2000 Mechanism of capacitative Ca^{2+} entry (CCE): interaction between IP3 receptor and TRP links the internal calcium storage compartment to plasma membrane CCE channels. Recent Prog Horm Res 55:127–161

Blanco G, Mercer RW 1998 Isozymes of the Na-K-ATPase: heterogeneity in structure, diversity in function. Am J Physiol 275:F633–F650

Blaustein MP, Golovina VA 2001 Structural complexity and functional diversity of endoplasmic reticulum Ca^{2+} stores. Trends Neurosci 24:602–608

Boyett MR, Hart G, Levi AJ 1986 Dissociation between force and intracellular sodium activity with strophanthidin in isolated sheep Purkinje fibres. J Physiol 381:311–331

Carafoli E, Santella L, Branca D, Brini M 2001 Generation, control, and processing of cellular calcium signals. Crit Rev Biochem Mol Biol 36:107–260

Chen Q, van Breemen C 1993 The superficial buffer barrier in venous smooth muscle: sarcoplasmic reticulum refilling and unloading. Br J Pharmacol 109:336–343

Clapham DE, Runnels LW, Strubing C 2001 The TRP ion channel family. Nat Rev Neurosci 2:387–396

Curtis TM, Scholfield CN 2001 Nifedipine blocks Ca^{2+} store refilling through a pathway not involving L-type Ca^{2+} channels in rabbit arteriolar smooth muscle. J Physiol 532:609–623

Devine CE, Somlyo AV, Somlyo AP 1972 Sarcoplasmic reticulum and excitation-contraction coupling in mammalian smooth muscles. J Cell Biol 52:690–718

Flynn ER, Bradley KN, Muir TC, McCarron JG 2001 Functionally separate intracellular Ca^{2+} stores in smooth muscle. J Biol Chem 276:36411–36418

Franzini-Armstrong C, Protasi F, Ramesh V 1998 Comparative ultrastructure of Ca^{2+} release units in skeletal and cardiac muscle. Ann N Y Acad Sci 853:20–30

Goldman WF, Bova S, Blaustein MP 1990 Measurement of intracellular Ca^{2+} in cultured arterial smooth muscle cells using Fura-2 and digital imaging microscopy. Cell Calcium 11:221–231

Golovina VA, Blaustein MP 1997 Spatially and functionally distinct Ca^{2+} stores in sarcoplasmic and endoplasmic reticulum. Science 275:1643–1648

Golovina VA, Blaustein MP 2000 Unloading and refilling of two classes of spatially resolved endoplasmic reticulum Ca^{2+} stores in astrocytes. Glia 31:15–28

Golovina V, Song H, James P, Robinson S, Lingrel J, Blaustein M 2002 Na^+ pump α2 subunit expression modulates Ca^{2+} signaling. Biophys J 82:570a–571a

Henkart M, Landis DMD, Reese TS 1976 Similarity of junctions between plasma membranes and endoplasmic reticulum in muscles and neurons. J Cell Biol 70:338–347

Iino M 2002 Molecular basis and physiological functions of dynamic Ca^{2+} signalling in smooth muscle cells. In: Role of the sarcoplasmic reticulum in smooth muscle. Wiley, Chichester (Novartis Found Symp 246) p 142–153

Inesi G, Sagara Y 1994 Specific inhibitors of intracellular Ca^{2+} transport ATPases. J Membr Biol 141:1–6

Jaconi M, Pyle J, Bortolon R, Ou J, Clapham D 1997 Calcium release and influx colocalize to the endoplasmic reticulum. Curr Biol 7:599–602

James PF, Grupp IL, Grupp G et al 1999 Identification of a specific role for the Na,K-ATPase α2 isoform as a regulator of calcium in the heart. Mol Cell 3:555–563

Janiak R, Wilson, SM, Montague S, Hume JR 2001 Heterogeneity of calcium stores and elementary release events in canine pulmonary arterial smooth muscle cells. Am J Physiol 280:C22–C33

Juhaszova M, Blaustein MP 1997a Na^+ pump low and high ouabain affinity α subunit isoforms are differently distributed in cells. Proc Natl Acad Sci USA 94:1800–1805

Juhaszova M, Blaustein MP 1997b Distinct distribution of different Na^+ pump alpha subunit isoforms in plasmalemma. Physiological implications. Ann NY Acad Sci 834:524–536

Lee C-H, Poburko D, Sahota P, Sandhu J, Ruehlmann DO, van Breemen C 2001 The mechanism of phenylephrine-mediated $[Ca^{2+}]_i$ oscillations underlying tonic contraction in the rabbit inferior vena cava. J Physiol 534:641–650

Lee C-H, Rahamian R, Szado T et al 2002a Requirement for the opening of the IP_3-sensitive Ca^{2+} channels and SOC in $α_1$-adrenergic receptor-mediated constriction of the rabbit inferior vena cava. Am J Physiol (Heart Circ Physiol), in press

Lee C-H, Poburko D, Kuo K-H, Seow C, van Breemen C 2002b Relationship between the sarcoplasmic reticulum and the plasma membrane. In: Role of the sarcoplasmic reticulum in smooth muscle. Wiley, Chichester (Novartis Found Symp 246) p 26–47

Levi AJ, Boyett MR, Lee CO 1994 The cellular actions of digitalis glycosides on the heart. Prog Biophys Mol Biol 62:1–54

McDaniel SS, Platoshyn O, Wang J et al 2001 Capacitative Ca^{2+} entry in agonist-induced pulmonary vasoconstriction. Am J Physiol 280:L870–L880

Meldolesi T, Pozzan T 1998 The endoplasmic reticulum Ca^{2+} store: a view from the lumen. Trends Biochem Sci 23:10–14

Miriel VA, Mauban JRH, Blaustein MP, Wier WG 1999 Local and cellular Ca^{2+} transients in smooth muscle of pressurized rat resistance arteries during myogenic and agonist stimulation. J Physiol 518:815–824

Parekh AB, Penner R 1997 Store depletion and calcium influx. Physiol Rev 77:901–930

Potocnik SJ, Hill MA 2001 Pharmacological evidence for capacitative Ca^{2+} entry in cannulated and pressurized skeletal muscle arterioles. Br J Pharmacol 134:247–256

Pozzan T, Rizzuto R, Volpe P, Meldolesi J 1994 Molecular and cellular physiology of intracellular calcium stores. Physiol Rev 74:595–636

Putney JW Jr, Broad LM, Braun FJ, Lievremont JP, Bird GS 2001 Mechanisms of capacitative calcium entry. J Cell Sci 114:2223–2229

Somlyo AV, Franzini-Armstrong C 1985 New views of smooth muscle structure using freezing, deep-etching and rotary shadowing. Experientia 41:841–856

Somlyo AP, Somlyo AV 1994 Signal transduction and regulation in smooth muscle. Nature 372:231–236

Stepien O, Marche P 2000 Amlodipine inhibits thapsigargin-sensitive Ca^{2+} stores in thrombin-stimulated vascular smooth muscle cells. Am J Physiol 279:H1220–H1227

Terasaki M 1989 Fluorescent labeling of endoplasmic reticulum. Methods Cell Biol 29:125–135

Tribe RM, Borin ML, Blaustein MP 1994 Functionally and spatially distinct Ca^{2+} stores are revealed in cultured vascular smooth muscle cells. Proc Natl Acad Sci USA 91:5908–5912

van Breemen C, Chen Q, Laher I 1995 Superficial buffer barrier function of smooth muscle sarcoplasmic reticulum. Trends Pharmacol Sci 16:98–105

Watanabe H, Burnstock G 1976 Junctional subsurface organs in frog sympathetic ganglion cells. J Neurocytol 5:125–136

Welsh DG, Morielli AD, Nelson MT, Brayden JE 2002 Transient receptor potential channels regulate myogenic tone of resistance arteries. Circ Res 90:248–250

Xu SZ, Beech DJ 2001 TrpC1 is a membrane-spanning subunit of store-operated Ca^{2+} channels in native vascular smooth muscle cells. Circ Res 88:84–87

Yamazawa T, Iino M, Endo M 1992 Presence of functionally different compartments of the Ca^{2+} store in single intestinal smooth muscle cells. FEBS Lett 301:181–184

DISCUSSION

Brading: A long time ago, we looked at the effects of removing ions on store filling and release. We found that in the guinea-pig taeniae, if you remove Na$^+$ from the system, you couldn't get store filling. This links up to a certain extent with what Mordy Blaustein is saying. I always wondered what was going on, and this type of effect—filling via Na$^+$/Ca^{2+} exchange from the outside—does fit in with the behaviour of whole tissues.

Burdyga: I was involved in these experiments, and we were also working on the Na$^+$/Ca^{2+} exchanger. When we used ouabain to block the Na$^+$ pump, there was a limited rise of Na$^+$ and then the Na$^+$/Ca^{2+} exchanger drives Na$^+$ out at the expense of Ca-ATPase activity. What I found later is that there is a massive load of the Ca^{2+}

in the SR, which you show nicely here. But at the same time Clair Aickin was measuring the membrane potential in these cells and had found that ouabain effect was concomitant with a strong hyperpolarization. Now I can see the increase in the frequency of Ca^{2+} sparks in these cells. Do the cells of yours generate sparks? If they do, then you should shift the membrane potential towards hyperpolarization and relax the contraction.

Blaustein: Victor Miriel, Gil Wier and I have done experiments looking at sparks in the small mesenteric arteries. They do generate sparks. We haven't had a chance yet to look at the effects of ouabain on sparks, although we plan to do this.

Paul: We have been working with Jerry Lingrel who has the α1 and α2 knockout mice, so we have a fair idea of what is happening in the living tissue. The α1 knockout is lethal: the homozygote never gets beyond the blastocyte stage. In the α2 knockout (the high-affinity oubain-sensitive isoform) animals are born but then die immediately. We have measured neonatal smooth muscle tissues. We have focused mainly on aorta and bladder. The interesting thing is that in our hands, when we measure by Western blot, we have 30% α2 and 70% α1. If we compare the neonatal α1 heterozygote, we reduce the Na^+/K^+ ATPase by 35%. If we compare this to the homozygous α2 knockout—the one that is born but then dies—we also reduce the Na^+/K^+ ATPase activity by about 30%. The answer seems to be that there is no change in the contractility to phenylephrine with the $α1^{+/-}$ smooth muscles, but you can reduce the α1 power by half and still not see any major changes. However, in the $α2^{-/-}$ contractility does change. The phenotype does not relax as well as wild-type, and it is more sensitive to phenylephrine. The KCl activation curves look identical, and this is very supportive for Mordy Blaustein's case. The postulated reason they are more sensitive to phenylephrine is that the Na^+/Ca^{2+} exchanger and the α2 isoform are localized in the areas near the plasma membrane SR junctions, which will have a higher Ca^{2+} level. The other thing I have to add is that perhaps it is easier for us to think of these things as being rigid and structured. But they are really diffusion–reaction processes. There was a fair difference—a shift of about 1 log unit on the phenylephrine scale—but it is still not absolute. They can relax just as well. In the adult heterozygote α2 knockout, which has about 15% less α2, there are minor differences, but the animals do pretty well.

Blaustein: We have been collaborating with Jerry Lingrel on these α2 knockout animals. We see the same thing. Unfortunately, the smooth muscle cells express α3, but not α2. We culture astrocytes prenatally; these cells express α1 and α2 Na^+ pumps. There is a little higher Ca^{2+} in the α2 heterozygote, but no difference in the Na^+ concentration compared with wild-type. What is really striking is that without an apparent change in cell Na^+, we see a change in the Ca^{2+} transient in the heterozygotes. I think this is due to the activity of the junctional complex.

Sanders: With the massive remodelling that is taking place in these cultured cells, do you not worry that there are also changes in the relationship between these proteins and the sarcoplasmic reticulum?

Blaustein: We see virtually identical effects of low-dose ouabain on the cultured cells and on the intact artery. My expectation is that there will be no change in Na$^+$ and yet we'll see large effects on signalling. Clearly, we would like to do knockout of the α3 Na$^+$ pump, to show that the knockout does the same thing as the ouabain. I expect that the results would be the same. Despite remodelling, we will still see these junctions. I think the behaviour is going to be very similar in cultured cells and in cells *in situ.*

Sanders: Let's consider TRP1. Burt Horowitz has looked at TRPs in gastrointestinal muscle. I am not saying that it is the same, but TRP4, 6 and 7 seem to be emerging as the channels that are expressed in smooth muscle cells. This is on individually selected smooth muscle cells. If you culture those cells for three days they start to express TRP1.

Blaustein: We know that these cells express TRP4 and TRP6. However, we also find TRP1 in the artery when we do Western blotting on the whole tissue.

Sanders: Is that from selecting myocytes, or from the whole artery?

Blaustein: It is whole artery. This is why we use cultured cells, because in arteries it is very difficult to do Western blotting of just myocytes. We want to look at the expressed proteins from the myocytes.

van Breemen: In terms of the compartmentalization, I was a little puzzled with the increase in Ca^{2+} that you saw in some areas when you added CPA. What is happening here?

Blaustein: Some areas empty and others fill when we put on CPA. This means that there are some SERCA Ca^{2+} pumps that are insensitive to CPA. A number of studies have demonstrated that there are some SERCA pumps that are insensitive. If you dump Ca^{2+} out of some regions and other Ca^{2+} pumps are still working, you are going to fill those regions.

van Breemen: Would that be PMR1?

Blaustein: I don't think it is PMR1 because it is a caffeine-sensitive store. I am not sure that Golgi is sensitive to caffeine.

Somlyo: The caffeine-sensitive store changes with time in culture.

Blaustein: There is no question that there will be developmental changes in culture, but I don't think that there is some fundamental change, and that some new phenomenon suddenly appears that wasn't there before.

Lompré: I would like to present some of our results describing the changes that occur to cells in culture. My concern is primarily with the organization of the SR when vascular smooth muscle cells proliferate in culture. This is of interest because of the importance of smooth muscle cell proliferation in atherosclerosis and restenosis. Freshly dissociated cells or those that have been cultured for just one day can release Ca^{2+} either by caffeine or ATP. If the stores have been emptied using one of the agonists, caffeine, then the other, ATP, cannot release Ca^{2+}, meaning that there is one Ca^{2+} store that contains both RyRs and $InsP_3$ receptors, or at least that these stores are interconnected. If we block the mitochondria by CCCP, we can still induce Ca^{2+} release by using caffeine or ATP. This indicates that the mitochondria and the SR are independent, at least after one day in culture. After one day in culture, 77% of the cells release Ca^{2+} by both caffeine and ATP. But after three days only 17% can release Ca^{2+} by both agonists, and after 5 days in culture you cannot release Ca^{2+} anymore by caffeine, but only by ATP. When cells stop proliferating when they are confluent, they regain a certain sensitivity to caffeine. Only 50% are sensitive to caffeine, and a higher dose is needed (40 mM as opposed to 10 mM in the earlier stage). When the cells are confluent, CCCP (carbonyl cyanide *m*-chlorophenylhydrazone) completely blocks caffeine-induced Ca^{2+} release but not ATP-induced Ca^{2+} release. This means that the mitochondria are involved in caffeine-induced Ca^{2+} release in some way when these cells have gone through the cell cycle. Freshly dissociated rat aortic smooth muscle cells have two SERCA isoforms (SERCA2a and SERCA2b) and two Ca^{2+} release channels (RyR and $InsP_3$ receptor). The $InsP_3$ receptors are around the nucleus. After three days cells start proliferating and they begin to lose SERCA2a. After five days in culture they no longer have SERCA2a or RyRs. When they stop proliferating and are confluent again, they regain both SERCA2a and RyR, but this is a different type of RyR. Freshly dissociated rat aortic SMCs have mainly RyR3, which gives two isoforms by differential splicing. One is constitutive and the other disappears in culture. When they stop proliferating they express RyR1. There are also two types of $InsP_3$ receptor: type 1 is the main isoform, but there is some type 2 also.

Eisner: So you are saying we shouldn't work on cultured cells!

Lompré: Cultured cells are a useful model for pathology. If you want to see what is happening in atherosclerosis, for example, it is very difficult to work on the atherosclerotic plaque, because there are many cell types involved. In culture you can reproduce some of the events. For example, we have done balloon injury in the rat, and we lose RyRs when cells proliferate *in situ.*

Sanders: You also lose most of the ion channels until you get confluence again.

Nelson: Still, you have to provide evidence that those cultured cells are similar to ones in atherosclerotic plaques.

Lompré: This is what we did. We can do immunofluorescence on atherosclerotic plaques, and we don't see any RyRs. But it is very difficult to work with plaques because they are very small.

Nixon: I don't think cultured cells are like atherosclerotic plaque cells, from what we have seen. In the data you showed on the expression of different channels, the expression and localization of $InsP_3$ receptors is very different. For example, the $InsP_3$ receptor expression around the nuclear envelope is not seen in tissue.

Lompré: Type 1 is localized mainly around the nucleus, but type 2 is more diffuse.

Blaustein: Clearly, we want to study things in intact tissue, preferably even *in situ*. But there are some things that we simply cannot do at the level that we would like to look in intact tissue. That is why we use model systems.

Paul: The evidence from the knockouts supports the idea of some kind of localized organization that is modulatory. They don't die, and they don't fail to respond to InsP$_3$ stimulation if they lack α2 isoform, but they are different.

Molecular basis and physiological functions of dynamic Ca^{2+} signalling in smooth muscle cells

Masamitsu Iino

Department of Pharmacology, Graduate School of Medicine, The University of Tokyo, Bunkyo-ku, Tokyo 113-0033, Japan

Abstract. We have visualized Ca^{2+} signals in smooth muscle cells mediated by the release of Ca^{2+} from intracellular Ca^{2+} stores and studied their underlying molecular basis. Ca^{2+} signals in smooth muscle cells within intact arterial tissues show diverse spatiotemporal patterns: Ca^{2+} waves and oscillations were induced by agonist stimulation or by sympathetic nerve stimulation. We also found spontaneous Ca^{2+} oscillations with low amplitudes (Ca^{2+} ripples) that were observed in the absence of extrinsic stimulation. These dynamic spatiotemporal patterns were generated by Ca^{2+} release via the inositol-1,4,5-trisphosphate (InsP$_3$) receptor (InsP$_3$R). We then studied the molecular basis of such complex Ca^{2+} signalling patterns. The activity of InsP$_3$R is regulated by the cytoplasmic Ca^{2+} concentration. The sensitivity of InsP$_3$R to Ca^{2+} provides feedback regulation of the Ca^{2+} release, which may be important for the generation of Ca^{2+} signalling patterns. A series of site-specific mutagenesis experiments in type 1 InsP$_3$R allowed us to identify glutamate at position 2100 as the Ca^{2+} sensor. Substitution of the amino acid by aspartic acid resulted in a 10-fold decrease in Ca^{2+} sensitivity. In cells expressing the mutant InsP$_3$R, Ca^{2+} release spikes and oscillations were inhibited, indicating the role of the Ca^{2+} sensitivity of InsP$_3$R in the generation of spatiotemporal patterns of Ca^{2+} signals.

2002 Role of the sarcoplasmic reticulum in smooth muscle. Wiley, Chichester (Novartis Foundation Symposium 246) p 142–153

An initial hint that Ca^{2+} stores are present and functional in smooth muscle cells came from earlier experiments revealing that agonist-induced contractions could be observed in the absence of extracellular Ca^{2+}. It is now known that smooth muscle Ca^{2+} stores express two types of Ca^{2+} release channels, the ryanodine receptor (RyR) and the inositol-1,4,5-trisphosphate (InsP$_3$) receptor (InsP$_3$R) (Somlyo & Somlyo 1994). Recent studies have shown that Ca^{2+} release from intracellular Ca^{2+} stores plays various important roles in the regulation of smooth muscle contraction. Local and transient releases of Ca^{2+} from RyR near the surface membrane, which are called Ca^{2+} sparks, activate Ca^{2+}-sensitive K$^+$

channels to induce hyperpolarization, and hence relaxation of smooth muscle cells (Nelson et al 1995). On the other hand, Ca^{2+} release via InsP$_3$R is the major source of Ca^{2+} associated with agonist-induced contractions in various smooth muscle cells, and the resulting Ca^{2+} signals have complex spatiotemporal patterns. Here, I will review our studies on the dynamic patterns of Ca^{2+} signals in smooth muscle cells, and discuss our recent work to understand the molecular basis of the complex spatiotemporal patterns of Ca^{2+} signals.

Ca^{2+} signals viewed in isolated smooth muscle cells

We measured changes in intracellular Ca^{2+} concentration ([Ca^{2+}]$_i$) in enzymatically isolated intestinal smooth muscle cells (Iino et al 1993). When the cells were activated by a muscarinic agonist, carbachol, we found Ca^{2+} waves, i.e. propagation of increase in [Ca^{2+}]$_i$ from one segment of the cell to another. Ca^{2+} waves can be observed in the absence of extracellular Ca^{2+} and are thought to be a phenomenon that involves positive feedback regulation of intracellular Ca^{2+} release. Furthermore, when we studied Ca^{2+} response to different concentrations of carbachol, we observed an all-or-none dose–response relationship. These results suggested that agonist-induced Ca^{2+} signalling in individual smooth muscle cells within intact tissues might be regulated not in a graded manner but in an all-or-none manner. However, such information would be lost when [Ca^{2+}]$_i$ averaged over many smooth muscle cells was measured. We therefore decided to image [Ca^{2+}]$_i$ in individual smooth muscle cells within intact vascular tissues by confocal microscopy.

Ca^{2+} signals viewed in individual smooth muscle cells in tissues

We carefully dissected rat tail arteries and loaded them with a Ca^{2+} indicator, Fluo-3. After a rectangular glass capillary was inserted into the lumen of the excised arteries, [Ca^{2+}]$_i$ in smooth muscle cells within the arterial wall was visualized using a confocal microscope. Brief electrical shocks were delivered at 5 Hz to the preparations to stimulate the sympathetic nerve network present in the adventitia. We found Ca^{2+} signals with diverse spatiotemporal patterns, Ca^{2+} waves and oscillations in individual smooth muscle cells during the sympathetic nerve stimulation (Iino et al 1994).

In another set of experiments, rat tail arteries were longitudinally cut open and viewed from the luminal face through the layer of endothelial cells using a wide-field fluorescence microscope. Ca^{2+} signals in individual vascular smooth muscle cells could be observed after loading Fluo-3. This imaging method allowed us to observe Ca^{2+} signals for a longer time with better signal-to-noise ratio than those observed using a confocal microscope. We found spontaneous Ca^{2+} oscillations with low amplitudes (Ca^{2+} ripples) in the absence of extrinsic stimulation (Asada et al 1999). These dynamic spatiotemporal patterns were generated by Ca^{2+} release via InsP$_3$Rs.

These results have raised several immediate questions. (1) What are the physiological roles of Ca^{2+} waves, oscillations and ripples? Since Ca^{2+} oscillation frequency increases with agonist concentration (Iino et al 1994), it is likely that the regulation of intensity of smooth muscle contraction is frequency coded. However, why does it have to be regulated by oscillation frequency rather than by graded change in $[Ca^{2+}]_i$? Are there any pathophysiological states in which frequency-coded regulation of smooth muscle contraction is disrupted? Nothing is known yet about the roles of Ca^{2+} ripples except for their possible contribution to the resting tension of the artery (Asada et al 1999). (2) What is the molecular basis of the dynamic spatiotemporal patterns of Ca^{2+} signals such as Ca^{2+} waves and oscillations? Although several mechanisms have been postulated, no definitive conclusion had been reached. One of the most straightforward and powerful ways to address these problems would be to use molecular genetic methods. In the following, I will describe our recent study using a model cell system to understand the molecular mechanism of complex Ca^{2+} signalling.

Ca^{2+} sensitivity of InsP$_3$R

The activity of InsP$_3$R is regulated not only by InsP$_3$ concentration but also by cytoplasmic Ca^{2+} concentration (Iino 1990, Bezprozvanny et al 1991, Finch et al 1991). Ca^{2+} sensitivity of the InsP$_3$R should provide feedback regulation of InsP$_3$R-mediated Ca^{2+} release, therefore, it may be important for the generation of complex Ca^{2+} signalling patterns (Berridge et al 2000). Although it has been suggested that calmodulin may be partly associated with Ca^{2+} sensitivity of InsP$_3$R, the molecular mechanism of the Ca^{2+} sensitivity has not been established thus far. We therefore decided to identify the Ca^{2+} sensor of InsP$_3$R.

Since every vertebrate cell line expresses intrinsic InsP$_3$R, it has been difficult to study the functions of extrinsic InsP$_3$R expressed in host cells infected or transfected with an expression vector. We used a mutant DT40 cell line in which all three InsP$_3$R genes were disrupted by homologous recombination (Sugawara et al 1997). Since this cell line does not express intrinsic InsP$_3$R, it can be used as a 'test-tube' to study the functions of extrinsic InsP$_3$Rs. The functions of thus-expressed InsP$_3$R can be studied using a luminal Ca^{2+} measurement method (Hirose & Iino 1994, Miyakawa et al 1999). Cells were first loaded with a low-affinity Ca^{2+} indicator, Furaptra, and then the surface membranes were permeabilized to allow removal of the cytoplasmic indicator while retaining the dye in the Ca^{2+} stores. Luminal Ca^{2+} concentration can be monitored and the

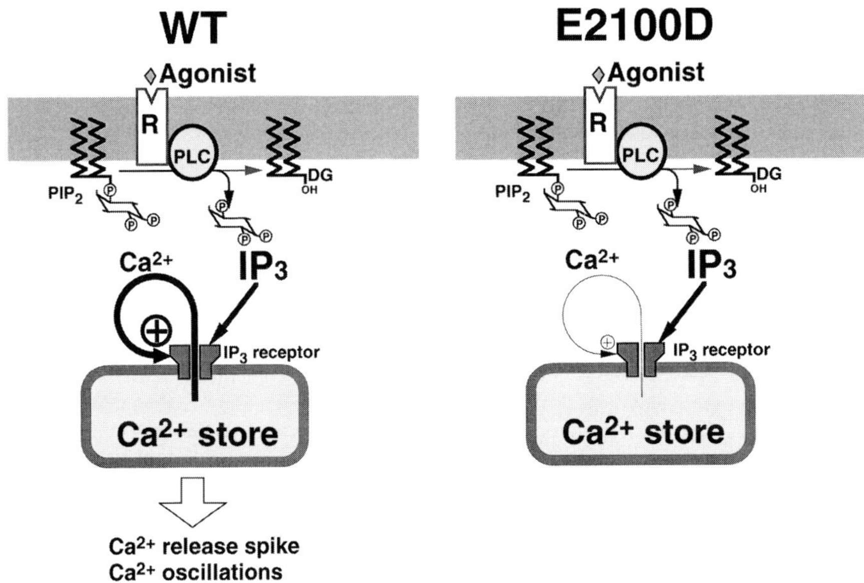

FIG. 1.   Schematic of the Ca^{2+}-mediated feedback regulation of Ca^{2+} release via InsP$_3$R in wild-type (WT) and E2100D mutant InsP$_3$Rs. In cells expressing WT InsP$_3$R, agonist-induced Ca^{2+} release spikes and Ca^{2+} oscillations were observed, while in cells expressing E2100D mutant InsP$_3$Rs, those Ca^{2+} signals were markedly inhibited.

time course of InsP$_3$-induced Ca^{2+} release from the Ca^{2+} store can be measured using this method.

A series of site-specific mutagenesis experiments in InsP$_3$R allowed us to identify glutamate at position 2100 of type 1 InsP$_3$R as the Ca^{2+} sensor. This amino acid corresponds to glutamate at position 4032 of type 1 RyR, which has been shown to be the Ca^{2+} sensor of this Ca^{2+} release channel. Substitution of glutamate 2100 by aspartic acid (E2100D) resulted in an approximately 10-fold decrease in the Ca^{2+} sensitivity of InsP$_3$R without any change in the InsP$_3$ sensitivity (Miyakawa et al 2001).

The E2100D mutant InsP$_3$R provided us with a unique opportunity to test the importance of Ca^{2+}-mediated feedback regulation of InsP$_3$R in the generation of Ca^{2+} signals in intact cells, because Ca^{2+} will have much less effect on the E2100D mutant InsP$_3$R (Fig. 1). Indeed, in cells expressing the E2100D mutant InsP$_3$R, agonist-induced Ca^{2+} release spikes and oscillations were markedly inhibited (Miyakawa et al 2001). These results clearly show that Ca^{2+} sensitivity of InsP$_3$R plays an important role in the generation of spatiotemporal patterns of Ca^{2+} signals.

Conclusions

The Ca^{2+} imaging technique has enabled us to monitor dynamic spatiotemporal patterns of Ca^{2+} signals in vascular smooth muscle cells. The Ca^{2+} sensor of $InsP_3R$ contributes to the generation of such dynamic spatiotemporal patterns of Ca^{2+} signals. In the future, we may be able to artificially control Ca^{2+} signalling patterns in smooth muscle cells using the mutant $InsP_3R$. When Ca^{2+} imaging and molecular genetic methods are combined, we will be able to understand the physiological roles of the dynamic Ca^{2+} signalling patterns.

Acknowledgements

This work was supported by CREST, Japan Science and Technology Corporation, and by grants from the Ministry of Education, Culture, Sports, Science and Technology of Japan.

References

Asada Y, Yamazawa T, Hirose K, Takasaka T, Iino M 1999 Dynamic Ca^{2+} signalling in rat arterial smooth muscle cells under the control of local renin-angiotensin system. J Physiol 521:497–505

Berridge MJ, Lipp P, Bootman MD 2000 The versatility and universality of calcium signalling. Nat Rev Mol Cell Biol 1:11–21

Bezprozvanny I, Watras J, Ehrlich BE 1991 Bell-shaped calcium-response curves of $Ins(1,4,5)P_3$- and calcium-gated channels from endoplasmic reticulum of cerebellum. Nature 351:751–754

Finch EA, Turner TJ, Goldin SM 1991 Calcium as a coagonist of inositol 1,4,5-trisphosphate-induced calcium release. Science 252:443–446

Hirose K, Iino M 1994 Heterogeneity of channel density in inositol-1,4,5-trisphosphate-sensitive Ca^{2+} stores. Nature 372:791–794

Iino M 1990 Biphasic Ca^{2+} dependence of inositol 1,4,5-trisphosphate-induced Ca release in smooth muscle cells of the guinea pig taeniae caeci. J Gen Physiol 95:1103–1122

Iino M, Kasai H, Yamazawa T 1994 Visualization of neural control of intracellular Ca^{2+} concentration in single vascular smooth muscle cells in situ. EMBO J 13:5026–5031

Iino M, Yamazawa T, Miyashita Y, Endo M, Kasai H 1993 Critical intracellular Ca^{2+} concentration for all-or-none Ca^{2+} spiking in single smooth muscle cells. EMBO J 12:5287–5291

Miyakawa T, Maeda A, Yamazawa T, Hirose K, Kurosaki T, Iino M 1999 Encoding of Ca^{2+} signals by differential expression of IP_3 receptor subtypes. EMBO J 18:1303–1308

Miyakawa T, Mizushima A, Hirose K et al 2001 Ca^{2+}-sensor region of IP_3 receptor controls intracellular Ca^{2+} signaling. EMBO J 20:1674–1680

Nelson MT, Cheng H, Rubart M et al 1995 Relaxation of arterial smooth muscle by calcium sparks. Science 270:633–637

Somlyo AP, Somlyo AV 1994 Signal transduction and regulation in smooth muscle. Nature 372:231–236

Sugawara H, Kurosaki M, Takata M, Kurosaki T 1997 Genetic evidence for involvement of type 1, type 2 and type 3 inositol 1,4,5-trisphosphate receptors in signal transduction through the B-cell antigen receptor. EMBO J 16:3078–3088

DISCUSSION

Nixon: Do the DT40 cells express any RyRs?

Iino: No. At least, they don't respond to 20 mM caffeine, an activator of the RyR.

Kotlikoff: The E2100A mutant had no response. Do you interpret that as an absolute requirement of Ca^{2+} binding for InsP$_3$ release?

Iino: We have not been able to activate the E2100A mutant InsP$_3$ receptor in DT40 cells. At the moment we can't go above 3 μM Ca^{2+} in our assay system because of a technical limitation. It may be that if we could go to very high Ca^{2+} concentration we could activate the E2100A channel. Ilya Bezprozvanny is collaborating with us and he is doing bilayer measurements. He says that it is difficult to get a Ca^{2+} channel recording from E2100A, but it is a functional channel.

Blaustein: This is not smooth muscle; it is using cultured cells. In relation to the previous discussion, does this have any relevance to real life? Are you doing the comparison to look at the role of PLC in activating smooth muscle? Clearly, you have shown that Ca^{2+} modulation of the InsP$_3$ receptor has profound effects. Relative to this, how about the effects of Ca^{2+} on the PLC cascade?

Iino: That is also an important issue. We need to do some equivalent experiments, knocking out the Ca^{2+} sensor region of PLC. In order to do this kind of experiment, however, we need to find a cell that has zero background for PLC.

Walsh: Do you have any information about the Ca^{2+} sensitivity of different PLC isozymes?

Iino: All PLC isozymes are sensitive to Ca^{2+}. PLCβ and γ require agonist stimulation, whereas the PLCδ isozymes do not require receptor activation. In other words PLCδ can be activated by Ca^{2+} alone.

Taylor: There has been quite a conflicting literature on whether accessory proteins might be involved in mediating some of these Ca^{2+} effects on InsP$_3$ receptors, and lots of bilayer recordings have failed to find inhibitory effects of Ca^{2+}. Can you comment on this?

Iino: The most simplistic interpretation of our results is that the glutamate 2100 constitutes a Ca^{2+} binding site. Although we cannot exclude the possibility that there is an accessory protein that binds to that position, or nearby, it is more likely that, with analogy to the RyR, the channel itself is binding Ca^{2+}. When we mutate the glutamate at position 2100 to aspartate we shift the entire biphasic response to the right. If there is an independent inhibitory mechanism we should see some depression of the peak activity at the optimal Ca^{2+} concentration when Ca^{2+} sensitivity is shifted toward the right, but we don't see this. It may be that this site is responsible for both activation and inhibition.

Taylor: However, there is evidence from us and other groups that the bivalent cation recognition properties of the sites that mediate stimulation and inhibition may be different.

Iino: If we just alter this single amino acid at position 2100 we shift everything towards the right. A simplistic view would be that the site, with some interaction between the four subunits, is responsible for both activation and inhibition.

van Breemen: It is nice that you showed that the Ca^{2+} sensitivity is responsible for the oscillations. It is easy to imagine that this could be responsible for the regenerative Ca^{2+} release. What are the complexities of the turning off and the downstroke of the oscillation, and in regard to this, do you know how much Ca^{2+} is released per oscillation from the SR?

Iino: I don't know exactly, but this is an important point. There is no conclusive mechanism known for explaining the Ca^{2+} oscillations. There has to be some sort of delay time within the system to generate oscillations. It has been proposed that there is a desensitization of the $InsP_3$ receptor, but in our hands we don't see any clear desensitization in terms of $InsP_3$ or Ca^{2+}. I think we have to think about a different mechanism. Yesterday, we saw results showing simultaneous measurement of the cytosolic Ca^{2+} concentration and luminal Ca^{2+} concentration in Susan Wray's paper (Wray et al 2002). I noticed that the decline in cytosolic Ca^{2+} was very rapid, whereas re-uptake was rather slow. It may be possible that luminal Ca^{2+} concentration has to reach some threshold level to fire a Ca^{2+} spike. Since $InsP_3$-induced Ca^{2+} release requires a positive feedback mechanism, regenerative Ca^{2+} release can be fired more easily at a higher Ca^{2+} content. This threshold level might become lower at high $InsP_3$ concentrations. This may explain the higher rate of Ca^{2+} oscillation at high $InsP_3$ concentrations.

Bolton: We flash-released $InsP_3$ from caged $InsP_3$ at different points in these oscillations in longitudinal muscle. We found that after a release, it was insensitive to $InsP_3$. Then, if you were at the trough and released $InsP_3$ you could advance the next oscillation. There seem to be two possible explanations. One is that there was a refractoriness to $InsP_3$ developing at the receptor level. The other is that the stores simply dump all their Ca^{2+}, so when we release $InsP_3$ at the peak of an oscillation there is no Ca^{2+} in the stores but at the trough they have restored it.

Blaustein: How do you visualize this in a cell in terms of the diffusion of Ca^{2+} and the diffusion of $InsP_3$? This has to occur at different sites along the cell. In order to get the regenerative response Ca^{2+} must diffuse to help activate the $InsP_3$ receptor and $InsP_3$ so that the process can be continued once it has started.

Iino: The initiation mechanism may be an event taking place in the vicinity of the ER.

Blaustein: How does the spread of the wave happen?

Iino: This could be through a Ca^{2+}-induced Ca^{2+} release (CICR) mechanism for the $InsP_3$ receptor.

Blaustein: Meaning that you also have to have $InsP_3$ as well as Ca^{2+} progressing along the cell.

Iino: Even if InsP$_3$ concentration is increased throughout the cell, InsP$_3$ alone is not capable of opening the InsP$_3$ receptors.

Blaustein: No, you need both.

Fry: That's an ideal mechanism. Essentially what you have is loose coupling, but the other way around, with the PLC mechanism. But you also have a rapid mechanism, which is autocatalytic on the receptor itself. You have a mechanism for producing something, but also a phase lag system that would generate a wave.

Blaustein: But both of them have to diffuse.

Fry: You need both to produce the wave, but you have a mechanism there for producing a wave. If only you could show that in the smooth muscle cell you had this loose coupling.

Bolton: It doesn't have to diffuse any distance at all. These waves have a duration of seconds; the diffusion will take only tens of milliseconds.

Blaustein: With diffusion you will lose the InsP$_3$ unless you are regenerating it some place.

Brading: But it will be regenerated. It is continually produced.

van Breemen: With respect to that explanation, what bothers me is that if you then increase the agonist dose, you will get a higher InsP$_3$ concentration which would lower your threshold, so therefore you would empty before the store got filled to the same level. You might expect that the amplitude of the oscillation would decrease with the increasing agonist concentration. We don't see this.

Iino: It seems possible that the filling level of Ca^{2+} stores at the peak of Ca^{2+} oscillation is also dependent on the InsP$_3$ concentration.

Blaustein: If I could return to the propagation, it could well be the Ca^{2+} dependence in the PLC cascade that also plays a role in this. As you progress along, you can generate more InsP$_3$ quickly.

Iino: We used green fluorescent protein (GFP)-tagged pleckstrin homology domain of PLCδ1 to monitor changes in the InsP$_3$ concentration within single MDCK cells (Hirose et al 1999). We saw the InsP$_3$ wave at the same time as the Ca^{2+} wave.

Blaustein: So you must be generating InsP$_3$ along the wave.

Sanders: You showed that there is a bell-shaped curve for the receptor to Ca^{2+}. Kevin Foskett has recently shown that at lower concentrations of InsP$_3$ the inhibitory part of this curve is shifted, which narrows the base quite a bit, so that lower concentrations of Ca^{2+} bring you over the top of the hump and start inhibiting the response (Mak et al 1998). Have you looked at the down slope for the Ca^{2+} sensitivity, and could you not interpret the results you showed not as a change in Ca^{2+} sensitivity, but bringing Ca^{2+} sensitivity to the right without moving Ca^{2+} inhibition to the right, which may cause you to go over the top of the hill quicker and therefore get into this inhibitory side?

Iino: We haven't yet analysed the descending limb very extensively using the DT40 cells. However, our results so far (and the bilayer measurements by Ilya Bezprozvanny) suggest that much higher Ca^{2+} concentration is required to go over the top of the hill in the E2100D mutant $InsP_3$ receptor than in the wild-type channel.

Somlyo: I was struck by your interesting observations when you activated the receptors and saw a very long lag phase with mutated $InsP_3$ receptors. What did you activate muscles with?

Iino: Those cells were B lymphocytes and their B cell receptors were activated.

Somlyo: What is upstream of this? How do we get to the $InsP_3$ receptor?

Iino: The B cell receptors were cross-linked by antibody against them, and the downstream pathway involves a series of tyrosine kinases.

Somlyo: Does it activate PLC?

Iino: Yes, the PLCγ is activated via tyrosine phosphorylation.

Somlyo: Presumably, then, most of that long lag phase involved the activation of PLC. Did you try to see what happens with $InsP_3$ itself, with regard to the lag phase? In normal smooth muscle we find a very short lag phase, whether we photo release $InsP_3$ or Ca^{2+} from caged Ca^{2+}. Why does that lag phase for PLC change so much?

Iino: We haven't compared the $InsP_3$ concentration time course in these two cell types (expressing either wild-type or E2100D mutant $InsP_3$ receptors). We presume that these were not so different. Again, we cannot rule out the possibility that Ca^{2+} is positively promoting the PLC activity. When the $InsP_3$ receptor is more sensitive to Ca^{2+}, it releases Ca^{2+} faster and gives off Ca^{2+} spikes, which may also feed back on PLC activity. This might have shifted the $InsP_3$ upstroke towards the earlier time in cells expressing the wild-type $InsP_3$ receptor.

Kotlikoff: Coming back to the context of RyRs coexisting with $InsP_3$ receptors, how do you conceive of this system working in isolation from the RyR Ca^{2+} sensor? In some ways it is easier to do the experiment to exclude the $InsP_3$ receptors because you can use something like heparin and completely block it, and have some confidence. The alternative experiment to block all RyRs is a little dirtier. What are your thoughts on how this system can exist in waves, and what is the wave transmission speed in the arteries?

Iino: The wave speed is about 20 μm per second at 25 °C. We haven't looked at whether or not the ryanodine receptor is involved in the Ca^{2+} waves in the arterial SMCs. However, we have looked at the Ca^{2+} waves in the intestinal SMCs in response to muscarinic stimulation. We applied carbachol, a muscarinic agonist, in the presence of ryanodine. Ryanodine had no effect on the Ca^{2+} response to repetitive stimulation by carbachol. On the other hand, if we gave caffeine in the presence of ryanodine, we observed Ca^{2+} response only once and the subsequent application of caffeine failed to induce a Ca^{2+} response. Ryanodine only binds to ryanodine receptors when the channels are open, so the absence of the effect of ryanodine during carbachol-induced Ca^{2+} waves is probably indirect evidence that the RyR is not involved in agonist-induced Ca^{2+} wave generation, at least in the intestinal SMCs.

Paul: Colin Taylor, I thought you said yesterday that unless InsP$_3$ was present, Ca^{2+} in your hands was always inhibitory. Is this correct?

Taylor: Yes.

Paul: How did that differ from what we have been told about here, with Ca^{2+} having a biphasic effect independent of InsP$_3$? Or did I just miss something?

Taylor: First, we are dealing with different InsP$_3$ receptor subtypes. Second, most of your measurements are steady-state measurements, where the InsP$_3$ receptor equilibrates Ca^{2+} with InsP$_3$. The net result is the effect of those kinetic observations that we were making; the same applies with Foskett's work, in which his patch-clamp measurements are all steady-state views of the kinetic properties that we are looking at. Broadly, many are consistent, although what wouldn't be consistent would be the notion that a single Ca^{2+}-binding site mediates the two effects. I don't think this sits comfortably with our observations.

Iino: A very low affinity Ca^{2+} site such as yours will require perhaps 100 μM Ca^{2+}.

Taylor: We don't require that; we just use this concentration to make it happen quickly, and lower concentrations will work.

Brading: Do your expression cells generate sparks? Do you get spontaneous release from the SR in these cells? You showed some of these cells where the SR was overloaded with Ca^{2+}, and I wondered whether there is a difference in the behaviour of the SR depending on the amount of intracellular Ca^{2+}. This is going back to this business of stretch-activated channels and whether if you stretch the SR by superfilling it with Ca^{2+} and getting some osmotic expansion it changes its behaviour.

Iino: Unfortunately, the DT40 cells we used have a very small size with a diameter of about 10 μm. The size of these cells is comparable to that of the Ca^{2+} puffs (spontaneous Ca^{2+} releases via the InsP$_3$ receptor) observed in *Xenopus* oocytes which are huge cells. Since we have not tried to look at subcellular Ca^{2+} transients in the DT40 cells, at the moment we can't say whether Ca^{2+} puffs are taking place or not.

Sanders: It is still amazing to me that there is positive feedback on the channel. What closes it? The single-channel records of InsP$_3$ receptors are very flickery. I thought that this was because there was a high concentration of Ca^{2+} outside the receptor, and this inhibits the channel and produces a flickeriness. Has anyone tried replacement of Ca^{2+} to see whether or not it changes the channel kinetics? You have this positive feedback that would seem to keep the channel open for long periods. This doesn't seem to match with the kinetics of single channels.

Iino: The ryanodine receptor has a square shape with each side being about 27 nm. The InsP$_3$ receptor should be slightly smaller than the RyR because the molecular weight is about half of the RyR's. We still don't know the topological relationship between the channel pore and the Ca^{2+} sensor on the InsP$_3$ receptor. It is possible that Ca^{2+} released via the channel pore does not have a direct access to the Ca^{2+} sensor and build-up of ambient Ca^{2+} concentration is required to influence the Ca^{2+} sensor. Elucidation of the three dimensional structure of the InsP$_3$ receptor may clarify the point.

Taylor: There is some evidence that speaks to this issue from Barbara Ehrlich, who has worked in bilayers on native cerebellar type 1 InsP$_3$ receptor. She varied what is in effect the luminal bivalent cation from Sr^{2+}, Ba^{2+} or Ca^{2+}. Their order of being able to inhibit the channel is Ca$^{2+} >$ Sr$^{2+} >$ Ba^{2+}. The mean open times of those channels varied in the same way. This suggests something is happening quite immediately.

Nixon: You showed that the glutamate residue was preserved through all three subtypes. Yet when you express each of the different subtypes and look at the Ca^{2+} oscillations, they are different. The type 3 InsP$_3$ receptor doesn't produce oscillations. Does that mean that this residue is not always involved in the same way? It is preserved there in the type 3 and you don't see Ca^{2+} oscillation.

Iino: The Ca^{2+} sensor regions are conserved in all the subtypes of InsP$_3$ receptors but are not identical. Both Ca^{2+} and InsP$_3$ sensitivities are different in all three subtypes. Type 3 is least sensitive to InsP$_3$; type 2 is most sensitive. The DT40 cells expressing only type 2 InsP$_3$ receptor show vigorous Ca^{2+} oscillation, but in cells expressing only type 3 it usually consists of just one initial spike. Such differences in the Ca^{2+} signalling pattern may be due to the difference in InsP$_3$ sensitivity. We may be able to do some swapping experiments, to see what happens with a mutant InsP$_3$ receptor having the InsP$_3$ sensitivity of type 2 and the Ca^{2+} sensitivity of type 3, for example.

Taylor: There is a much-quoted paper (Hagar et al 1998) that tends to suggest that type 3 InsP$_3$ receptors are not inhibited by cytosolic Ca^{2+}. Every subsequent attempt to examine this has produced contradictory results.

Iino: Ilya Bezprozvanny's unpublished results also suggest that type 3 InsP$_3$ receptor has a biphasic Ca^{2+} dependence.

Taylor: There are countless papers now that suggest this is the case, not least Kevin Foskett's work on recombinant type 3 InsP$_3$ receptors in *Xenopus* oocyte nuclei. The original notion that type 3 is not inhibited by cytosolic Ca^{2+} is pretty much dead.

References

Hagar RE, Burgstahler AD, Nathanson MH, Ehrlich BE 1998 Type III InsP$_3$ receptor channel stays open in the presence of increased calcium. Nature 396:81–84

Hirose K, Kadowaki S, Tanabe M, Takeshima H, Iino M 1999 Spatio-temporal dynamics of inositol 1,4,5-trisphosphate that underlies complex Ca^{2+} mobilization patterns. Science 284:1527–1530
Mak DO, McBride S, Foskett JK 1998 Inositol 1,4,5-trisphosphate activation of inositol trisphosphate receptor Ca^{2+} channel by ligand tuning of Ca^{2+} inhibition. Proc Natl Acad Sci USA 95:15821–15825
Wray S, Kupittayanant S, Shmigol A 2002 Role of the sarcoplasmic reticulum in uterine smooth muscle. In: Role of the sarcoplasmic reticulum in smooth muscle. Wiley, Chichester (Novartis Found Symp 246) p 6–25

Calcium release events in excitation–contraction coupling in smooth muscle

T. B. Bolton, D. V. Gordienko, V. Pucovský, S. Parsons and O. Povstyan

Department of Pharmacology & Clinical Pharmacology, St George's Hospital Medical School, Cranmer Terrace, London SW17 0RE, UK

Abstract. Although smooth muscle cells are not organized in sarcomeres, as are striated muscles, nevertheless Ca^{2+} for contraction is released from the sarcoplasmic reticulum (SR) at certain preferred sites. These sites commonly discharge packets of Ca^{2+} spontaneously and have been called frequent discharge sites (FDSs). Each spontaneous release of a Ca^{2+} packet usually leads to a burst of openings of Ca^{2+}-activated K^+ channels in the cell membrane which produces a spontaneous transient outward current (STOC) in smooth muscle cells under voltage clamp. When fluorescent Ca^{2+} indicators such as Fluo-3 became available, the spontaneous transient increases in $[Ca^{2+}]_i$ produced by Ca^{2+} packets released from the SR were also detected in cardiac muscle as flashes of fluorescence or 'sparks'. Sparks in smooth muscle consist of smaller Ca^{2+} packets that can give rise to 'microsparks'. In some smooth muscles which have Ca^{2+}-activated Cl^- channels, STICs (spontaneous transient inward currents) are also found to be associated with sparks. FDSs have been found to be important initiating sites for a Ca^{2+} wave in response to an action potential or in response to receptor activation and possibly other stimuli, such as stretch. In both cases Ca^{2+}-induced Ca^{2+} release seems to be crucially involved.

2002 Role of the sarcoplasmic reticulum in smooth muscle. Wiley, Chichester (Novartis Foundation Symposium 246) p 154–173

Smooth muscles lack the regular sarcomeric arrangement of the contractile filaments and of the Ca^{2+} storage, release, and re-uptake processes that are found in striated muscles. Studies of subcellular morphology and the use of Ca^{2+} indicator dyes such as Fluo-3 have revealed that in skeletal and cardiac striated muscles the movements of Ca^{2+} during excitation–contraction coupling occur within a sarcomere, each of which possesses all the necessary mechanisms. In smooth muscle cells the consensus is that Ca^{2+} is stored and released from the sarcoplasmic reticulum (SR) but it is only recently that this process has been amenable to investigation in more detail due to the use of the laser scanning confocal microscope and various organelle-specific fluorescent stains. However, since smooth muscle lacks sarcomeres, it has been unclear from where exactly Ca^{2+} is released within the smooth muscle cell in response to the different stimuli which lead to contraction; the SR has other important functions besides Ca^{2+} storage, for example protein synthesis, and there is every indication that different parts of the SR are specialised for particular functions (e.g. Golovina & Blaustein 1997).

Phasic and tonic tension generation in smooth muscle

Many if not most smooth muscle tissues will generate and propagate action potentials under at least some conditions. The propagation of an action potential from one cell to another through a, usually limited, part of a smooth muscle tissue synchronizes contraction of a portion of the tissue and this can generally be seen as a small phasic rise in tension generated by the whole tissue. A phasic contraction is characterized by tension rising and falling on a second timescale. There are some smooth muscles, such as those in the wall of large blood vessels and stomach fundus which do not readily generate action potentials and so, rather than phasic, characteristically show tonic changes in tension (Golenhofen 1981). These take place over a longer timescale and are associated with pharmacomechanical coupling rather than electromechanical coupling (Somlyo & Somlyo 1968) which is found in smooth muscles which contract phasically. Many tonic smooth muscles will generate action potentials under the influence of agents such as tetraethylammonium (TEA) showing that they do also possess voltage-dependent Ca^{2+} channels which are responsible for the upstroke of the action potential but that the contribution of these is normally masked by currents through various delayed rectifier and Ca^{2+}-activated K$^+$ channels (Bolton et al 1988).

Because excitation–contraction coupling is normally different in phasic and tonic smooth muscles it is possible that intracellular Ca^{2+} release events which lead to contraction may be different in these two types of smooth muscle. So far most studies have concentrated on smooth muscles which can readily generate action potentials and so are well-endowed with voltage-dependent Ca^{2+} channels through which Ca^{2+} may enter into the cell to activate tension generation either directly, or indirectly through Ca^{2+}-induced Ca^{2+} release (CICR). Tonic smooth muscles are probably less well endowed with such Ca^{2+} channels (Bolton et al 1988) and external Ca^{2+} entering the cell is likely to be of lesser importance for tension generation; the supposition would be, especially as such smooth muscles generally exhibit pharmacomechanical coupling and small changes in membrane potential, that Ca^{2+} store release via the receptor-activated G protein/ phospholipase C/inositol-1,4,5-trisphosphate (InsP$_3$) system would be important.

Effects of receptor activation on tension

Smooth muscles not only alter their tension in response to changes in the frequency of action potential discharge, but also in response to activation of various receptors

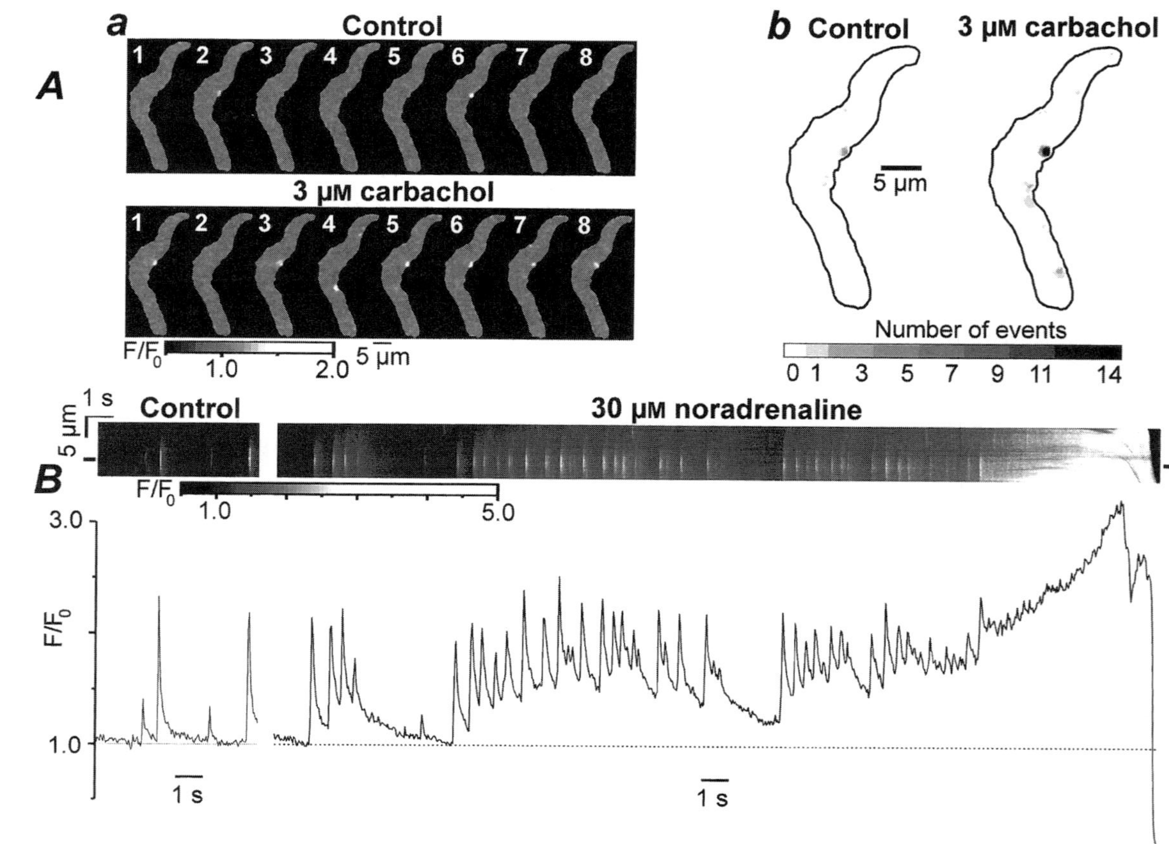
A
a Control
1 2 3 4 5 6 7 8
3 μM carbachol
1 2 3 4 5 6 7 8
F/F0
1.0 2.0 5 μm
b Control   3 μM carbachol
5 μm
Number of events
0 1 3 5 7 9 11 14
B
1 s
5 μm
Control   30 μM noradrenaline
F/F0
1.0 5.0
3.0
F/F0
1.0
1 s   1 s

on their surface. Some receptors may be activated by neurotransmitters released from adjacent nerves. Activation of receptors may alter tension both by altering the frequency of action potential discharge and also through biochemical pathways. These include, for example, release of Ca^{2+} by InsP$_3$ from stores (causing tension generation, e.g. Bramich & Hirst 1999) and β-adrenoceptor mediated increases in cAMP formation (causing relaxation, Butcher et al 1965, Harbon & Clauser 1971). Ca^{2+} release events in several smooth muscles have been studied in some detail and the arrangement of Ca^{2+} release sites and the mechanism involved in Ca^{2+} wave formation and subsequent contraction in response to various methods of excitation in these smooth muscles are now beginning to be understood. Both the action potential and biochemical mechanisms involved in tension changes likely exert their effects through alterations of the mechanisms of Ca^{2+} release and re-uptake in the cell, even though other effects such as sensitization of the contractile proteins have been described (Somlyo & Somlyo 2000).

Spontaneous transient increases in [Ca^{2+}]$_i$ or 'sparks'

Transient spontaneous increases in intracellular ionized Ca^{2+} concentration, [Ca^{2+}]$_i$, were first detected in smooth muscle as bursts of openings of Ca^{2+}-activated K$^+$ channels (Benham & Bolton 1986). Several years later, laser scanning confocal imaging showed that similar spontaneous transient increases in

---

FIG. 1. (*Opposite*)   Effect of stimulation of the receptors linked to InsP$_3$ formation on frequent discharge sites (FDSs) in freshly isolated smooth muscle cells. (A) Carbachol potentiates Ca^{2+} spark discharge in myocyte from guinea-pig ileum. (a) Eight sequential fluorescence confocal images taken as a part of a series of 40 images (acquired at 2 Hz) from Fluo-3 loaded myocyte in control and after 1 min exposure to carbachol. The normalized fluorescence intensity is grey-scale coded as indicated by bar (F/F$_0$). (b) Frequency of occurrence of Ca^{2+} sparks at each site within the confocal optical slice in control and in the presence of carbachol was computed and coded as shown by bar. Sparks were discharged mostly (5 out of 9 discharges) at a single site (FDS) in control; in carbachol this site discharged sparks more frequently (14 discharges) and other FDSs occurred. Modified from Gordienko et al (1999) with permission of the publishers. (B) Noradrenaline increases Ca^{2+} spark frequency at FDS leading to summation of individual sparks and initiation of propagating Ca^{2+} wave in rabbit portal vein myocyte. The line-scan images shown were obtained from Fluo-4 loaded myocyte and were formed by aligning (from left to right) the successive fluorescence images of the scan line (acquired at 14 Hz) positioned close (within 1.5 μm) and parallel to the cell membrane in the central region of the myocyte. Plots below the images show the time course of the normalized fluorescence signal at site indicated by black bars on the images. Note that application of noradrenaline leads to an increase in spark frequency at FDS and summation of individual sparks triggering Ca^{2+} wave and myocyte contraction (the latter can be seen as a displacement at the end of the line-scan image). To minimize possible contribution of voltage-gated Ca^{2+} channels, experiment was performed in the presence of 1 μM nicardipine.

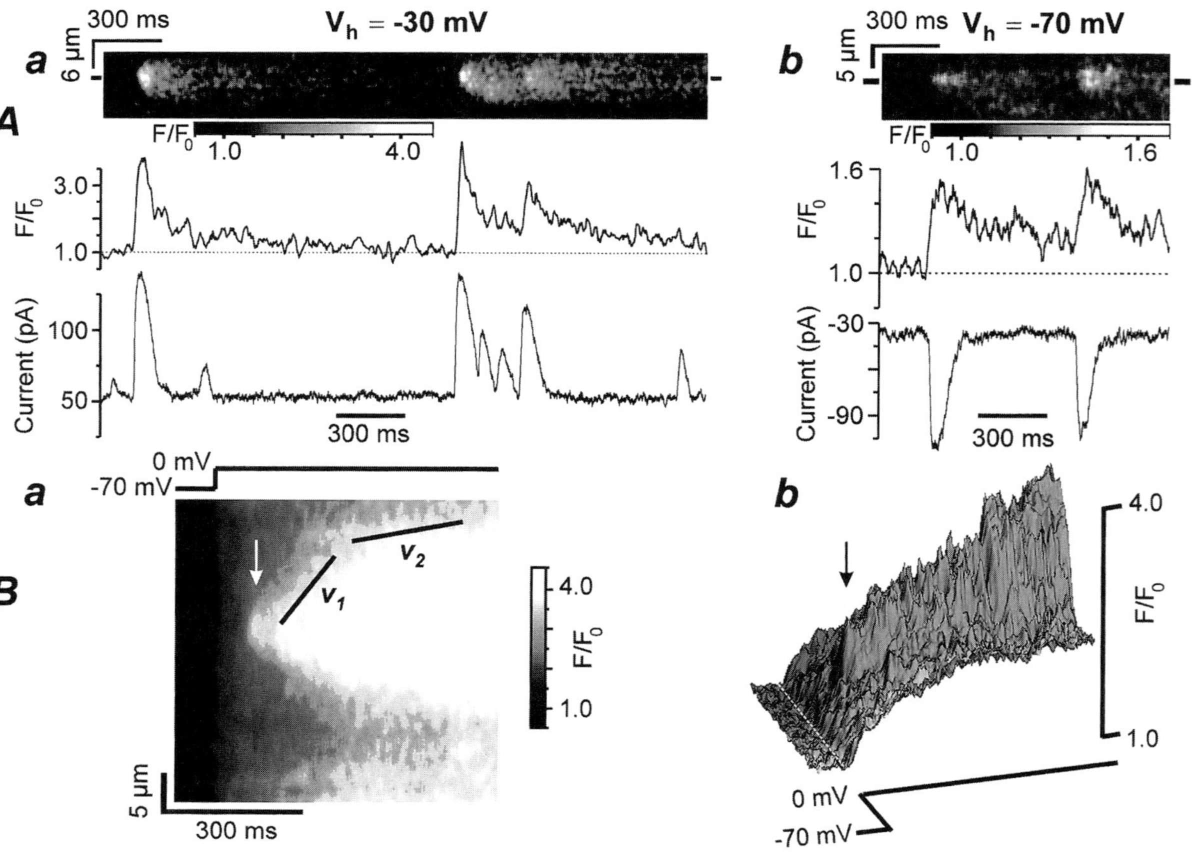
A
a
V_h = -30 mV
300 ms
6 μm
F/F_0
1.0
4.0
F/F_0
3.0
1.0
Current (pA)
100
50
300 ms
b
V_h = -70 mV
300 ms
5 μm
F/F_0
1.0
1.6
F/F_0
1.6
1.0
Current (pA)
-30
-90
300 ms
B
a
0 mV
-70 mV
v_1
v_2
F/F_0
4.0
1.0
5 μm
300 ms
b
F/F_0
4.0
1.0
0 mV
-70 mV

[Ca^{2+}]$_i$ occur in cardiac (Cheng et al 1993) and skeletal muscles (Klein et al 1996). These were observed as flashes of fluorescence when cells were loaded with Ca^{2+}-sensitive indicators (see Figs 1, 2 and 4), hence the name 'sparks' (Cheng et al 1993). In single cells of smooth muscles Ca^{2+} sparks generally trigger a burst of openings of Ca^{2+}-activated K$^+$ channels. These bursts are seen under voltage-clamp as spontaneous transient outward currents (Fig. 2A panel a) or STOCs (Nelson et al 1995, Bolton & Gordienko 1998, ZhuGe et al 2000). The suggestion was made that the spark is the fundamental building block of larger Ca^{2+} transients and Ca^{2+} waves (Cheng et al 1993). Sparks have been most frequently seen in single smooth muscle cells isolated by enzyme treatment but have been also recorded in the smooth muscle cells of the media of isolated segments of small arteries (e.g. Jaggar 2001).

In smooth muscle cells from guinea-pig small mesenteric arteries and longitudinal muscle of the small intestine, 'microsparks' have been recorded. These are more rapid and smaller increases in [Ca^{2+}]$_i$. In some cells fluctuations in baseline fluorescence exceed recording noise and a low frequency peak in the power spectrum of the fluorescent signal was found (Bolton & Gordienko 1998). These observations, and others we have made, suggest that sparks in smooth muscle consist of smaller elementary events which, when they occur alone, give rise to microsparks (Gordienko et al 1998, Bolton & Gordienko 1998). These

FIG. 2. (*Opposite*) Local Ca^{2+} signalling in relation to membrane events in voltage-clamped myocytes from small mesenteric arteries of guinea-pig. (A) Imaging of [Ca^{2+}]$_i$ with simultaneous recordings of net membrane current in voltage-clamped myocytes. The myocytes were dialysed with solution containing 0.1 Fluo-3 (pentapotassium salt). Normalized line-scan images (acquired at 500 Hz) were obtained from two different myocytes at two different holding potentials: (a) -30 mV and (b) -70 mV. Plots of the time course of the normalized fluorescence signal (F/F$_0$) at sites indicated by black bars on the images and corresponding records of the net membrane current are shown below the images. Note that at -30 mV each Ca^{2+} spark leads to a spontaneous transient outward current (STOC) resulting from a burst of openings of Ca^{2+}-activated K$^+$ channels in the cell membrane. At -70 mV, Ca^{2+} sparks increase the probability of opening of Ca^{2+}-activated Cl$^-$ channels triggering spontaneous transient inward currents (STICs). Modified with permission of the publishers (a) from Bolton & Gordienko (1998), and (b) from Gordienko et al (1999). (B) Ca^{2+} wave triggered by step-like depolarization of the myocyte to 0 mV. (a) Normalized line-scan image was obtained during voltage step from the holding potential -70 mV to 0 mV. The Ca^{2+} wave propagation velocities are: $v_1 = 60\,\mu$m s^{-1}; $v_2 = 10\,\mu$m s^{-1}. (b) Image from panel *a* is shown as shaded surface plot. Note that voltage step initially causes relatively uniform rise in [Ca^{2+}]$_i$ (probably due to Ca^{2+} entering the cell) which eventually triggers Ca^{2+} spark at a single point along the scan line. This, in turn, gives rise to propagating Ca^{2+} wave. Note also that fluctuations in the fluorescence signal during the Ca^{2+} wave are significantly larger than those before the voltage step was applied and those at sites where the wave had not invaded. This suggests that Ca^{2+} wave consisted of numerous, local Ca^{2+}-release events varying in amplitude and size. Modified with permission of Blackwell Science Ltd from Bolton & Gordienko (1998).

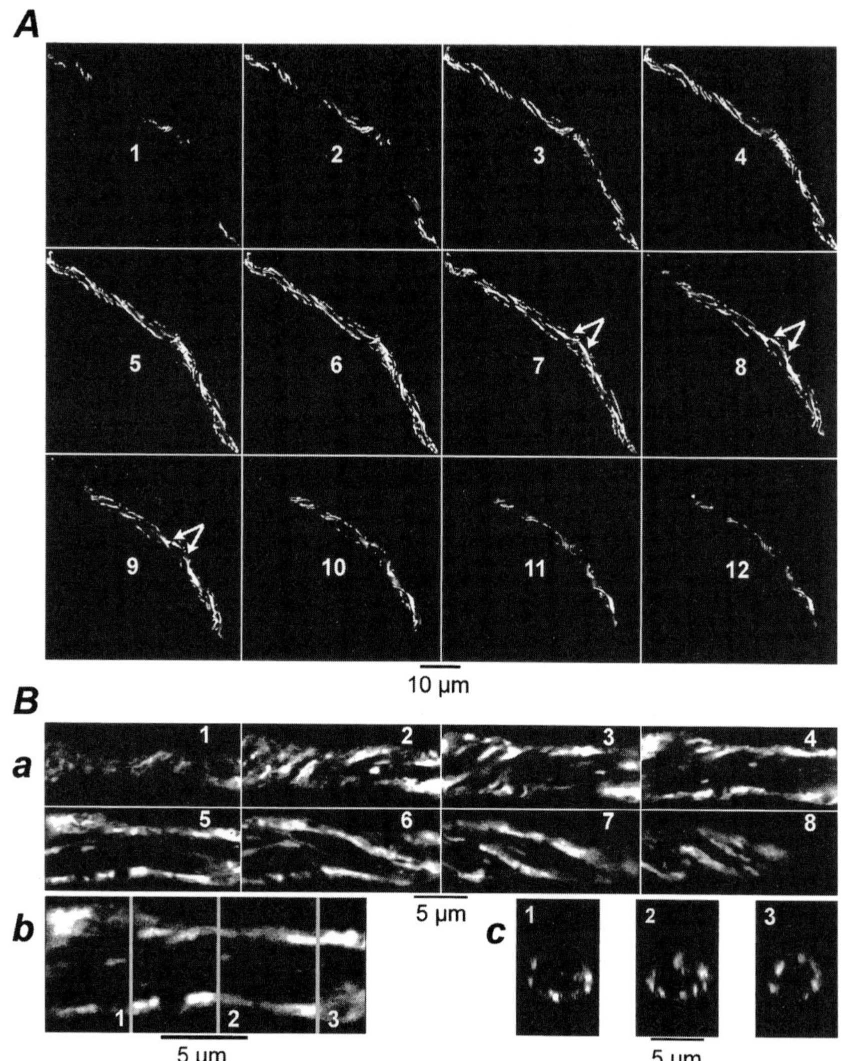

A
1
2
3
4
5
6
7
8
9
10
11
12
10 μm
B
a
1
2
3
4
5
6
7
8
5 μm
b
1
2
3
5 μm
c
1
2
3
5 μm

observations argue that a spark is not the fundamental event in smooth muscle cells, nor does it represent the opening of a single SR Ca^{2+} channel, but instead it is composed of smaller components which may themselves represent the opening of one, or a few, SR Ca^{2+} channels. Further evidence has been obtained by high-resolution recordings of the point of origin of sparks. This shows submicron 'jitter' consistent with the hypothesis that sparks may be initiated by different SR Ca^{2+} channels in a tight cluster. Channels in the cluster would recruit each other and variations in the initiating channel (or conceivably initiating 'subcluster', an even tighter group of a few channels) would be the cause of this jitter (Fig. 4D panel b, Gordienko et al 2001, Pucovský et al 2002). Similar possibilities have been considered for cardiac and skeletal muscle (Shirokova et al 1999).

Frequent discharge sites

It has become abundantly clear that spontaneous sparks occur at certain preferred locations in single smooth muscle cells loaded with Fluo-3 or Fluo-4 and imaged by laser scanning confocal microscopy (Figs 1 and 4); these preferred locations were called 'frequent discharge sites' (FDSs; Gordienko et al 1998). They were often but not exclusively located close to the cell membrane. In some smooth muscle cell types such as longitudinal muscle of guinea-pig small intestine (Fig. 1A and Gordienko et al 1998), feline oesophageal muscle (Kirber et al 2001), or the muscle of guinea-pig small mesenteric arteries (Pucovský et al 2002), a number of FDSs are seen in each confocal plane (about 1 μm thick). In rabbit portal vein muscle only one, or less commonly two, such FDSs are found (Fig. 4 and Gordienko et al 2001).

Our studies of the structures giving rise to sparks suggest that the SR is the source of Ca^{2+} release through ryanodine receptors which act as Ca^{2+} release channels, but that adjacent InsP$_3$ receptors can influence the behaviour of the

---

FIG. 3. (*Opposite*) Three-dimensional distribution of the sarcoplasmic reticulum (SR) visualized by confocal z-sectioning in single rabbit portal vein myocyte stained with 0.1 μM DiOC$_6$. (A) The gallery shows images 7, 9, 11, 13, 14, 15, 17, 21, 23, 25, 27 and 28 from z-sectioning protocol comprising 35 individual x–y confocal optical slices taken with a step of 0.35 μm. Arrows on images 7, 8 and 9 emphasise the continuity of the SR and the nuclear envelope. (B) z-sectioning protocol was similar to that in (A) but imaging of a small portion of another myocyte was performed at higher magnification. (a) Images 11, 13, 15, 17, 19, 21, 23, 25 are shown in the gallery. z-cuts of 3D reconstructed image providing a transverse section of the myocyte in z–y plane at three different x coordinates depicted by vertical lines in (b) are presented in (c). Note that the superficial SR had a helical arrangement: twisted longitudinal strands lining the internal side of the plasmalemma appeared in x–y images obtained during z-sectioning (A and Ba); these could be seen as discrete sub-plasmalemmal bands in transverse (z–y) sections of the myocyte (c). Note also that strands of weaker staining could be detected deeper within the cell. Modified with permission of the publishers from Gordienko et al (2001).

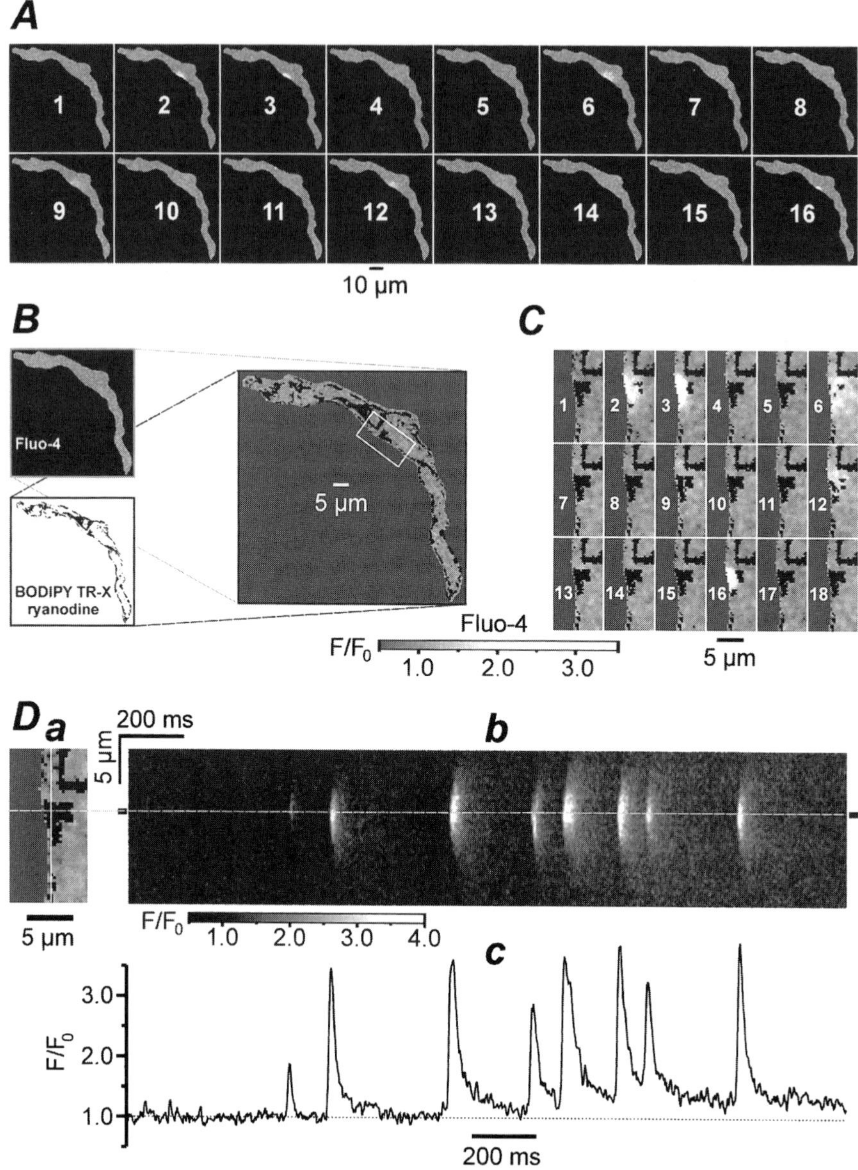
A
1 2 3 4 5 6 7 8
9 10 11 12 13 14 15 16
10 µm
B
Fluo-4
BODIPY TR-X
ryanodine
5 µm
Fluo-4
F/F_0
1.0 2.0 3.0
C
1 2 3 4 5 6
7 8 9 10 11 12
13 14 15 16 17 18
5 µm
D a
200 ms
5 µm
b
F/F_0
1.0 2.0 3.0 4.0
c
3.0
2.0
1.0
F/F_0
200 ms

ryanodine receptor channels probably by allowing the escape of Ca^{2+} from the SR when they are activated by InsP$_3$ and increasing [Ca^{2+}]$_i$ in the vicinity of the ryanodine receptors (Boittin et al 1998). In rabbit portal vein the three dimensional structure of the SR has been studied in some detail by reconstruction from confocal z-sections made at submicron intervals (Gordienko et al 2001). The SR when stained with DiOC$_6$ (Fig. 3) or fluorescently labelled ryanodine (Fig. 4B) is revealed to be well developed in the perinuclear region and generally forms a lozenge-shaped envelope around the nucleus; such staining almost certainly does not differentiate the Golgi apparatus which at least in other cell types behaves as an InsP$_3$-sensitive Ca^{2+} store (Pinton et al 1998). From this lozenge-shaped volume strands of SR emanate and spiral down the cell towards its ends (Fig. 3), mainly but not exclusively in a superficial (subplasmalemmal) location (Gordienko et al 2001).

The FDSs are important for the initial release of Ca^{2+} which occurs following an action potential as has been elegantly shown by the work of Imaizumi's group (Imaizumi et al 1998, Ohi et al 2001). The influx of Ca^{2+} during the upstroke of an action potential in vas deferens or urinary bladder smooth muscle triggers Ca^{2+} release from FDSs close under the membrane and from these a Ca^{2+} wave spreads from under the membrane to all parts of the cell, eliciting contraction. Careful study showed that FDSs from which spontaneous sparks were discharged were also the sites (or 'hot spots') where Ca^{2+} was first released during an action potential. The Ca^{2+} entering during an action potential or step-like membrane depolarization (Fig. 2B, panel b) causes further Ca^{2+} release from the FDSs by a

---

FIG. 4. (*Opposite*) FDS coincides with the SR portion showing a high density of ryanodine receptors (RyRs). (A) Imaging was performed on isolated rabbit portal vein myocytes loaded with Ca^{2+}-sensitive indicator Fluo-4 AM. The gallery shows 16 sequential fluorescence confocal images taken as a part of time series comprising 100 x–y images (acquired at 2 Hz). Note that sparks (images 2, 3, 6, 9, 12, 16) were discharged at a single site within myocyte (frequent discharge site, FDS). (B) Following [Ca^{2+}]$_i$ imaging the same myocyte was stained with 1 μM Bodipy TR-X ryanodine and each confocal image of the normalized Fluo-4 fluorescence from the x–y time series was overlaid with Bodipy TR-X ryanodine confocal image. (C) The gallery shows 18 sequential images of the boxed region (B) after enlargement and rotation by 54°. Note that Ca^{2+} sparks originate from a prominent portion of the SR enriched with RyRs. Also note that the initiating point of an individual spark cannot be identified at such a low (2 Hz) acquisition rate. (D) To improve temporal resolution of [Ca^{2+}]$_i$ imaging, a single line depicted by white line in enlarged image of the boxed region (a) was repetitively scanned every 7 ms. Corresponding line-scan image and a time course plot of the normalised fluorescence signal at site indicated by black bars on the image are shown in (b) and (c), respectively. Grey dashed line in (a–b) projects the position of the initiating site of Ca^{2+} sparks in the line-scan image (b) onto the SR portion in the boxed region (a). Note that the precise position of the Ca^{2+} spark initiating point varies slightly but detectably from one event to another (b) which suggests that a spark results from the opening of more than one Ca^{2+}-release channel. Modified with permission of publishers from Gordienko et al (2001).

process of CICR which is inhibited by treatment of the cell with caffeine, ryanodine or cyclopiazonic acid to deplete Ca^{2+} stores (Imaizumi et al 1998). Under voltage clamp the rate of spark and STOC discharge in single cells is increased by depolarization (Bolton & Gordienko 1998).

During the action potential in vas deferens or urinary bladder the rise in $[Ca^{2+}]_i$ close under the cell membrane is responsible, in combination with the depolarization, for the repolarization phase as it causes the opening of Ca^{2+}-activated K^+ channels through which a large repolarizing outward current flows (Arnaudeau et al 1997, Imaizumi et al 1998, Ohi et al 2001). This may lead to a transient period of hyperpolarization (an 'afterhyperpolarization') following the action potential (Imaizumi et al 1998).

Receptor-evoked Ca^{2+} waves are initiated at FDSs

The FDSs are also important for the effects of activation of receptors which act through the formation of $InsP_3$ because they serve as initiation points for Ca^{2+} waves. In longitudinal muscle of guinea-pig ileum, activating muscarinic receptors with carbachol (leading to $InsP_3$ formation) stimulates the discharge of Ca^{2+} sparks from FDSs (Fig. 1A). One of the FDSs is dominant and acts as a pacemaker, giving rise to a Ca^{2+} wave which spreads to recruit other FDSs. At least over short periods of time, one dominant FDS determines the rate at which Ca^{2+} waves are discharged during muscarinic receptor activation and increased production of $InsP_3$ (Gordienko et al 1999). Stimulation of adrenoreceptors in myocytes of rabbit portal vein (Fig. 1B, D.V. Gordienko & T. B. Bolton, unpublished results) or of guinea-pig small mesenteric artery (Pucovský et al 2002) leads to an increase in spark frequency at FDSs and summation of individual sparks triggering a Ca^{2+} wave and myocyte contraction. Caffeine has also been reported to initiate Ca^{2+} waves from FDSs (Arnaudeau et al 1996).

Action of agents causing relaxation

Agents which relax smooth muscle such as SNAP (S-nitroso-N-acetyl-penicillamine) and SNP (sodium nitroprusside) and which are nitric oxide donors (Hughes 1999) also exert actions on FDSs. In recent studies from this laboratory it was found that these agents inhibited spark discharge and Ca^{2+} wave formation in response to noradrenaline in single cells from small guinea-pig mesenteric arteries (Pucovský et al 2002). However, in coronary artery smooth muscle cells SNP was found to increase the frequency of spark discharge and of STOCs (Porter et al 1998).

Activation of membrane ion channels

Besides activating Ca^{2+}-activated K$^+$ channels, a rise in [Ca^{2+}]$_i$ can also increase the probability of opening of Ca^{2+}-activated Cl$^-$ channels (Fig. 2A panel b) and cation channels, the latter often requiring previous gating by activation of G protein-linked membrane receptors. In some smooth muscles, such as guinea-pig tracheal smooth muscle, a spark may activate a burst of openings of both Ca^{2+}-activated Cl$^-$ channels (giving rise to a STIC—spontaneous transient inward current) and K$^+$ channels (giving rise to a STOC); sometimes a biphasic transient (outward–inward) current response is produced when activation of both types of channel overlaps, which has been termed a STOIC. Whether Cl$^-$ or K$^+$ channels are activated seems to depend on their sensitivity to [Ca^{2+}]$_i$, the membrane potential (K$^+$ channels are much more voltage-sensitive) and the proximity of the channel type to ryanodine receptors (ZhuGe et al 1998, Gordienko et al 1999).

Studies on the relationship between sparks and STOCS have shown the proportion of sparks giving rise to STOCs was 73% (feline oesophageal smooth muscle, Kirber et al 2001), 79% (toad gastric muscle, ZhuGe et al 2000) and 96% (rat cerebral arteries, Pérez et al 1999). Where FDSs are numerous, as in longitudinal muscle of guinea-pig ileum, they occur most frequently in a position close beneath the cell membrane; however, in these smooth muscle cells in the relaxed state the diameter of the cell may not exceed 3 μm for much of their length, so any sparks occurring in such narrow regions of the cell are likely to raise [Ca^{2+}]$_i$ close to the internal mouths of Ca^{2+}-activated K$^+$ channels (Gordienko et al 1998). If the cell is normally much wider than this, as for example during contraction, then sparks distant from the cell membrane would be unlikely to evoke a STOC. Other factors have also been suggested to operate. For example, parts of the cell membrane act as anchorage for the contractile filaments and it seems likely that these areas lack channels (Bolton et al 1999). In rabbit portal vein single smooth muscle cells there was usually only one site discharging sparks which was within 1–2 μm of the SR nuclear envelope and close beneath the cell membrane (Fig. 4 and Gordienko et al 2001). It is apparent that Ca^{2+} release sites in smooth muscle cells from different tissues are organized in different ways for reasons that are not yet understood and which provide important functional relationships to other specific structures in the cell, including cell membrane ion channels.

An interesting idea is that in small arteries such as the rat basilar STOCs elicited by sparks can exert a negative feedback on the membrane potential of the smooth muscle syncytium (cells are linked by low resistance pathways); the hyperpolarization so produced will reduce Ca^{2+} entry through voltage-dependent Ca^{2+} channels. Thus, in pressurized arteries where the membrane potential is lower than in the relaxed state and there is significant Ca^{2+} influx into the cell through open voltage-dependent Ca^{2+} channels leading to loading of SR Ca^{2+} stores and STOC discharge, the influx of Ca^{2+} into the cell would be expected to be reduced by increased STOC discharge (Nelson et al 1995, Knot et al 1998).

Summary

Preferred sites of Ca^{2+} release—FDSs—are found in smooth muscle which lacks the regular sarcomeric arrangement of the contractile proteins, Ca^{2+} release mechanisms and uptake sites found in striated muscles. Filamentous and irregular strands of SR which are more frequently found in a superficial location in the cell, at least in phasic smooth muscles, show one or more FDSs at which Ca^{2+} is released spontaneously. In some smooth muscle cells which are up to 500 μm or more long, there may be several of these FDSs in a single confocal plane and perhaps as many as 20 in a single cell; in other smooth muscle cells there may only be a single FDS. An FDS seems to represent a differentiated portion of the SR specialized for Ca^{2+} release. Such FDSs release packages of Ca^{2+} spontaneously in many smooth muscles. The FDSs also act as trigger sites for Ca^{2+} waves in response to action potentials or to activation of G protein-coupled receptors which increase the formation of $InsP_3$. Thus they are responsive to Ca^{2+} entering through voltage-dependent channels in the cell membrane or to Ca^{2+} released in response to $InsP_3$ generated by activation of various excitatory receptors. They may also serve as initiator sites of Ca^{2+} waves in response to other types of stimuli such as stretch.

The model described above is based on observations described so far on only a few smooth muscles and will need to be tested for its applicability to other or all smooth muscles.

Acknowledgements

This work was supported by The Wellcome Trust (grant numbers 042293, 060659 and 051162) and a programme grant from the British Heart Foundation (RG 99001).

References

Arnaudeau S, Macrez-Lepretre N, Mironneau J 1996 Activation of calcium sparks by angiotensin II in vascular myocytes. Biochem Biophys Res Commun 222:809–815

Arnaudeau S, Boittin FX, Macrez N, Lavie JL, Mironneau C, Mironneau J 1997 L-type and Ca^{2+} release channel-dependent hierarchical Ca^{2+} signalling in rat portal vein myocytes. Cell Calcium 22:399–411

Benham CD, Bolton TB 1986 Spontaneous transient outward currents in single visceral and vascular smooth muscle cells of the rabbit. J Physiol 381:385–406

Boittin FX, Coussin F, Macrez N, Mironneau C, Mironneau J 1998 Inositol 1,4,5-trisphosphate- and ryanodine-sensitive Ca^{2+} release channel-dependent Ca^{2+} signalling in rat portal vein myocytes. Cell Calcium 23:303–311

Bolton TB, Gordienko DV 1998 Confocal imaging of calcium release events in single smooth muscle cells. Acta Physiol Scand 164:567–575

Bolton TB, Large WA 1986 Are junction potentials essential? Dual mechanism of smooth muscle cell activation by transmitter released from autonomic nerves. Q J Exp Physiol 71:1–28

Bolton TB, Aaronson PI, MacKenzie I 1988 Voltage-dependent calcium channel in intestinal and vascular smooth muscle cells. Ann NY Acad Sci 522:32–42

Bolton TB, Prestwich SA, Zholos AV, Gordienko DV 1999 Excitation–contraction coupling in gastrointestinal and other smooth muscles. Annu Rev Physiol 61:85–115

Bramich NJ, Hirst GDS 1999 Sympathetic neuroeffector transmission in the rat anococcygeus muscle. J Physiol 516:101–115

Butcher RW, Ho RJ, Meng HC, Sutherland EW 1965 Adenosine 3′,5′-monophosphate in biological materials. II. The measurement of adenosine 3′,5′-monophosphate in tissues and the role of the cyclic nucleotide in the lipolytic response of fat to epinephrine. J Biol Chem 240:4515–4523

Cheng H, Lederer WJ, Cannell MB 1993 Calcium sparks: elementary events underlying excitation–contraction coupling in heart muscle. Science 262:740–744

Golenhofen K 1981 Differentiation of calcium activation processes in smooth muscle using selective antagonists. In: Bülbring E, Brading AF, Jones AW, Tomita T (eds) Smooth muscle: an assessment of current knowledge. Edward Arnold, London, p 157–170

Golovina VA, Blaustein MP 1997 Spatially and functionally distinct Ca^{2+} stores in sarcoplasmic and endoplasmic reticulum. Science 275:1643–1648

Gordienko DV, Bolton TB, Cannell MB 1998 Variability in spontaneous subcellular calcium release in guinea-pig ileum smooth muscle cells. J Physiol 507:707–720

Gordienko DV, Zholos AV, Bolton TB 1999 Membrane ion channels as physiological targets for local Ca^{2+} signalling. J Microsc 196:305–316

Gordienko DV, Greenwood IA, Bolton TB 2001 Direct visualization of sarcoplasmic reticulum regions discharging Ca^{2+} sparks in vascular myocytes. Cell Calcium 29:13–28

Harbon S, Clauser H 1971 Cyclic adenosine 3′,5′ monophosphate levels in rat myometrium under the influence of epinephrine, prostaglandins and oxytocin. Correlations with uterus motility. Biochem Biophys Res Commun 44:1496–1503

Hughes MN 1999 Relationships between nitric oxide, nitroxyl ion, nitrosonium cation and peroxynitrite. Biochim Biophys Acta 1411:263–272

Imaizumi Y, Torii Y, Ohi Y et al 1998 Ca^{2+} images and K$^+$ current during depolarization in smooth muscle cells of the guinea-pig vas deferens and urinary bladder. J Physiol 510:705–719

Jaggar JH 2001 Intravascular pressure regulates local and global Ca^{2+} signaling in cerebral artery smooth muscle cells. Am J Physiol 281:C439–C448

Kirber MT, Etter EF, Bellve KA et al 2001 Relationship of Ca^{2+} sparks to STOCs studied with 2D and 3D imaging in feline oesophageal smooth muscle cells. J Physiol 531:315–327

Klein MG, Cheng H, Santana LF, Jiang YH, Lederer WJ, Schneider MF 1996 Two mechanisms of generalised calcium release in skeletal muscle. Nature 379:455–458

Knot HJ, Standen NB, Nelson MT 1998 Ryanodine receptors regulate arterial diameter and wall [Ca^{2+}] in cerebral arteries of rat via Ca^{2+}-dependent K$^+$ channels. J Physiol 508:211–221

Nelson MT, Cheng H, Rubart M et al 1995 Relaxation of arterial smooth muscle by calcium sparks. Science 270:633–637

Ohi Y, Yamamura H, Nagano N et al 2001 Local Ca^{2+} transients and distribution of BK channels and ryanodine receptors in smooth muscle cells of guinea-pig vas deferens and urinary bladder. J Physiol 534:313–326

Pérez GJ, Bonev AD, Patlak JB, Nelson MT 1999 Functional coupling of ryanodine receptors to K$_{Ca}$ channels in smooth muscle cells from rat cerebral arteries. J Gen Physiol 113:229–238

Pinton P, Pozzan T, Rizzuto R 1998 The Golgi apparatus is an inositol 1,4,5-trisphosphate-sensitive Ca^{2+} store, with functional properties distinct from those of the endoplasmic reticulum. EMBO J 17:5298–5308

Porter VA, Bonev AD, Knot HJ et al 1998 Frequency modulation of Ca^{2+} sparks is involved in regulation of arterial diameter by cyclic nucleotides. Am J Physiol 274:C1346–C1355

Pucovský V, Gordienko DV, Bolton TB 2002 Effect of nitric oxide donors and noradrenaline on Ca^{2+} release sites and global intracellular Ca^{2+} in myocytes from guinea-pig small mesenteric arteries. J Physiol 539:25–39

Shirokova N, Gonzalez A, Kirsch WG et al 1999 Calcium sparks: release packets of uncertain origin and fundamental role. J Gen Physiol 113:377–384

Somlyo AV, Somlyo AP 1968 Electromechanical and pharmacomechanical coupling in vascular smooth muscle. J Pharmacol Exp Ther 159:129–145

Somlyo AP, Somlyo AV 2000 Signal transduction by G-proteins, Rho-kinase and protein phosphatase to smooth muscle and non-muscle myosin II. J Physiol 522:177–185

ZhuGe R, Sims SM, Tuft RA, Fogarty KE, Walsh JV 1998 Ca^{2+} sparks activate K^+ and Cl^- channels, resulting in spontaneous transient currents in guinea-pig tracheal myocytes. J Physiol 513:711–718

ZhuGe R, Fogarty KE, Tuft RA, Lifshitz LM, Sayar K, Walsh JV 2000 Dynamics of signaling between Ca^{2+} sparks and Ca^{2+}-activated K^+ channels studied with a novel image-based method for direct intracellular measurement of ryanodine receptor Ca^{2+} current. J Gen Physiol 116:845–864

DISCUSSION

Wray: Can you clarify the difference, if any, between a hotspot and an FDS? Is an FDS just a 'tickled' hotspot, to use your terminology, or is there no difference?

Bolton: Initially we were just looking at spontaneous events and mapping where these occur. Then Imaizumi started to look at what happens at the Ca^{2+} wave initiation point if you depolarize. These initiation points turned out to be the same positions as the FDSs, suggesting that they are specialized areas from which there is preferential Ca^{2+} release from the SR.

Wray: So we can assume they are synonymous?

Bolton: Essentially, yes.

Hirst: You started by telling us that this phenomenon was seen in tissues and has physiological significance. I think one of the reasons that I ran into disagreement with Mike Kotlikoff was because I didn't believe it occurred physiologically. Under the conditions of your experiment I don't believe that you would see STOCs and sparks in tissues. You made recordings from a single arteriole cell and an intestinal muscle cell from the longitudinal layer not under stretch or stimulation, and in both cases there were STOCs and sparks. I don't regard those as likely to occur physiologically. Instead, you have to provide some sort of physiological stimulus to the tissue, and then those phenomena will occur.

Bolton: You are right. At first we thought these sparks were a Ca^{2+} overload phenomenon. This is because we saw them in single cells. Also, when we depolarized the cell the frequency went up. However, since then they have been described in whole tissues.

Hirst: Only under specific conditions.

Bolton: Yes, but even in your own work you don't normally stretch the cell. With pacemaker cells which are driving the intestine the idea is that these things may be taking place spontaneously. (This is certainly true in the sinoatrial node; Lipsius et al 2001.) Presumably you are not stretching there.

Hirst: No, but here we are talking about smooth muscle cells.

Bolton: I call interstitial cells of Cajal (ICCs) a type of smooth muscle.

Wier: The specific conditions under which sparks were seen in intact mesenteric arteries was when they were pressurized at 37 °C. I think there can be little doubt that these are occurring spontaneously in arteries. This is as close to physiological conditions as can be obtained outside the living animal.

Sanders: Did you typically see sparks, or waves in these arteries?

Wier: We see both to certain extents. I think that Dr Iino would agree that we see waves in a certain fraction of these cells under this normal pressurized condition.

Nelson: Only a small percentage (6%) of the smooth muscle cells in pressurized cerebral arteries exhibit Ca^{2+} waves, whereas all cells have sparks. Under normal conditions only a small fraction of cells have waves.

Hirst: You can easily see spontaneous synaptic potentials if you make long recordings from a non-pressurized arteriole. The clamps are big enough to see this property *in situ*. I have never seen a STOC in an arteriole or a piece of longitudinal muscle of the intestine unless I have applied a non-physiological stimulant.

Kotlikoff: It also requires a sufficient number of those stochastic events to occur in time, such that you can observe it. I can imagine a situation where there are stochastic events that are below the threshold of resolution.

Somlyo: What about microsparks? How does one differentiate a microspark?

Bolton: By the time course.

Somlyo: Given the point spread function of the instrument how do you differentiate it from a large spark in another, slightly different focal plane?

Bolton: If it happens in another confocal plane then it is diffusing from there.

Somlyo: It is not a question of diffusion, it is because the confocal microscope does not have a 1 Å confocal plane. At best, given a ×40 objective with a numerical aperture of 1.3, it will have a z-axis resolution of about 0.7 μm.

Bolton: This problem has been investigated by us and others. My understanding is that if you have a spark which is of the same size, duration and spatial spread as the spark in your confocal plane, and it is in another confocal plane then, if you see it, it will have a slower time course because diffusion will have occurred. If it is not in your confocal plane you will not see it at the beginning, so it has to spread to be visible. As it diffuses it will take longer, so the timecourse is very critical for deciding which confocal plane it is in (Gordienko et al 1998).

Somlyo: We are talking about what the optics will see. If something fires at the same time as a large round source in one plane it will appear as a 'micro' object in a plane above and below (e.g. Ho & Shao 1991).

Wier: Perhaps I can add a few things. The critical point is that the differences will be subtle in time course and difficult to predict on a theoretical basis, without knowing the exact geometrical relationship among the different sources. In striated muscle where we have better information on where sparks might arise, the calculations have shown that it is hard to distinguish a small, in-focus spark, from a large out of focus one. They may have the same amplitude.

Bolton: You think the time-course is going to be similar?

Wier: In theory the time course will be different for the reasons you state, but the signal:noise ratio and other practical aspects prevent it from being a reliable way to distinguish the two.

Bolton: I don't think signal:noise ratio is a problem in our experiments.

Kotlikoff: You implied that the $InsP_3$ receptors were clustered at the same spot as the FDSs and that somehow they were acting specifically to trigger this process. I'm not sure how you can distinguish that idea from the suggestion that they are just diffusely raising cytosolic Ca^{2+}, doing the same thing you see with CICR, and that this is getting you to a threshold where you have enough stochastic gating of ryanodine receptors to generate observable sparks. Do you think you can distinguish between those two possibilities?

Bolton: There are different initiation sites for sparks. The point at which we can first see a rise in fluorescence shows jitter; these points correspond to where we see the microsparks.

Kotlikoff: If you take that point as an FDS and evoke it with an $InsP_3$ stimulus, you seem to imply that this means that the $InsP_3$ is acting preferentially or locally at that FDS. I don't see how you can distinguish that from $InsP_3$ generally raising global cytosolic Ca^{2+} and producing the same kind of CICR that you see either as a spontaneous spark or a triggered CICR.

Bolton: When we apply an agonist that activates a G protein-coupled receptor and increases $InsP_3$ in the cell, we don't normally see a small general rise in the fluorescence over the whole cell (indicating a rise in ionized Ca^{2+} concentration) until it reaches some point at which sparks start to go off, as you suggest. There is virtually no detectable change in the general Ca^{2+} level. Nevertheless, some of these sites do start to fire-off. I would therefore argue that it is not the general level of cytoplasmic Ca^{2+} that is rising, but it is the Ca^{2+} immediately outside the SR. We can't image this because it is in a domain that is too small to resolve, but this Ca^{2+} level is increasing and that is tickling the FDSs.

Kotlikoff: It is also common not to be able to resolve the rise in cytosolic Ca^{2+} from a short, triggered Ca^{2+} current. But you can see the release. Presumably this triggered depolarization is as homogenous a stimulus as can be imagined for the cell. It can't be resolved, yet we see the CICR.

Fry: In the traces that you showed of a general spark discharge, I'm unsure of the calibration you used, but they would fit with the notion that there was a small generalized increase of Ca^{2+}, which is then followed by the discharge.

Nelson: One of the things that happens when cells are isolated is that the Ca^{2+} channels run down or become non-functional. For this reason it is important to ensure that you have a robust Ca^{2+} current. Dr Isenberg has done this. If you have good Ca^{2+} current you should see a nice change in global Ca^{2+} as you depolarize. In your experiment, did you measure the Ca^{2+} currents under voltage clamp to see what type of Ca^{2+} current was being delivered to the cell?

Bolton: Those experiments were by Imaizumi and as I recollect the inward Ca^{2+} current was about 100 picoamps in that cell. One problem is that the Ca^{2+}-activated K$^+$ current is a lot bigger, in the order of nanoamps. Therefore the Ca^{2+} current is not very obvious.

Nelson: Why doesn't the Ca^{2+} current go across the whole cell?

Bolton: It probably does, but it doesn't go up as much as you would expect because of buffering in the cell. Ca^{2+} entry by itself, unless it continues for a period, is difficult to detect. Using the colour-coded presentation of [Ca^{2+}]$_i$ changes we get more of a qualitative picture. If you wanted to detect Ca^{2+} entry you would have to look carefully at the threshold fluorescence. I'm sure you will detect responses that look as if Ca^{2+} is coming in, but I think you have to look under different conditions to the ones we used. The rise in [Ca^{2+}]$_i$ is relatively much less. For example, if you block or discharge the Ca^{2+} stores with ryanodine or CPA, then you do get a generalized rise in [Ca^{2+}]$_i$, but it takes longer for the general [Ca^{2+}]$_i$ to rise and it rises much less.

Brading: But you can easily load Ca^{2+} stores with Ca^{2+} from outside without any contraction in lots of different smooth muscle cells.

Sanders: I think the importance of this mechanism is underestimated when we are looking at single cells. With a lot of single cells you have to voltage clamp positive to -50 mV to really get sparking going. Yet at the resting membrane potential of the gut you can see Ca^{2+} waves in cells in the intact gut. This suggests that this is a more excitable phenomenon in the intact gut than in single cells. One reason you don't see STOCs in the intact gut is because of the resting potential. You don't get any substantial coupling to the BK channels.

Hirst: That's simply not true. If you take a piece of the muscle that Tom Bolton is talking about and add a low concentration of acetylcholine, you will activate an InsP$_3$ pathway. Then you can see STOCs clearly. You can then use something non-physiological for the preparation.

Sanders: But when you add acetylcholine you are raising Ca^{2+}.

Hirst: You are suggesting that there is a resolution problem. I don't think this is a resolution problem: we simply don't see STOCs in the absence of a stimulation.

Sanders: That preparation has a resting potential of -65 mV. If you add charybdotoxin to that tissue you are not going to depolarize it because there are very few BK channels being activated.

Hirst: But you can see a STOC if you artificially activate a STOC pathway by applying acetylcholine, so the resolution is there. Even if you look in short segments, or at 1000 cells, you don't see a STOC until you pressurize it.

Bolton: I think there could be a terminology problem here. What is the resting state of smooth muscle?

Nelson: It never rests.

Bolton: Exactly. The blood vessels are always subject to a degree of pressure, so what is physiological? When you take it out and put it in an organ bath, if it's a piece of vascular muscle, then it is not under pressure.

Hirst: Just like your single smooth muscle cells, so they shouldn't be doing anything.

Bolton: They are not if they are at the resting membrane potential, but they are, if depolarized.

Hirst: You showed STOCs and sparks in a resting preparation.

Bolton: But these are single cells, they are not resting. If you look at the membrane potentials they are very likely to be active. If you don't voltage clamp them, they can be spontaneously active.

Sanders: In the preparations that you use, you are already getting waves before depolarization. Haven't you actually started to deactivate the process somewhat in these single cells?

Bolton: If you are saying that single cells are not a good paradigm for whole tissues, that may be true: they have been damaged in the preparation process. But, equally when you take the preparation out of the body and put it in a physiological salt solution, it still isn't physiological.

Blaustein: When you used bodipy ryanodine did you try to compete with ryanodine? We found that this is not as specific as Fluo-3 FF or ER-Tracker.

Bolton: We also tried with the $DiOC_6$ which is not specific for RyRs, but bodipy ryanodine binds specifically and with relatively high affinity to these (Zhang et al 1997, Holz et al 1999, Cifuentes et al 2001).

Blaustein: This is true; $DiOC_6$ also stains the reticulum loop amongst other things. I think we need to be careful about equating the sites that you look at with RyRs.

Paul: If we accept that there are FDSs, what is special about these sites? Is there anything structurally different about them? There is the danger that the preparations are damaged when you treat them with enzymes to get them out and that these are just damaged sites. Do they move in time?

Bolton: I don't know. Over the period of observation most people find that certain sites will discharge. But no one has looked into whether when you look over another period of time at the same cell, say an hour later, if they will have changed. These sorts of experiments are not being done. How fixed might they be in time? I would suspect that the cell produces them and they are used for a period and then, like all other things in the cell, there is a turnover and some stop and others start, and this is how the cell operates. However, this is just speculation on my part.

Brading: Could these sites be where either a nerve varicosity is close or a process from an ICC is present? These are normally where they start in terms of physiological activation.

Bolton: The SR normally underlies the rows of caveolae. Between those are the dense bars where the contractile proteins are attached. It seems likely that the caveolae contain receptors and channels, and that the SR occurs under the voltage-dependent channels.

References

Cifuentes F, González CE, Fiordelisio T, Guerrero G, Lai FA, Hernández-Cruz A 2001 A ryanodine fluorescent derivative reveals the presence of high affinity ryanodine binding sites in the Golgi complex of rat sympathetic neurons, with possible functional roles in intracellular Ca^{2+} signalling. Cell Signal 13:353–362

Gordienko DV, Bolton TB, Cannell MB 1998 Variability in spontaneous subcellular Ca^{2+} release in guinea-pig ileum smooth muscle cells. J Physiol 507:707–720

Ho R, Shao Z 1991 Axial resolution of confocal microscopes revisited. Optiik 88:147–154

Holz GG, Leech CA, Heller RS, Castonguay N, Habener JF 1999 cAMP-dependent mobilization of intracellular Ca^{2+} stores by activation of ryanodine receptors in pancreatic β cells. A Ca^{2+} signaling system stimulated by the insulinotropic hormone glucagon-like peptide-1-(7-37). J Biol Chem 274:14147–14156

Lipsius SL, Hüser J, Blatter LA 2001 Intracellular Ca^{2+} release sparks atrial pacemaker activity. News Physiol Sci 16:101–106

Zhang X, Wen J, Bidasee KR, Besch HR Jr, Rubin RP 1997 Ryanodine receptor expression is associated with intracellular Ca^{2+} release in rat parotid acinar cells. Am J Physiol 273:C1306–C1314

Sarcoplasmic reticulum, calcium waves and myometrial signalling

Roger C. Young

Department of Obstetrics and Gynecology, 96 Jonathan Lucas Street, Suite 634, Medical University of South Carolina, Charleston, SC 29425, USA

Abstract. Ca^{2+} waves are rises of intracellular free Ca^{2+} that occur in a temporally and spatially coordinated manner, such that a leading edge of a wave front can be clearly discerned. Waves that occur within the bounds of a single cell are termed intracellular Ca^{2+} waves. Generation of a Ca^{2+} wave may be the end result of multiple cellular signalling events and, consequently, is a mechanism of signal integration or information processing. Passage of a Ca^{2+} wave is a mechanism for signalling within a cell, or across a cell. This paper reviews intracellular Ca^{2+} waves and their relationship with the sarcoplasmic reticulum (SR) in uterine smooth muscle. Wave speeds are the rate at which the leading front of the Ca^{2+} wave travels, and are measured by time-lapse imaging of Ca^{2+}-dependent fluorescent dyes. Waves can be experimentally generated in cultured cells by agonist or mechanical stimulation. Wave speeds are unaffected by the removal of Ca^{2+} in the bathing solution, indicating that the source of the Ca^{2+} is the SR. The mechanisms of intracellular wave propagation can be investigated by modulating the Ca^{2+} release mechanisms of the SR. The mechanism most consistent with our observations is that propagation of intracellular Ca^{2+} waves in cultured human uterine myocytes is SR Ca^{2+} release that can utilize $InsP_3$ receptors alone, Ca^{2+}-induced Ca^{2+} release receptors alone, or both together. The rate-determining step for Ca^{2+} wave propagation is diffusion of Ca^{2+} through a highly buffered cytoplasm. Passage of a Ca^{2+} wave also results in capacitative Ca^{2+} entry, which links deep cytoplasmic Ca^{2+} changes to subplasmalemmal Ca^{2+} concentrations.

2002 Role of the sarcoplasmic reticulum in smooth muscle. Wiley, Chichester (Novartis Foundation 246) p 174–188

Myometrium of pregnancy is a unique example of visceral smooth muscle. Essentially quiet for the majority of gestation, a burst of rhythmic contractile activity is responsible for delivery of the human fetus. Endocrine signals originating from the fetus are likely responsible for the events that begin labour (McLean & Smith 2001), but the mechanisms of intra- and intercellular signalling during the course of labour are not fully resolved.

The long duration of an individual uterine contraction in human (60–90 s) is difficult to explain using a single signalling mechanism. Myometrium clearly demonstrates action potential propagation, but this rapid (5 cm/s) mechanism suggests that the entire contraction should last no more than 26 s (6 s for action potential propagation and 20 s that each myocyte remains contracted). Additionally, if action potential propagation recruited all the cells in the uterus over 6 s, then the majority of the myocytes would be contracting simultaneously. This is not observed during normal human labour, but is seen when oxytocin overstimulation results in an iatrogenic tetanic contraction. This condition results in intrauterine pressures more than double the normal 50 to 75 torr of spontaneous labour, and clearly has adverse effects on the fetus. These discrepancies, as well as others (Young 2000), suggest that action potential propagation alone cannot easily explain the uterine physiology seen in labouring women.

Another signalling mechanism to consider in myometrium is Ca^{2+} wave propagation. Ca^{2+} waves (Young & Hession 1996) are rises of intracellular free Ca^{2+} that occur in a temporally and spatially coordinated manner, such that a leading edge of a wave front can be clearly discerned (Fig. 1). Waves that occur within the bounds of a single cell are termed intracellular Ca^{2+} waves, and waves that cross cell boundaries are intercellular Ca^{2+} waves. This paper will focus on the mechanism of intracellular Ca^{2+} wave propagation in short-term cultured human myometrium. Specifically, the relationships among Ca^{2+} waves, intracellular Ca^{2+} metabolism and the sarcoplasmic reticulum (SR) will be presented.

The general mechanism

Intracellular Ca^{2+} waves have been described in a number of cell types. Intracellular Ca^{2+} waves are expressed in the absence of Ca^{2+}, therefore stored intracellular Ca^{2+} must be one source of the Ca^{2+} responsible for the Ca^{2+} wave. Early work in *Xenopus* oocytes demonstrated that Ca^{2+} waves can be observed with expression of only ryanodine receptors (Clapham et al 1993).

The general mechanism of Ca^{2+} wave propagation is believed to be as follows. An event occurs that results in a small rise of intracellular Ca^{2+} near an SR release receptor (either ryanodine receptor or inositol-1,4,5-trisphosphate [InsP$_3$] receptor). The release of SR Ca^{2+} results in a local increase of cytoplasmic Ca^{2+}. The nascent cytoplasmic Ca^{2+} then diffuses to adjacent release sites, causing further SR release. This process is a regenerative process and will remain so as long as the receptors are functional and the SR Ca^{2+} stores are sufficient. The Ca^{2+} wave can then be envisioned as a series of SR Ca^{2+} release events that are dependent upon local Ca^{2+} concentrations, receptor function, receptor density and diffusion of cytoplasmic Ca^{2+}. The rate of propagation of the Ca^{2+} wave is possibly dependent upon any of these processes.

Observing Ca^{2+} waves

Since Ca^{2+} waves are rises of intracellular free Ca^{2+}, Ca^{2+} imaging is required to visualize them. Generally fast, high-affinity Ca^{2+}-sensitive fluorescent dyes such

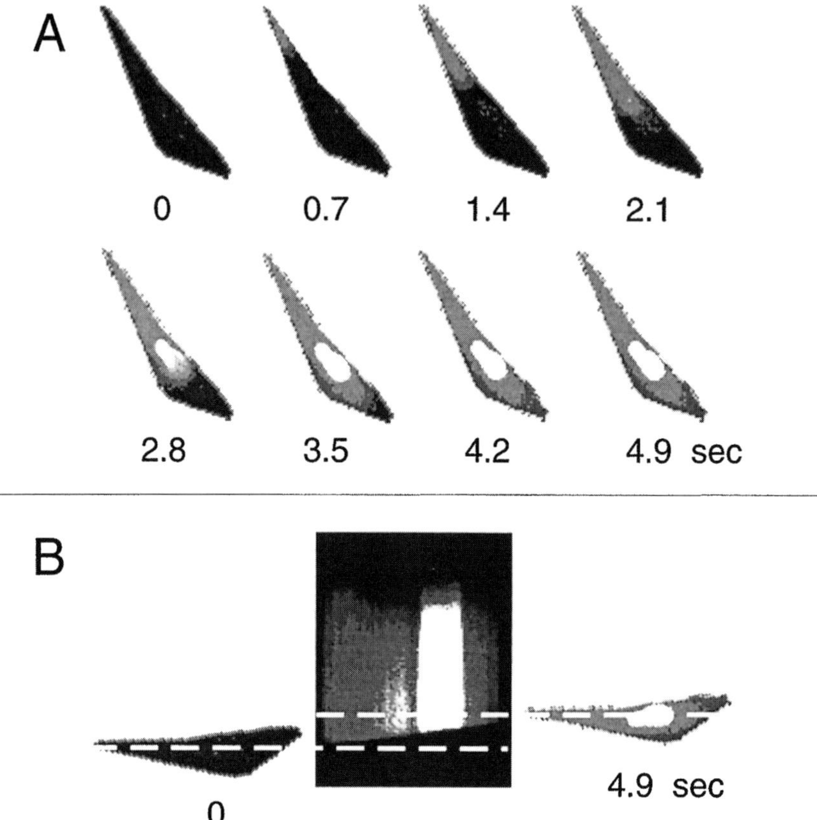

FIG. 1.   Intracellular Ca^{2+} wave in cultured human myometrium. Greyscale images of a cultured myocyte, where black represent low intracellular Ca^{2+} concentrations and white represents high intracellular Ca^{2+} concentrations. The wave begins at the upper left and spreads to the lower right. The reconstructed line scan (see text) demonstrates a sharp line that represents motion of the leading edge of the Ca^{2+} wave front. The slope of the leading edge represents the wave speed. (Reprinted from Young & Zhang 2001).

as Ca^{2+} green 1 or Fluo-3 are used. The inherent advantages of these dyes are the relatively large dynamic range and the good quantum yield of the fluorescence. As with all 'single-wavelength' dyes, close estimation of absolute concentrations of intracellular free Ca^{2+} is not possible. Ratiometric dyes such as Fura-2 generally allow more accurate estimation of concentration changes. However, the dynamic range of ratiometric dyes is less than single-wavelength dyes, which results in less sensitivity.

Classically, laser scanning confocal microscopy has been used to investigate Ca^{2+} waves. For relatively flat, cultured cells, less technically demanding epifluorescence also yields good results, as long as the detection camera allows for rapid enough image acquisition.

To be considered a Ca^{2+} wave, the wave front should traverse a significant portion of the cell. Generally this is not a problem, since waves are usually all-or-nothing phenomena. In short-term culture (fewer than 10 passages), uterine smooth muscle cells obtained from pregnant women remain elongated. The initiation point of a wave can be in the centre of the cell, but most agonist-induced waves are initiated at one of the elongated ends.

The quintessential element of a Ca^{2+} wave is the leading edge. If the intracellular Ca^{2+} rises are so fast that the leading edge cannot be discerned, then the entire cell appears to exhibit global rises of Ca^{2+}. Since this phenomenon is dependent upon the measuring equipment, it is important to delineate the time frame of the waves studied in each protocol. Since most reported intracellular Ca^{2+} waves in smooth muscle are in the range of 5–30 μm/s, we will limit our discussions to this range for the remainder of the paper.

Wave speeds are measured by the technique of line scanning. With the laser scanning confocal microscope, a true line of analysis can be probed. For example, the scan of the laser line is positioned across the long axis of a cell, and repetitive images of only that line are obtained. When those images are stacked on top of each other, a distance versus time image is produced. This image is called a line scan image. As described, this technique presents an experimental problem since it requires making a decision regarding the location of the wave before experiment is begun.

To overcome this limitation, a series of images of an entire field of cells can be obtained, then the waves can be visualized when the images are played back in rapid sequence. Specific lines of analysis can then be chosen after the experiment is completed. The line is selected to be perpendicular to the wave front (that is, along the direction of propagation). As before, the lines are stacked on top of one another, yielding a reconstructed line scan that contains all the information of a true line scan. The leading edge of the Ca^{2+} wave appears on the line scan as a sharp line with a constant slope. Since the slope is distance/time, appropriate conversion factors allow true speeds to be obtained in μm/s.

Intracellular Ca^{2+} stores

Intracellular Ca^{2+} waves have been observed both in the absence of Ca^{2+} in the bathing solution and in the presence of non-specific blockers of transmembrane Ca^{2+} entry, such as La^{3+}. Thus, the primary source of Ca^{2+} for intracellular Ca^{2+} waves is intracellular, specifically the SR Ca^{2+} stores. In human myocytes, these stores can be visualized (Young & Mathur 1999) using the low-affinity Ca^{2+}-dependent fluorescent dye Fluo-3-FF (Fig. 2). Interestingly, the Ca^{2+} stores

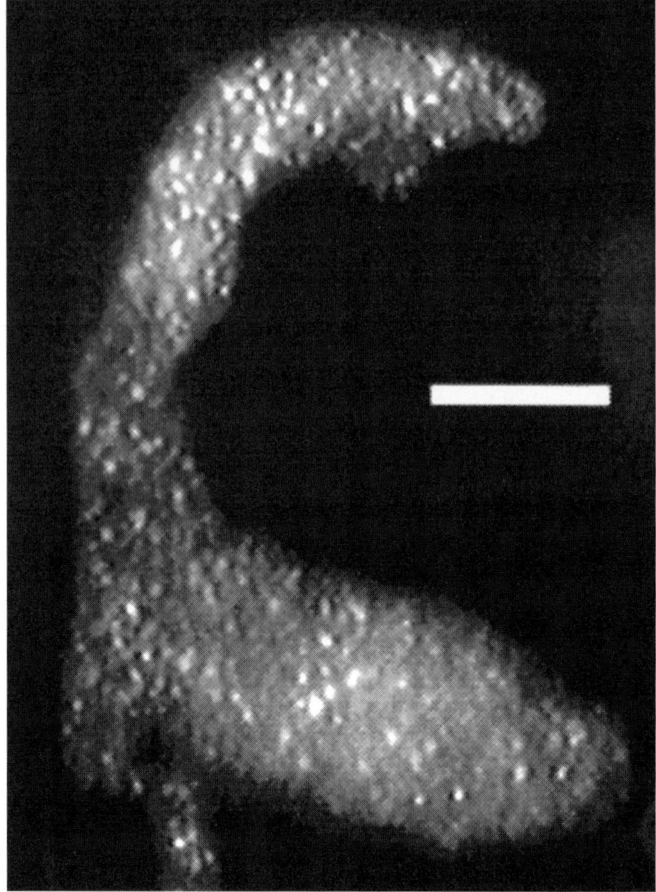

FIG. 2.   Imaging of Ca^{2+} within the SR stores using the low-affinity Ca^{2+}-sensitive fluorescent dye Fluo-3-FF. This cell is a myocyte freshly dispersed from term pregnant myometrium, but similar results are seen with cultured cells. Scale bar = 10 μm (reprinted from Morgan et al 1996).

appear to be focally distributed within the SR in both cultured and freshly dispersed human uterine myocytes.

SR receptors in human myometrium

Human myometrium has been demonstrated to contain the mRNA of all three isoforms of the ryanodine receptor, although type 1 and 3 mRNA are the most abundant (Martin et al 1999). Similarly, mRNA encoding the three types of InsP$_3$ receptor are present in human myometrium (Morgan et al 1996). In cultured human myocytes, a lacy network can be outlined by using antibodies against both ryanodine receptors and InsP$_3$ receptors, suggesting the presence of these release receptors on the SR of cultured human myocytes (Young & Mathur 1999).

Involvement of InsP$_3$ receptors in Ca^{2+} wave propagation

In myometrium, oxytocin (OT) exposure increases InsP$_3$. By comparing Ca^{2+} wave speeds that occur spontaneously (low InsP$_3$) with those that occur following OT exposure (high IP$_3$), we can determine the effect of increasing activity of the InsP$_3$ receptors (Young & Zhang 2001). In other words, if activation of the InsP$_3$ receptors were rate limiting, then the OT-induced waves should be faster than the spontaneously occurring waves. No difference was found (10.3±3.4 μm/s vs. 9.6±2.6 μm/s). These data suggest that InsP$_3$ receptors are not necessary for propagation of intracellular Ca^{2+} waves and that they are not rate limiting even if activated.

Involvement of ryanodine receptors in Ca^{2+} wave propagation

Ryanodine receptors can be blocked by exposing the cells to 10 μM ruthenium red. Under these conditions, oxytocin exposure still elicits Ca^{2+} waves, and the observed wave speeds are the same as waves measured in the absence of ruthenium red (Young & Zhang 2001). Thus, similarly to InsP$_3$ receptors, functional ryanodine receptors are not rate limiting in the mechanism of Ca^{2+} wave propagation.

Diffusion of Ca^{2+}

Since neither the ryanodine nor the InsP$_3$ receptors appear to be rate limiting for Ca^{2+} wave propagation, the diffusion of Ca^{2+} along the surface of the SR may be the rate-limiting step in Ca^{2+} wave propagation. It is therefore important to consider this process in some detail.

The cytoplasm of virtually all cells is highly buffered with regard to Ca^{2+}, and diffusion is highly dependent upon buffering:

$$D' = D \times K_\mathrm{s}/(K_\mathrm{s} + B)$$

Where D' is the apparent Ca^{2+} diffusion coefficient, D is the true diffusion coefficient in water, K_s is the buffer dissociation constant, and B is the concentration of the buffer. For a detailed analysis of Ca^{2+} diffusion in the presence of fixed and mobile buffers, see Wagner & Keizer (1994).

The essential aspect of Ca^{2+} buffering is that the Ca^{2+} diffusion rate is inversely dependent upon the concentration of Ca^{2+} buffer. Since virtually all organic Ca^{2+} buffers are pH dependent, the intracellular pH greatly influences the intracellular buffering capacity. Therefore, if the rate-limiting step of Ca^{2+} wave propagation is Ca^{2+} diffusion, then the wave speeds should be dependent upon pH. Indeed this was found (Young & Zhang 2001), with acidification (using 20 nM sodium butyrate) resulting in a 44% increase in wave speed, and alkalinization (using 20 nM ammonium chloride) resulting in a 35% decrease in wave speed.

Lastly, it is important to consider the possibility that the buffering capacity of the cytoplasm is exceeded during rises of intracellular Ca^{2+}. If the local Ca^{2+} concentration change is rapid (for example, with the opening of an SR receptor), then it is possible that the local Ca^{2+} concentration could exceed the buffering capacity of the medium. This would result in D' approaching D, and possibly a large increase in the observed Ca^{2+} wave speed. In our experiments in cultured human myometrium, the speed of the wave did not increase with the intensity of the fluorescent signal, implying that the buffering capacity of the cytoplasm was not exceeded. Nonetheless, Ca^{2+} buffering capacity is clearly a critical parameter in Ca^{2+} wave propagation, and should be considered in detail when investigating time-dependent intracellular Ca^{2+} signalling processes.

The superficial buffer barrier

The Ca^{2+} concentration at the inner surface of the plasma membrane can be qualitatively monitored using the Ca^{2+}-activated K^+ channel as a reporter. By simultaneously monitoring the deep cytoplasmic Ca^{2+} with a fluorescent dye, the relationship between the Ca^{2+} concentrations at the two locations can be determined. Using mechanical stimulation to initiate a Ca^{2+} wave, the timing of the opening of the K^+ channels was compared to the passage of the Ca^{2+} wave (Fig. 3). A significant delay was found, with the K^+ channels opening 6.3±4.7 s after the wave reached the pipette location (Young et al 2001). These data suggest that the Ca^{2+} concentration at the inner surface of the plasma membrane did not reflect the concentration of the deep cytoplasm. Furthermore, the data suggest that Ca^{2+} released from the deep SR does not freely diffuse to the plasma membrane. These observations are consistent with the superficial buffer barrier hypothesis developed by van Breemen (van Breemen et al 1995).

Cell-to-cell information transmission

An intracellular Ca^{2+} wave in one cell can initiate an intracellular Ca^{2+} wave in an adjacent cell. This wave that crosses cell boundaries is then termed an *inter*cellular Ca^{2+} wave. Two distinctly different mechanisms appear likely for this process: gap junctional (Paemeleire et al 2000) or paracrine (Young & Hession 1997). It is the ability of Ca^{2+} waves to communicate relatively slowly (5 to 30 μm/s) over short ranges (hundreds of microns) that distinguishes them from action potential

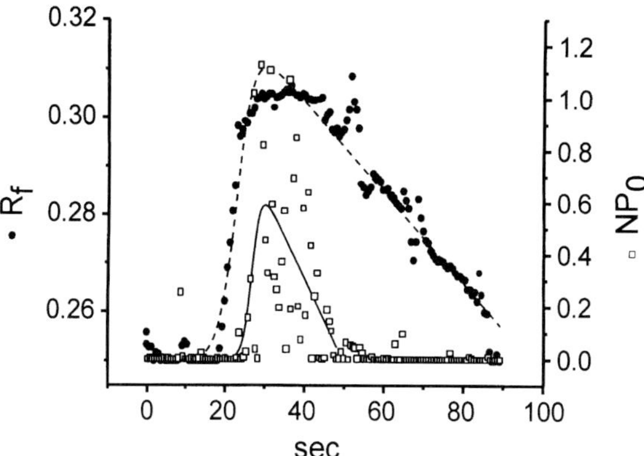

FIG. 3.   By observing the opening of Ca^{2+}-dependent K^{+} channels using patch clamp technology simultaneously with fluorescence imaging of intracellular Ca^{2+}, we can compare the rise of the subplasmalemmal Ca^{2+} with passage of Ca^{2+} wave. Circles are relative fluorescence (R_f) and squares are the open probability of the K^{+} channel (NP_0). Of specific importance is the time delay (in this case five seconds) between the passage of the Ca^{2+} wave and the opening of the K^{+} channels (reprinted from Young et al 2001).

propagation, which communicates rapidly (5 cm/s) over long distances (many centimetres).

In summary, cultured human myocytes exhibit intracellular Ca^{2+} waves that may assist with communication within and between adjacent cells. The primary mechanism of intracellular Ca^{2+} wave propagation is regenerative release of intracellular SR Ca^{2+} stores through ryanodine or InsP$_3$ receptors, or both together (Fig. 4). As such, the SR Ca^{2+} store/release system can be considered to

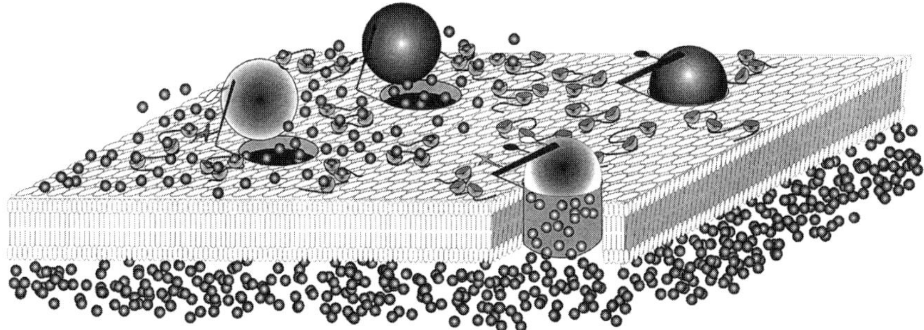

FIG. 4.   Pictorial representation of the mechanism of intracellular Ca^{2+} wave propagation in cultured human uterine smooth muscle cells. Ryanodine (dark spheres) and InsP$_3$ receptors (white spheres) are present in the SR membrane. Within the SR is a high concentration of Ca^{2+}. As the leading edge of the Ca^{2+} wave reaches each receptor, the channels open, releasing additional Ca^{2+}. Diffusion of Ca^{2+} is retarded by Ca^{2+} buffers (represented by small cups), and is the rate-determining step in the mechanism of propagation.

be an excitable system that complements the electrical excitable system of the plasma membrane. In cultured human myocytes, the rate-limiting step of intracellular Ca^{2+} wave propagation is diffusion of Ca^{2+} through a highly buffered medium. Since this mechanism requires functional SR release receptors, both the ryanodine and $InsP_3$ receptors appear to function more like all-or-nothing switches rather than modulators of function.

There appears to be some restriction on the diffusion of the intracellular free Ca^{2+} between the deep cytosol and the plasma membrane. If true, then there is an inherent physiological difference between the ryanodine receptor (which is primarily responsive to Ca^{2+}) and the $InsP_3$ receptor (which is primarily responsive to $InsP_3$) in the mechanism of communication with the plasma membrane.

References

Clapham DE, Lechleiter JD, Girard S 1993 Intracellular waves observed by confocal microscopy from Xenopus oocytes. Adv Second Messenger Phosphoprot Res 28:161–165

Martin C, Chapman KE, Thornton S, Ashley RH 1999 Changes in the expression of myometrial ryanodine receptor mRNAs during human pregnancy. Biochim Biophys Acta 1451:343–352

McLean M, Smith R 2001 Corticotrophin-releasing hormone and human parturition. Reproduction 121:493–501

Morgan JM, De Smedt H, Gillespie JI 1996 Identification of three isoforms of the $InsP_3$ receptor in human myometrial smooth muscle. Pflüger's Arch 431:697–705

Paemeleire K, Martin PE, Coleman SL et al 2000 Intercellular calcium waves in HeLa cells expressing GFP-labeled connexin 43, 32, or 26. Mol Biol Cell 11:1815–1827

van Breemen C, Chen Q, Laher I 1995 Superficial buffer barrier function of smooth muscle sarcoplasmic reticulum. Trends Pharmacol Sci 16:98–105

Wagner J, Keizer J 1994 Effects of rapid buffers on Ca^{2+} diffusion and Ca^{2+} oscillations. Biophys J 67:447–456

Young RC 2000 Tissue-level signaling and control of uterine contractility: the action potential-calcium wave hypothesis. J Soc Gynecol Investig 7:146–152

Young RC, Hession RO 1996 Intra-and intercellular calcium waves in cultured human myometrium. J Muscle Res Cell Motil 17:349–355

Young RC, Hession RO 1997 Paracrine and intracellular signaling mechanisms of calcium waves in cultured human uterine myocytes. Obstet Gynecol 90:928–932

Young RC, Mathur SP 1999 Focal sarcoplasmic reticulum calcium stores and diffuse inositol 1,4,5-trisphosphate and ryanodine receptors in human myometrium. Cell Calcium 26:69–75

Young RC, Schumann R, Zhang P 2001 Intracellular calcium gradients in cultured human uterine smooth muscle: a functionally important subplasmalemmal space. Cell Calcium 29:183–189

Young RC, Zhang P 2001 The mechanism of propagation of intracellular calcium waves in cultured human uterine myocytes. Am J Obstet Gynecol 184:1228–1234

DISCUSSION

Fry: We were talking earlier about restricted spaces. Working from the calculation that there is half a Ca^{2+} ion in a sampling space, the laws of diffusion don't apply; so looking at diffusion models does not seem to be the best way to look at movement of ions into and out of sampling spaces. Alternatively, we can become more esoteric about it and to look at a quantum mechanical model where the only thing that determines the rate at which you will get an ion into or out of a space will be the activation energy of something it binds to. The concentration would be irrelevant, as is the dimension of the space, and it is only dependant on what it binds to. In a way yours and Professor Blaustein's approaches are quite useful, because you can change the thing that it potentially binds to and see if you can change the diffusion sites. Could you comment on this?

Young: I don't know what the subplasmalemmal Ca^{2+} concentration is, but I can take a guess. You will only get no ions if you assume it is 100 nM. But in all likelihood, even in a cell without any oscillations — what we would call 'at rest' — just thinking about where the sitting Ca^{2+} exchanger would reverse, it is in the order of 1 μM, somewhere nearer the concentration of the SR. It may be slightly less, but it is certainly not 100 nM.

Fry: So you are saying there are 10 ions there then. You still have the same problem.

Young: But these are 10 free ions. There is protein buffering and so even though there are a huge number of ions there, only 10 are free.

Fry: There are a lot of Ca^{2+} ions present, just not many are free. The question is still the same — diffusion will not now determine any rates; it will be the speed at which the Ca^{2+} will come off or go onto something that will bind to it. This is a different way of looking at how things will move into spaces and determine rates of reactions.

Young: I have done tunnelling calculations on excited state transitional complexes. These start to fall apart at 10–15 Å. I don't think that tunnelling or anything quantum mechanical, when you are talking about single nanometers, is really going to be of any significance.

Bolton: Did I understand correctly that you are getting Ca^{2+} waves, but you don't believe there are any ryanodine receptors (RyRs) or InsP$_3$ receptors involved?

Young: Exactly the opposite. I think you can have either one or the other, or both. My cells had both. I can set the conditions so either one or the other is functional and you still see the same rate of waves.

Blaustein: Did you ever knock them both out?

Young: I have done that; I didn't show those data.

Bolton: How do you knock out RyRs in uterus?

Young: I don't. I added excess ryanodine and then either locked them open or depleted them — that was the best I could do.

van Breemen: If you witness Ca^{2+} release and you don't see activation of the large conductance Ca^{2+}-activated K$^+$(BK) channels, does that mean that the Ca^{2+} release is only towards the centre of the cell and not towards the membrane?

Young: When we image the cells it looks like the Ca^{2+} release goes all the way throughout the cell, it doesn't look like there are restricted spaces. When we put the patch on the plasma membrane to obscure the BK channel, then block the Ca^{2+} entry through the membrane, it appears that Ca^{2+} released from the SR doesn't reach it.

Kotlikoff: Don't you worry about the restricted space of the omega of the patch?

Young: That's another problem, but then when I don't have the nifedipine it is effective, and I can see the BK channels.

Wier: What is the spatial resolution of your Ca^{2+} measurements? In other words, are you picking up the edges of the wave before it actually arrives at your patch pipette, just reflecting limited spatial resolution?

Young: I would think that is possible for one or two seconds, not 6–10.

Brading: You said that you were trying to solve a problem in a real human uterus. How does this fit in with Sue Wray's work, which shows that there is no involvement of the stores in Ca^{2+} waves in human uterus (Wray et al 2002, this volume)?

Young: I certainly believe both of our sets of experiments. I think we need to sit down and resolve this apparent discrepancy.

Hellstrand: What is the role of the action potentials here? The textbook idea would be that there are trains of action potentials that maintain the contraction. Is this wrong?

Young: Parkington (Monash, Australia) has taken microelectrodes and measured contractility simultaneously (Parkington et al 1999). Half the experiments showed the cells experienced an overshoot action potential and the other half did not. The ones that do only have the overshoot in human, which is contrary to mouse and other animals. These cells have a single spike at the beginning and then have a depolarization to about $-25\,mV$. For example a Ca^{2+} activated Cl^- channel might somehow be activated, but the action potential itself is only the initiation which fits nicely with my hypothesis that the pacing of the uterine contraction in the order of minutes is by propagation of the action potential, but the recruitment of the cells is by the movement of the long distance Ca^{2+} waves.

Raeymaekers: Mechanical stimulation of cells in culture has been shown to release ATP which could diffuse to neighbouring cells and activate purinergic receptors. Can you exclude this possibility in the interpretation of your data?

Young: I can explain ATP. I did these experiments looking at intercellular Ca^{2+} waves and apyrase, and there is no effect. It turns out that the presence of either a prostaglandin transporter blocker, which is the same class as a Cl^- channel blocker, or a prostaglandin synthesis inhibitor, interrupts intracellular Ca^{2+} that is far away. It does not interrupt the intercellular Ca^{2+} through the gap junctions; there are two different mechanisms.

Paul: What does Mg^{2+} do?

Young: I don't have any direct experiments on that.

Somlyo: Why do you need the apyrase? Doesn't the human uterus have enough ecto-ATPase? There is a lot of ecto-ATPase in most smooth muscles. Is it not high in human uterus?

Young: I was merely trying to find a way to remove the possibility of having ATP in my signal. I don't have any other data on that.

Wray: Can you explain more about when you think gap junctions are involved and when they are not? Do you think that gap junctions are involved to get a spread over, say, 20 cells, and then you need another mechanism? How much have you manipulated them?

Young: I think gap junctions are very important in the initiation of labour. It has been shown that they increase the conduction rate of the action potential prior to the onset of labour. They seem to be induced prior to the initiating event in labour. You can have the gap junctions and the rapid action potential propagation, yet no spontaneous labour. Therefore they are necessary, but not sufficient. If I look at intercellular waves, as I flow the solution across, there is about a five cell communication that goes against the flow. In other words it has an intracellular mechanism, and it seems to be about four or five cells that go against the flow and the rest go with the flow.

Fry: In agreement with that would be the observation that acidosis increases the rate and alkalinity does the opposite. Acidosis would increase gap junction resistance, so if it was going to have to be through cell–cell coupling then you would expect the opposite result from what you got. To a certain extent that fits in with your supposition. What was the magnitude of the pH changes?

Young: I used identical conditions as for Dr Wray's work; I think it went down to pH 6.7 and up to about pH 7.6.

Wray: During contractions *in vivo* there are small acidifications. This also occurs *in vitro*, but to a smaller extent.

Eisner: My concern about your pH experiment would be that acid would affect the RyR, and decrease its open probability. The one time we did try this in cardiac muscle, to look at the effect of pH on Ca^{2+} buffering, we saw nothing, but that's a different tissue.

Young: I have tried to exclude that possibility by isolating conditions when I thought that the RyR really wasn't participating, and by looking only at the InsP$_3$ component as best I could. I know it's not perfect when you add excess ryanodine for 10 min. But wouldn't alkylinization have recruited RyRs, or increased the open probability with Ca^{2+}?

Eisner: In the heart acidification decreases the RyR opening probability, the SR Ca^{2+} content goes up, you get bigger waves, and they tend to propagate faster. It sounds the wrong way round but that is the way it happens.

Somlyo: Are these intracellular waves over long distances occurring in the absence of action potentials?

Young: Yes, that was the reason I did this. In these cells under these conditions, the waves are occurring in the absence of action potential because these cells do not express any voltage-activated Ca^{2+} channels. I have tried ruptured and nystatin patches, but I cannot see any inward current secondary to Ca^{2+}. Additionally, waves are observed under depolarizing high K^+ conditions.

McHale: What is your view of action potential propagation in the uterus *in vivo*, rather than in cultured cells?

Young: My view is that it travels through the fasciculus, it requires gap junctions and it travels at 5 cm/s.

McHale: Are you saying that this goes too fast for contraction?

Young: It goes from the top of the uterus to the bottom of the uterus within 5–7 s.

McHale: But it doesn't make it contract?

Young: It has a circuitous pathway through all the fasciculus cells, so every fasciculus in every bundle has one favourite pathway possibly, a way of least resistance. Each bundle would therefore stimulate each pathway in each group of cells or small fraction of cells and then from there the intercellular waves within the bundle will recruit the other cells in order to form that typical contraction that we see on labour and delivery.

Eisner: I'm confused. If Ca^{2+} can spread between those cells, why can't the action potential?

Young: It can, but we are looking at a tortuous pathway and electrically it should take the least resistance. The most current is going to go through the pathway of the least resistance.

Eisner: Yes, but if the other pathways are well enough coupled for Ca^{2+} to spread, I would have thought that electrons would use them as well.

Bolton: So it's actually more favourable for action potential spread than it is for Ca^{2+} spread?

Young: But for a 500 μm bundle, the action potential spread would be completed in milliseconds.

Eisner: Have you, or anyone else, used a voltage-sensitive dye to look at this?

Young: That is what we are trying to do now, in whole tissue. I have done some of these experiments and the only thing I can really say is that the intracellular Ca^{2+} waves are generally the same as they are in cultured cells.

Nixon: Why do you get that spread of Ca^{2+} in the opposite direction in some of the cells?

Young: I think it is just agonist stimulation. In the intercellular wave propagation the sequential cells are producing prostaglandin, transporting Ca^{2+} from the inside to the outside and diffusing to the adjacent cell. This is the mechanism for the fan wave: each individual cell all of a sudden sees the bath of prostaglandin on top of it and it is going to wave from its most likely hotspot position to the other direction.

Blaustein: You started by talking about the uterus in labour and uterine contraction, but then you talked about the gravid uterus, which is not the same as a uterus in labour. A gravid uterus is hardly going to want to spread contraction or be excitable, rather the opposite. Will all that change dramatically when labour begins, or just before labour begins? Is there a way to treat this with hormones to make that change, and how will that change what you are looking at?

Young: In the clinic when we face the problem, we have a patient in unwanted labour so we start from the opposite side of this equation: we want to stop the labour. The key question here is how do we stop a labouring uterus?

Blaustein: Why is that gravid uterus in labour unless something happened prematurely and something else has gone wrong to make that uterus different from a typical gravid uterus?

Young: There are some answers provided by Roger Smith's work. He looked at corticotropin-releasing hormone (CRH) of the fetus, placental conversion of steroids and up-regulation of a whole lot of machinery in myocytes. But I have no way to address these questions with my work.

Sanders: Did you ever look to see whether indomethacin (i.e. prostaglandin inhibition) decreases coupling between these Ca^{2+} waves and activation of BK channels? Perhaps this is your time delay. Production of prostaglandin should activate protein kinase A because of cAMP production, and this will increase BK open probability.

Young: That's a good point. I have seen that indomethacin not only does prostaglandin inhibition but there also are other mechanisms, possibly as simple as intracellular binding of Ca^{2+} in terms of buffer. However, what you are saying is entirely possible.

Somlyo: Do your calculations measure length of time over time travelled for say human uterus, do they assume linear distance or more zig-zag movement such as you showed here.

Young: In any calculation you allow for a number of variables in terms of trying to model the system of action potential propagation. We know that action potentials have to travel within a fasciculus, which is a macroscopic 1–2 mm connective tissue grouping of bundles. Those do not follow a very tortuous route, at the most they wind once or twice over 30 cm, but not much larger than that. The best simulation that's been done using action potential alone used around nine parameters and was totally unrealistic. I probably didn't point it out but the dashed line that overlayed my clinical contraction was where I actually assumed a 350 μm size bundle. I assumed an intracellular Ca^{2+} wave and cell activity for 20 s, and I was able to fit it except for the foot at the end (Young 1997).

References

Parkington HC, Tonta MA, Brennecke SP, Coleman HA 1999 Contractile activity, membrane potential, and cytoplasmic calcium in human uterine smooth muscle in the third trimester of pregnancy and during labor. Am J Obstet Gynecol 181:1445–1451

Wray S, Kupittayanant S, Shmigol A 2002 Role of the sarcoplasmic reticulum in uterine smooth muscle. In: Role of the sarcoplasmic reticulum in smooth muscle. Wiley, Chichester (Novartis Found Symp 246) p 6–25

Young RC 1997 A computer model of uterine contractions based on action potential propagation and intercellular calcium waves. Obstet Gynecol 89:604–608

Sarcoplasmic reticulum and membrane currents

Gerald M. Herrera and Mark T. Nelson[1]

Department of Pharmacology, University of Vermont, Given Medical Building, 89 Beaumont Avenue, Burlington, VT 05405-0068, USA

Abstract. Local and global Ca^{2+} signals from voltage-dependent Ca^{2+} channels (VDCCs) and ryanodine-sensitive Ca^{2+} release (RyRs) channels in the sarcoplasmic reticulum (SR) encode information to different Ca^{2+}-sensitive targets including the large- (BK) and small-conductance (SK) Ca^{2+}-activated K^+ channels in the surface membrane. In smooth muscle, unlike cardiac muscle, Ca^{2+} signalling to RyRs is not local, exhibiting a significant lag between VDCC activation and subsequent RyR stimulation, measured as Ca^{2+} sparks and associated BK currents. However, Ca^{2+} signalling from RyRs (Ca^{2+} sparks) to BK channels appears to be local in arterial (ASM) and urinary bladder smooth muscle (UBSM), consistent with a close proximity of SR RyRs to BK channels. The response of BK channels in ASM and UBSM depends on the tuning of the Ca^{2+}/voltage sensitivity of the BK channel by its accessory subunit, the β_1 subunit. UBSM, in contrast to ASM, has both BK and SK channels. SK channels in UBSM are solely activated by Ca^{2+} signals from VDCCs, whereas BK channels are activated by Ca^{2+} from both VDCCs and RyRs. The differential regulation of BK and SK channels by Ca^{2+} signals underlies their roles in regulating action potential duration and membrane potential (BK channels) and after-hyperpolarizations (SK channels) in smooth muscle.

2002 Role of the sarcoplasmic reticulum in smooth muscle. Wiley, Chichester (Novartis Foundation Symposium 246) p 189–207

In smooth muscle, the sarcoplasmic reticulum (SR) plays an important role in regulating cell excitability by communicating intimately with ion channels in the surface membrane. In most cases, Ca^{2+} release from ryanodine-sensitive Ca^{2+} release channels (RyRs) in the SR leads to a paradoxical decrease in smooth muscle cell excitability due to activation of plasma membrane K^+ channels (Fig. 1; Nelson et al 1995). This is in stark contrast to cardiac muscle, where Ca^{2+} release from RyRs supplies the majority ($> 90\%$) of Ca^{2+} required for contraction (Cheng et al 1993, Cannell et al 1995). In this paper, we will briefly review the basic

---

[1]This paper was presented at the symposium by Mark T. Nelson, to whom correspondence should be addressed.

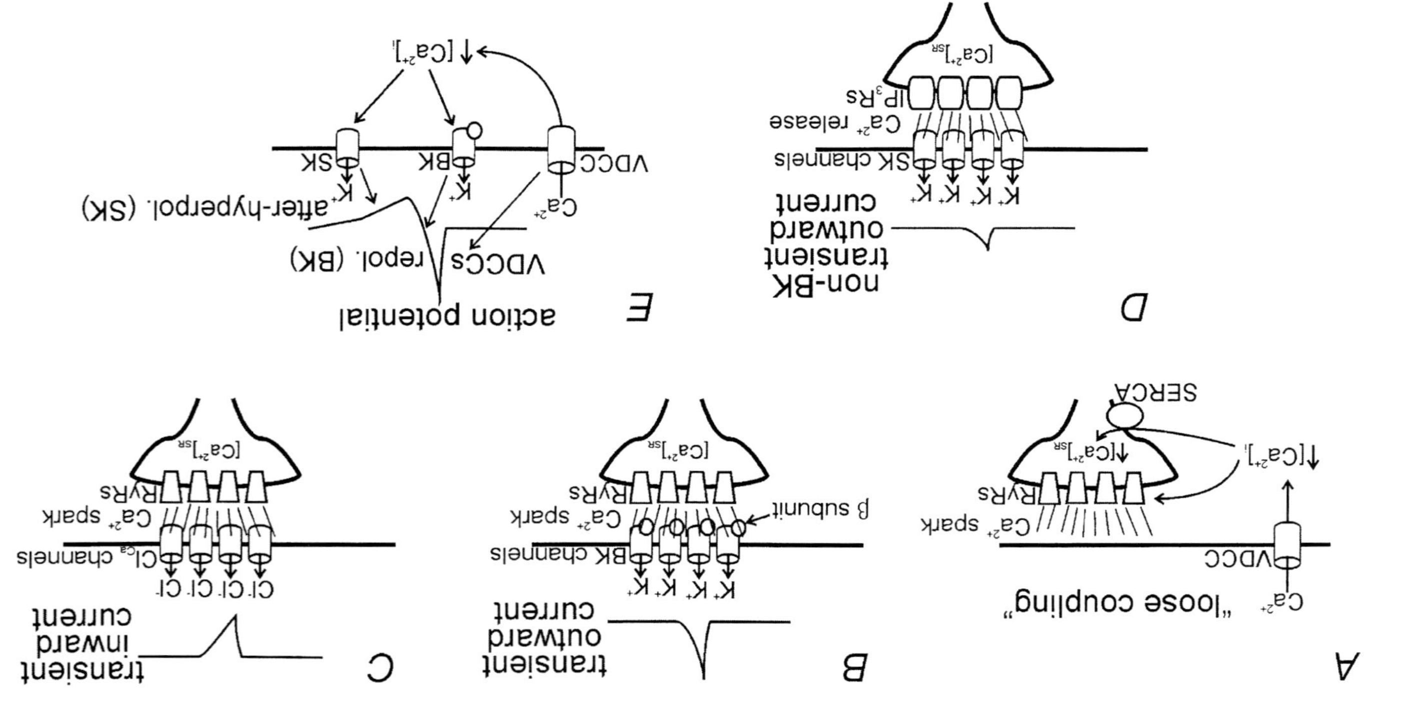

A
Ca²⁺
VDCC
"loose coupling"
Ca²⁺ spark
RyRs
↑[Ca²⁺]ᵢ
↑[Ca²⁺]_SR
SERCA
B
transient outward current
BK channels
Ca²⁺ spark
K⁺ K⁺ K⁺ K⁺
RyRs
β subunit
[Ca²⁺]_SR
C
transient inward current
Cl_Ca channels
Ca²⁺ spark
Cl⁻ Cl⁻ Cl⁻ Cl⁻
RyRs
[Ca²⁺]_SR
D
non-BK transient outward current
SK channels
Ca²⁺ release
K⁺ K⁺ K⁺ K⁺
IP₃Rs
[Ca²⁺]_SR
E
action potential
K⁺ repol. (BK)
K⁺ after-hyperpol. (SK)
VDCCs
SK
BK
Ca²⁺
↑[Ca²⁺]ᵢ
VDCC

signalling mechanisms between the SR and the surface membrane that have been described, and then discuss our current data relating to the nature of SR Ca^{2+} release and the discrimination of different types of Ca^{2+} signals by large (BK)- and small (SK)-conductance Ca^{2+}-dependent K^+ (K_{Ca}) channels.

Ca²⁺ sparks and BK channels in smooth muscle

To date, all smooth muscle types examined display characteristic Ca^{2+} release events known as Ca^{2+} sparks (Fig. 1; see Jaggar et al 2000 for review). Ca^{2+} sparks are localized Ca^{2+} release events through RyRs in the SR located very close (10–20 nm) to the surface membrane. Ca^{2+} sparks can activate Ca^{2+}-dependent ion channels in the cell surface, and the physiological response of the particular cell type will reflect the type of ion channels involved (Fig. 1).

In most cases, Ca^{2+} release through RyRs acts to paradoxically limit contractility of smooth muscle. This relaxing influence can by unmasked by blocking RyRs with the plant alkaloid ryanodine. Ryanodine increases myogenic activity in both arterial smooth muscle (ASM) (myogenic tone) and urinary bladder smooth muscle (UBSM) (spontaneous phasic contractions) (Fig. 2) (see e.g. Nelson et al 1995, Knot et al 1998, Herrera et al 2000). The predominant role of Ca^{2+} sparks in tissues where blocking RyRs causes a contractile response, such as in ASM and UBSM, is to activate BK channels in the surface membrane causing a transient outward current (Fig. 1B; Nelson et al 1995, Bolton & Imaizumi 1996, Perez et al 1999, Herrera et al 2001). This results in cell membrane hyperpolarization and decreased entry of Ca^{2+} through voltage-dependent Ca^{2+} channels (VDCCs), which opposes contraction (Nelson et al 1995, Jaggar et al 2000, Jaggar 2001).

FIG. 1. (*Opposite*)   Communication between the SR and ion channels in the surface membrane. (A) As opposed to cardiac muscle, VDCCs in the surface membrane of smooth muscle and RyRs in the SR appear to be spatially separated and loosely coupled (see Collier et al 2000). Ca^{2+} entry through VDCCs in the plasma membrane raises the cytosolic Ca^{2+} concentration ($[Ca^{2+}]_i$). The rise in $[Ca^{2+}]_i$ activates RyRs through two ways: (1) activation of RyRs by cytosolic Ca^{2+}, and (2) increasing SR Ca^{2+} content ($[Ca^{2+}]_{SR}$). SERCA, SR Ca^{2+} ATPase. (B) Transient outward currents appear ubiquitously in smooth muscle. BK channels in the cell surface are activated by a Ca^{2+} spark originating from a cluster of RyRs 10–20 nm beneath the cell surface. The K_{Ca} channel accessory $\beta 1$ subunit tunes the Ca^{2+} sensitivity of the channel. (C) Some smooth muscle types express Ca^{2+}-activated Cl^- channels which can be activated by a Ca^{2+} spark to cause transient inward currents (Wang et al 1992, Hogg et al 1993, ZhuGe et al 1998). (D) In gastrointestinal smooth muscle, SK channels can be activated by Ca^{2+} release through InsP$_3$Rs in the SR, giving rise to small non-BK transient outward currents that are blocked by the SK inhibitor apamin (Bayguinov et al 2000). (E) Ca^{2+} signalling from VDCCs to K_{Ca} channels also plays an important role in smooth muscle excitability. In UBSM, Ca^{2+} entry through VDCCs causes the upstroke of the action potential and activates both BK and SK channels. BK channels cause action potential repolarization, while SK channels cause the after-hyperpolarization.

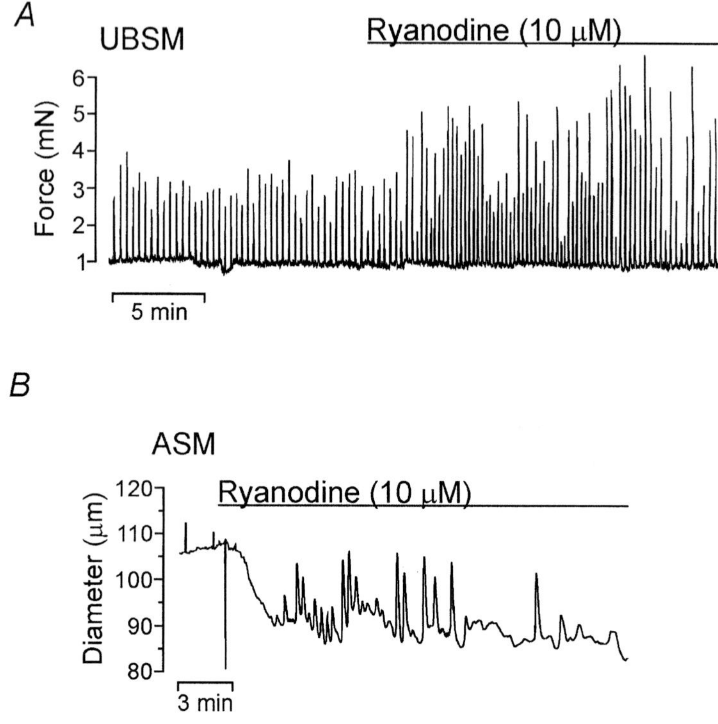

FIG. 2.   Ca^{2+} release through RyRs opposes smooth muscle contraction. (A) Original record of spontaneous phasic contractions in a strip of guinea-pig UBSM. Blocking SR Ca^{2+} release through RyRs with ryanodine increases contraction frequency, and results in a twofold increase in force production (see Herrera et al 2000). (B) Representative trace illustrating the effect of ryanodine on the diameter of a mouse (129/svj) cerebral artery. The artery was pressurized to 60 mmHg, and exhibited myogenic tone. Blocking RyRs with ryanodine causes vasoconstriction (see Nelson et al 1995, Knot et al 1998). The passive diameter for this artery at 60 mmHg was 134 µm. (Data courtesy of Dr D.M. Eckman, unpublished observations.)

Furthermore, Ca^{2+} spark activity in smooth muscle increases with membrane depolarization (Jaggar et al 1998, Knot et al 1998, Herrera et al 2001, Jaggar 2001). RyRs are activated by cytosolic Ca^{2+} ([Ca^{2+}]$_i$) (Rousseau et al 1987). Depolarization raises [Ca^{2+}]$_i$ due to increased activity of VDCCs. Thus, it has been suggested that Ca^{2+} which enters the smooth muscle cell can activate RyRs to cause Ca^{2+} release (Ganitkevich & Isenberg 1992, Kamishima & McCarron 1997, Imaizumi et al 1998), analogous to Ca^{2+}-induced Ca^{2+} release (CICR) originally described in cardiac muscle (Fabiato 1983, Cheng et al 1993, Cannell et al 1995).

'Loose coupling' of VDCCs and RyRs

Indeed, Kotlikoff and co-workers (Collier et al 2000) recently showed that RyRs are activated by Ca^{2+} entry through VDCCs in UBSM. However, this CICR is markedly different from that in cardiac muscle, where Ca^{2+} entry through a single VDCC can trigger a spark within milliseconds. This fast response reflects the local communication between VDCCs and RyRs in cardiac muscle (Cheng et al 1993, Cannell et al 1995). In smooth muscle, the communication between VDCCs and RyRs is much slower, and Ca^{2+} sparks are activated only after a substantial lag following activation of VDCCs (Collier et al 2000). This process has been termed 'loose coupling' (Fig. 1A) (Collier et al 2000), because activation of RyRs requires a significant accumulation in $[Ca^{2+}]_i$, rather than opening of a single VDCC. In smooth muscle, activation of RyRs subsequent to a rise in $[Ca^{2+}]_i$ likely involves two mechanisms: (1) activation of RyRs by cytosolic Ca^{2+}, and (2) an increase in SR Ca^{2+} content ($[Ca^{2+}]_{SR}$) (Fig. 1A). In the present study, we use transient spark-activated BK currents as indicators of SR Ca^{2+} release (Fig. 1B, 3, 4) to assess the communication between VDCCs and RyRs in smooth muscle.

Ca^{2+} sparks and Ca^{2+}-activated Cl^- channels

Some types of smooth muscle express a second type of Ca^{2+} activated ion channel, the Ca^{2+}-activated Cl^- (Cl_{Ca}) channel. Thus, in tissues such as portal vein and airway, Ca^{2+} sparks activate Cl_{Ca} channels, causing transient inward currents (Wang et al 1992, ZhuGe et al 1998) (Fig. 1C). These spark-activated inward currents would tend to depolarize smooth muscle cells at resting membrane potentials, and they have been implicated in triggering spontaneous action potentials and rhythmic contractions in tissues such as portal vein (Wang et al 1992) and trachea (Janssen & Sims 1994). In some cases, spark-activated inward currents are seen following transient spark-activated BK currents. These currents have been termed spontaneous transient outward–inward currents or 'STOICs' referring to the fact that the outward phase is always seen preceding the inward phase (ZhuGe et al 1998). Interestingly, the kinetics of the outward and inward currents are markedly different (Wang et al 1992, ZhuGe et al 1998), especially with regard to the decay phase. In portal vein myocytes, transient outward currents last about 100 ms, whereas the inward currents are much slower, lasting 400 ms (Hogg et al 1993). In tracheal myocytes, Ca^{2+} sparks and the resulting STOICs have been recorded simultaneously (ZhuGe et al 1998). In this study, it was apparent that the decay of the inward transient Cl_{Ca} currents parallels that of a Ca^{2+} spark, whereas the outward transient BK current decays much faster (ZhuGe et al 1998). These observations lend insight into the nature of the Ca^{2+} signals that activate Cl_{Ca} channels and BK channels yielding STOICs. The relatively low-affinity of BK channels for Ca^{2+} means that these channels require large increases in $[Ca^{2+}]_i$ for significant activation. We have estimated that, in order to account for the activity of BK channels during a spark, Ca^{2+} must increase to $10–100\,\mu M$ during a spark (Perez et al 2001). Thus, BK channels are activated briefly at the peak of a spark, then channel activity declines, relating to the kinetics of channel deactivation, as the $[Ca^{2+}]_i$ dissipates rapidly from the local vicinity of BK channels. Cl_{Ca} channels, on the other hand, have a higher Ca^{2+} sensitivity, and their activity tracks more closely the change in average cytosolic $[Ca^{2+}]_i$ in the vicinity of a Ca^{2+} spark (see ZhuGe et al 1998).

SR Ca^{2+} release and SK channels

Yet another type of communication between the SR and the surface membrane involves activation of SK channels in the sarcolemma by Ca^{2+} release through inositol-1,4,5-trisphospate (InsP$_3$) receptors (InsP$_3$Rs) in the SR (Bayguinov et al 2000) (Fig. 1D). In gastrointestinal smooth muscle, spontaneous or evoked release of Ca^{2+} through InsP$_3$Rs, called Ca^{2+} 'puffs', can activate K_{Ca} channels, resulting in transient outward currents that are insensitive to the BK channel blocker charybdotoxin (Bayguinov et al 2000). These non-BK transient currents are blocked by the peptide toxin apamin, which selectively inhibits SK channels. These InsP$_3$R-mediated transient SK currents are especially predominant following purinergic stimulation, and have been hypothesized to underlie inhibitory neurotransmission in response to purinergic stimulation in the gut (Bayguinov et al 2000).

BK and SK channels and the configuration of the UBSM action potential

In the present study, we have utilized electrophysiological recordings and confocal fluorescence imaging to study the activation of BK and SK channels by Ca^{2+} entry through VDCCs and Ca^{2+} release through RyRs in guinea-pig UBSM cells (see Fig. 1E). The techniques for studying whole-cell currents and confocal fluorescence imaging of Ca^{2+} sparks have been described in detail (Perez et al 1999, Herrera et al 2001). Our most interesting findings relate to how BK and SK channels can be differentially activated by Ca^{2+} influx through VDCCs and SR Ca^{2+} release from RyRs to control different phases of the UBSM action potential. Other implications from this work emphasize the 'loose coupling' relationship between VDCCs and RyRs (Fig. 1A) (Collier et al 2000) characteristic of smooth muscle.

Results and discussion

Relaxation of smooth muscle by Ca^{2+} release through RyRs

One of the most perplexing findings regarding the role of the SR in smooth muscle contraction has been that blocking Ca^{2+} release through RyRs with ryanodine causes smooth muscle contraction, rather than relaxation (Fig. 2). If SR Ca^{2+} release through RyRs contributed to raising $[Ca^{2+}]_i$ leading to contraction, then blocking RyRs should relax smooth muscle. The observation that ryanodine causes smooth muscle contraction indicates that Ca^{2+} release events through RyRs (Ca^{2+} sparks) elicit a response which opposes contraction. Recently, we showed that Ca^{2+} spark–BK channel communication also occurs in UBSM (Herrera et al 2001), and blocking RyRs or BK channels results in UBSM contraction (Fig. 2A) (Herrera et al 2000). UBSM undergoes repeated fluctuations in membrane potential resulting from the spontaneous action potentials in this tissue (Creed 1971). Since BK channel activity is sensitive to both Ca^{2+} and voltage (Cox et al 1997, Cui et al 1997), we sought to determine how the communication between Ca^{2+} sparks and BK channels changes at membrane potentials UBSM cells experience during normal action potentials.

Voltage-dependence of the coupling of RyRs (Ca^{2+} sparks) to BK channels

Using simultaneous measurements of whole-cell membrane currents with amphotericin-perforated patch (Horn & Marty 1988) in conjunction with confocal measurements of Ca^{2+} sparks using the fluorescent indicator Fluo-3, we found that every Ca^{2+} spark causes a transient BK current in UBSM cells (Fig. 3A; Herrera et al 2001). Furthermore, larger Ca^{2+} sparks were associated with larger BK currents (Fig. 3A). A correlation was found between the amplitude of sparks and the spark-induced activity of BK channels at all membrane potentials tested from $-50\,mV$ to $-20\,mV$ (Fig. 3B). Interestingly, the slope of the relationship between spark amplitude and BK channel activity increases at depolarized potentials, such that a given size Ca^{2+} spark causes a greater increase in BK channel activity at $-20\,mV$ compared to $-50\,mV$ (Fig. 3B). This finding could be quantitatively explained by the known voltage-dependence of the apparent Ca^{2+}-sensitivity of the BK channel (Cox et al 1997, Cui et al 1997).

Membrane potential depolarization does not change the affinity with which Ca^{2+} binds to the BK channel, *per se*. Instead, the BK channel is thought to exist in one of at least two conformational states having different affinities for Ca^{2+}. The transition from low- to high-affinity Ca^{2+} binding is voltage dependent (Cui et al 1997). The apparent dissociation constant of the BK channel for Ca^{2+} decreases roughly sixfold from $-50\,mV$ to $-20\,mV$ (Cui et al 1997). Our estimates indicated that the Ca^{2+} concentration during a spark in

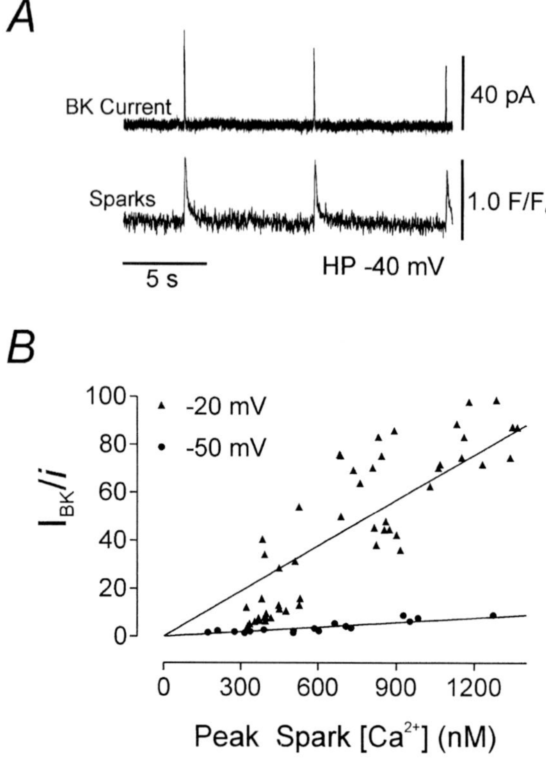

FIG. 3.   Ca^{2+} sparks activate BK channels in smooth muscle. (A) Simultaneous measurements of Ca^{2+} sparks and whole-cell BK currents in a guinea-pig UBSM cell. In this cell, three sparks occurred from a single spark site, and each spark activated BK channels causing a transient outward current. HP, holding potential. (B) Voltage-dependence of the coupling of RyRs to BK channels in UBSM cells. BK channel activity is measured as the amplitude of the transient BK current (I_{BK}) divided by the unitary BK current (i) at a given membrane potential. I_{BK}/i is plotted as a function of the peak [Ca^{2+}] during the associated spark at two different membrane potentials (-50 mV and -20 mV). Depolarization from -50 mV to -20 mV increases the slope of the relationship between spark Ca^{2+} and I_{BK}/i approximately threefold, indicating stronger coupling between sparks and BK channels at depolarized potentials (see Herrera et al 2001).

the vicinity of BK channels increases from 100 nM to at least 10 μM (Perez et al 2001). For a spark size of 10 μM, we found that BK channel activity is sixfold higher at -20 mV than at -50 mV (Fig. 3B), which is precisely the increase one would expect based on the voltage-dependence of the apparent Ca^{2+} sensitivity (Cox et al 1997, Cui et al 1997). The increased coupling strength of RyRs (Ca^{2+} sparks) to BK channels with depolarization indicates that Ca^{2+} sparks would have a more substantial impact on UBSM membrane potential at depolarized

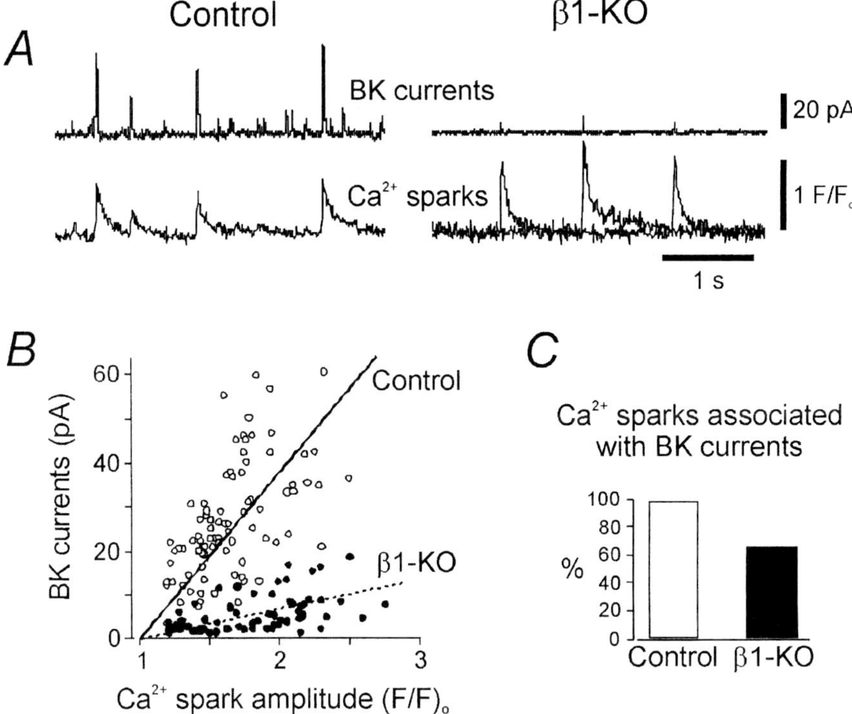

FIG. 4. β1 subunit of the BK channel increases coupling strength of RyRs (Ca^{2+} sparks) to BK channels in ASM. (A) Simultaneous recordings of Ca^{2+} sparks and BK currents in cerebral arterial smooth muscle cells from a control mouse and a β_1 knockout (KO) mouse. Holding potential was -40 mV for both cells. (B) Correlation between Ca^{2+} spark amplitude and BK current amplitude in control and β1 KO myocytes ($n=6$ cells and 7 cells, respectively). The slope of the relationship is greater in control myocytes. D, Percentage of Ca^{2+} sparks associated with transient BK currents in control and β_1 KO cells. In control myocytes, virtually all sparks caused BK currents, whereas in β_1 KO ASM cells, there were several Ca^{2+} sparks which occurred with no detectable currents. Modified from Brenner et al (2000).

potentials. This effect of membrane potential is very much like the one seen with the BK channel accessory β1 subunit (Fig. 1B), which also increases the apparent Ca^{2+}-sensitivity of the BK channel (Wallner et al 1999, Brenner et al 2000). We have found that a given size Ca^{2+} spark causes a larger transient BK current in mice with the β1 subunit expressed compared to mice lacking β1-subunit expression (Fig. 4) (Brenner et al 2000). Thus, with agonist-evoked UBSM membrane potential depolarization, or during the peak of a UBSM action potential, Ca^{2+} spark-induced activation of BK channels would be a very effective means of repolarizing the membrane potential.

Transient BK currents in UBSM: evidence in support of
'loose coupling' of VDCCs and RyRs

Measurements like those in Fig. 3 and 4 of SR Ca^{2+} release (Ca^{2+} sparks) and simultaneous membrane currents are technically very difficult. In particular, when holding a cell at a depolarized potential positive to -20 mV, the average cytosolic $[Ca^{2+}]$ rises to such an extent that detecting localized increases in $[Ca^{2+}]$, such that occur during a Ca^{2+} spark, becomes problematic due to poor contrast. However, under physiological conditions, UBSM cells frequently experience membrane potentials that are positive to -20 mV, especially during an action potential which peaks around $+15$ mV (Heppner et al 1997). To determine how SR Ca^{2+} release changes at a potential close to the peak of a UBSM action potential, we developed a voltage-step protocol that enables us to use BK channel activity as an indicator of SR Ca^{2+} release. This protocol consists of holding the cell at -70 mV and stepping the cell to $+10$ mV for 100 ms. During this voltage step, the current recorded was biphasic, consisting of inward and outward phases. The inward phase is due to VDCCs, and the outward phase consists of current conducted by BK channels, SK channels, and voltage-dependent K^+ (K_V) channels.

To determine the contribution of BK channels to this mixed outward current, the portion of the current that was sensitive to the BK channel blocker tetraethylammonium (TEA^+) was measured (see Nelson & Quayle 1995). TEA^+ was applied at a concentration (1 mM) which should have very little effect on K_V currents. BK current was found by subtracting the current in the presence of TEA^+ from the control current. Similar results were seen when BK channels were blocked with the highly selective peptide inhibitor of BK channels iberiotoxin (200 nM) (Galvez et al 1990). To ensure that K_V channels did not significantly contribute to the current attributed to BK channels, a subset of experiments were performed in which iberiotoxin (200 nM) was applied to block BK channels, followed by TEA^+ (1 mM) in the continued presence of iberiotoxin. The mean current was reduced by iberiotoxin (200 nM). In the continued presence of iberiotoxin, TEA^+ application was without further effect. Thus, 1 mM TEA^+ appears to be a selective blocker of BK channels in UBSM.

BK currents consisted of transient and steady-state components. The transient currents were abolished by ryanodine (10 μM), indicating that they are activated by Ca^{2+} sparks. The lag from the onset of depolarization (activation of VDCCs) to the first transient BK current was around 50 ms. This delay is much too long to be attributed to local signalling from VDCCs to RyRs such as occurs in heart, and instead is consistent with the idea of 'loose coupling' between VDCCs and RyRs (Fig. 1A) (Collier et al 2000). Spark-activated transient BK currents remained in the presence of the VDCC antagonist diltiazem (50 μM). However, the frequency of transient BK currents, which is an index of Ca^{2+} spark probability, was dependent upon Ca^{2+} entry through VDCCs, as the VDCC blocker diltiazem (50 μM) reduced the frequency of transient currents substantially (Herrera & Nelson 2001). These findings suggest that RyRs (Ca^{2+} spark sites) are activated not by Ca^{2+} influx through VDCCs directly, but by the elevation in cytosolic $[Ca^{2+}]$ or SR $[Ca^{2+}]$ which occurs following Ca^{2+} entry through VDCCs during depolarization (Fig. 1A).

Steady-state BK currents and their dependence on Ca^{2+} entry through VDCCs

In contrast to transient BK currents, activation of steady-state BK currents were almost entirely attributed to Ca^{2+} entry through VDCCs. This steady-state BK current is measured as the BK current remaining in the presence of 10 μM ryanodine, which inhibits transient BK currents. Roughly 50% of the BK current is steady state. The VDCC blocker diltiazem (50 μM) inhibits this steady-state BK current by 96% (Herrera & Nelson 2001). The nature of the communication between VDCCs and BK channels, leading to steady-state BK currents during depolarization pulses, is not known. Activation of BK channels by Ca^{2+} entry through VDCCs likely plays an important role in repolarizing the action potential (Heppner et al 1997).

Communication between VDCCs and SK channels

The SK channel is an important K_{Ca} channel that regulates the action potential after-hyperpolarization (Creed et al 1983) and spontaneous contractile activity (Herrera et al 2000) in UBSM. To determine the nature of the Ca^{2+} signal that activates SK channels in UBSM, whole-cell SK currents were recorded during 100 ms depolarizations from -60 mV to $+10$ mV (Fig. 5). SK currents were measured by applying the potent peptide inhibitor of SK channels apamin (Köhler et al 1996). Apamin (100 nM) was applied after obtaining a control recording (Fig. 5). SK currents are defined as the current in the presence of apamin subtracted from the control current (Fig. 5B). We found that blocking RyRs with ryanodine (10 μM) did not affect the size of the mean SK current, whereas blocking VDCCs with diltiazem (50 μM) completely abolished SK currents (Herrera & Nelson 2001).

We have also found that blocking SK channels with apamin has no effect on the amplitude or kinetics of transient Ca^{2+} spark-activated currents in UBSM cells (Herrera et al 2001). Based on these observations, SK channels in UBSM cells do not appear to be in sufficient density above a spark site to give rise to measurable currents during a Ca^{2+} spark. SK channels seem to be activated by Ca^{2+} which enters the cell through VDCCs. Thus, during a UBSM action potential, Ca^{2+}

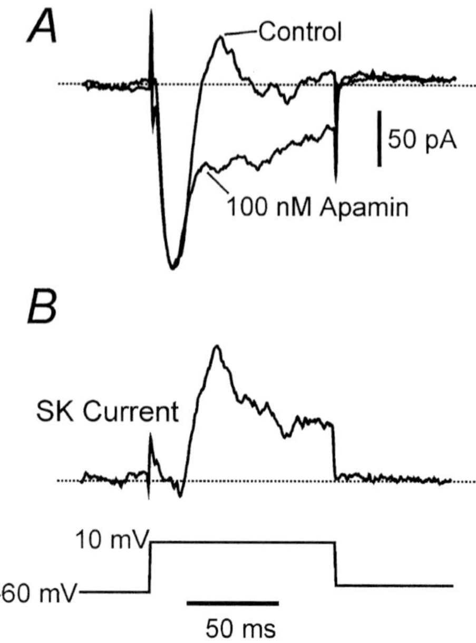

FIG. 5.   SK currents in UBSM cells. (A) Original records of whole-cell currents recorded from a guinea-pig UBSM cell during a 100 ms depolarization from $-60\,mV$ to $+10\,mV$. Current is shown under baseline conditions (control), and after addition of the SK channel blocker apamin (100 nM). (B) The apamin-sensitive portion of the current in A is shown (SK current). Dotted lines indicated zero current. (From G.M. Herrera & M.T. Nelson, unpublished observations.)

entry through VDCCs during the upstroke raises $[Ca^{2+}]_i$, which then activates SK channels to cause the after-hyperpolarization (Fig. 1E).

Summary

The SR plays an important role in regulating the excitability, and thus contractility, of smooth muscle. In addition to the stereotypic role of SR Ca^{2+} release through InsP$_3$ receptors which contributes to contractile responses, SR Ca^{2+} release through RyRs (Ca^{2+} sparks) also plays an important role to limit contractility by activating plasma membrane ion channels such as BK and SK channels. VDCCs serve as the major pathway for Ca^{2+} entry during myogenic contractions in smooth muscle (Knot et al 1998, Herrera et al 2000). Ca^{2+} entry through VDCCs can activate RyRs in smooth muscle through a process that is fundamentally different from the local control of CICR in cardiac muscle. In smooth muscle, RyRs are activated only after sufficient Ca^{2+} accumulation has occurred, which takes tens of milliseconds, as opposed to local control in cardiac muscle which occurs in less than 5 ms.

We have found that the communication between Ca^{2+} sparks and BK channels can be tuned by membrane potential (this study, Herrera et al 2001) and by the accessory β_1 subunit (Brenner et al 2000, Petkov et al 2001). This effect can be accounted for based on the allosteric relationship between the BK channel voltage and Ca^{2+} sensors, whereby membrane potential depolarization favours a channel conformation that has a higher Ca^{2+} sensitivity. The β_1 subunit has a similar role to increase the apparent Ca^{2+} sensitivity of the α subunit.

BK channels can also be activated by Ca^{2+} influx through VDCCs. Whereas the communication between RyRs and BK channels is local, it is not clear whether this is the case for the communication between VDCCs and BK channels. Communication between VDCCs and BK channels likely plays an important role in repolarizing the UBSM action potential (Fig. 1E).

SK channels are not activated by Ca^{2+} release through RyRs (Herrera et al 2001, this study), however SK channels are activated by $InsP_3R$-mediated Ca^{2+} release in gastrointestinal smooth muscle (Bayguinov et al 2000). In UBSM, SK channels are activated by Ca^{2+} entry through VDCCs, and this likely contributes to the action potential after-hyperpolarization (Fig. 1E).

Acknowledgements

The authors would like to thank Dr D.M. Eckman for supplying the data used for Fig. 2B, and Dr A.D. Bonev for providing the data used for Fig. 4. This study was supported by National Institutes of Health Grants DK-53832, HL-44455, HL-63722 and NS-39405, and by a Training Grant from the National Institutes of Health T32 HL/AR 07944.

References

Bayguinov O, Hagen B, Bonev AD, Nelson MT, Sanders KM 2000 Intracellular calcium events activated by ATP in murine colonic myocytes. Am J Physiol 279:C126–C135

Bolton TB, Imaizumi Y 1996 Spontaneous transient outward currents in smooth muscle cells. Cell Calcium 20:141–152

Brenner R, Perez G J, Bonev AD et al 2000 Vasoregulation by the beta1 subunit of the calcium-activated potassium channel. Nature 407:870–876

Cannell MB, Cheng H, Lederer W J 1995 The control of calcium release in heart muscle. Science 268:1045–1049

Cheng H, Lederer W J, Cannell MB 1993 Calcium sparks: elementary events underlying excitation-contraction coupling in heart muscle. Science 262:740–744

Collier ML, Ji G, Wang Y-X, Kotlikoff MI 2000 Calcium-induced calcium release in smooth muscle: loose coupling between the action potential and calcium release. J Gen Physiol 115:653–662

Cox DH, Aldrich RW 2000 Role of the $\beta1$ subunit in large-conductance Ca^{2+}-activated K^+ channel gating energetics: mechanisms of enhanced Ca^{2+} sensitivity. J Gen Physiol 116:411–432

Cox DH, Cui J, Aldrich RW 1997 Allosteric gating of a large conductance Ca-activated K^+ channel. J Gen Physiol 110:257–281

Creed KE 1971 Membrane properties of the smooth muscle of the guinea-pig urinary bladder. Pflügers Arch 326:115–126

Creed KE, Ishikawa S, Ito Y 1983 Electrical and mechanical activity recorded from rabbit urinary bladder in response to nerve stimulation. J Physiol 338:149–164

Cui J, Cox DH, Aldrich RW 1997 Intrinsic voltage dependence and Ca^{2+} regulation of *mslo* large conductance Ca-activated K^+ channels. J Gen Physiol 109:647–673

Fabiato A 1983 Calcium-induced release of calcium from the cardiac sarcoplasmic reticulum. Am J Physiol 245:C1–C14

Galvez A, Gimenez-Gallego G, Reuben JP et al 1990 Purification and characterization of a unique, potent, peptidyl probe for the high conductance calcium-activated potassium channel from venom of the scorpion *Buthus tumulus*. J Biol Chem 265:11083–11090

Ganitkevich VY, Isenberg G 1992 Contribution of Ca^{2+}-induced Ca^{2+} release to the $[Ca^{2+}]_i$ transients in myocytes from guinea-pig urinary bladder. J Physiol 458:119–137

Heppner TJ, Bonev AD, Nelson MT 1997 Ca^{2+}-activated K^+ channels regulate action potential repolarization in urinary bladder smooth muscle. Am J Physiol 273:C110–C117

Herrera GM, Nelson MT 2001 Remote sensing of Ca^{2+} influx through voltage-dependent Ca^{2+} channels (VDCCs) by ryanodine receptors (Ca^{2+} sparks) in the sarcoplasmic reticulum (SR) of urinary bladder smooth muscle (UBSM). FASEB J 15:A1116

Herrera GM, Heppner TJ, Nelson MT 2000 Regulation of urinary bladder smooth muscle contractions by ryanodine receptors and BK and SK channels. Am J Physiol 279:R60–R68

Herrera GM, Heppner TJ, Nelson MT 2001 Voltage dependence of the coupling of Ca^{2+} sparks to BK_{Ca} channels in urinary bladder smooth muscle. Am J Physiol 280:C481–C490

Hogg RC, Wang Q, Large WA 1993 Time course of spontaneous calcium-activated chloride currents in smooth muscle cells from the rabbit portal vein. J Physiol 464:15–31

Horn R, Marty A 1988 Muscarinic activation of ionic currents measured by a new whole-cell recording method. J Gen Physiol 92:145–159

Imaizumi Y, Torii Y, Ohi Y et al 1998 Ca^{2+} images and K^+ current during depolarization in smooth muscle cells of the guinea-pig vas deferens and urinary bladder. J Physiol 510:705–719

Jaggar JH 2001 Intravascular pressure regulates local and global Ca^{2+} signaling in cerebral artery smooth muscle cells. Am J Physiol 281:C439–C448

Jaggar JH, Stevenson AS, Nelson MT 1998 Voltage dependence of Ca^{2+} sparks in intact cerebral arteries. Am J Physiol 274:C1755–C1761

Jaggar JH, Porter VA, Lederer WJ, Nelson MT 2000 Calcium sparks in smooth muscle. Am J Physiol 278:C235–C256

Janssen LJ, Simms SM 1994 Spontaneous transient inward currents and rhythmicity in canine and guinea-pig tracheal smooth muscle cells. Pflügers Arch 427:473–480

Kamishima T, McCarron JG 1997 Regulation of the cytosolic Ca^{2+} concentration by Ca^{2+} stores in single smooth muscle cells from rat cerebral arteries. J Physiol 501:497–508

Knot HJ, Standen NB, Nelson MT 1998 Ryanodine receptors regulate arterial diameter and wall $[Ca^{2+}]$ in cerebral arteries of rat via Ca^{2+}-dependent K^+ channels. J Physiol 508:211–221

Köhler M, Hirschberg B, Bond CT et al 1996 Small-conductance, calcium-activated potassium channels from mammalian brain. Science 273:1709–1714

Nelson MT, Quayle JM 1995 Physiological roles and properties of potassium channels in arterial smooth muscle. Am J Physiol 268:C799-C822

Nelson MT, Cheng H, Rubart M et al 1995 Relaxation of arterial smooth muscle by calcium sparks. Science 270:633–637

Pérez GJ, Bonev AD, Patlak JB, Nelson MT 1999 Functional coupling of ryanodine receptors to K_{Ca} channels in smooth muscle cells from rat cerebral arteries. J Gen Physiol 113:229–238

Pérez GJ, Bonev AD, Nelson MT 2001 Micromolar Ca^{2+} from sparks activates Ca^{2+}-sensitive K^+ channels in rat cerebral artery smooth muscle. Am J Physiol Cell Physiol 281: C1769–C1775

Petkov GV, Bonev AD, Heppner TJ, Brenner R, Aldrich RW, Nelson MT 2001 β1-subunit of the Ca^{2+}-activated K^+ channel regulates contractile activity of mouse urinary bladder smooth muscle. J Physiol 537:443–452
Rousseau E, Smith JS, Meissner G 1987 Ryanodine modifies conductance and gating behavior of single Ca^{2+} release channel. Am J Physiol 253:C364–C368
Wallner M, Meera P, Toro L 1999 Molecular basis of fast inactivation in voltage and Ca^{2+}-activated K^+ channels: a transmembrane β-subunit homolog. Proc Natl Acad Sci USA 96:4137–4142
Wang Q, Hogg RC, Large WA 1992 Properties of spontaneous inward currents recorded in smooth muscle cells isolated from the rabbit portal vein. J Physiol 451:525–537
ZhuGe R, Sims SM, Tuft RA, Fogarty KE, Walsh JV 1998 Ca^{2+} sparks activate K^+ and Cl^- channels, resulting in spontaneous transient currents in guinea-pig tracheal myocytes. J Physiol 513:711–718

DISCUSSION

Young: Are you saying that you find RyRs deep in the cell? If so, and ryanodine is found deep in the cytosol, how do you have the ryanodine only affecting the subplasmalemmal RyRs as opposed to throughout the whole cell?

Nelson: I assume we are affecting RyRs throughout the whole cell. It seems that the RyRs that are obviously communicating with the BK channels are very close to the cell membrane.

Young: My point is that you would be completely perturbing the Ca^{2+} concentration in the SR by adding ryanodine.

Nelson: Under voltage-clamp conditions, if we add ryanodine and block the sparks, we see no rise in Ca^{2+}. Why would you think it would go up?

Young: Not go up: I thought the stores would be depleted by added ryanodine.

Nelson: The stores might be. Caffeine has no effect in the presence of ryanodine. We haven't looked at whether or not we observe $InsP_3$-induced Ca^{2+} release from the stores in the presence of ryanodine. One of the issues is that in bilayer experiments, ryanodine from cardiac muscle can induce a subconductance state of the RyR. This is thought to deplete the SR. I should point out that the few bilayer studies looking at RyRs from smooth muscle did not show this. We don't know whether ryanodine is blocking the release channel or inducing a subconductance state. In either case, Ca^{2+} sparks would be inhibited.

Isenberg: I would like to ask about the time relationship between the Ca^{2+} currents and the SK channel activity. Is there a delay?

Nelson: That's a good question. We hope that there would be a delay, or we wouldn't have an action potential. I guess the delay is about 30 ms, which is when we start seeing the prominent after-hyperpolarization.

Isenberg: You didn't say this explicitly, but I got the feeling you would like to say that Ca^{2+} release does not activate SK channels. Did I misinterpret?

Nelson: All I can say is that when we add ryanodine and block sparks, as well as presumably the RyRs, we saw no effect in the SK currents. There could be a couple of reasons for that. First, the density of SK channels: if we take the conductance published by John Adelman and calculate the number of channels, it is only a few hundred channels per cell. If these channels are uniformly distributed, then the number of channels (about 1) above a spark will be too small in order to observe a current. Second, the Ca^{2+} sensor for the SK channel is calmodulin. So the issue here is what is the on-rate of Ca^{2+} on the SK–calmodulin complex. If we take Mike Walsh's data from different preparations of myosin light-chain kinase (MLCK), they had an exceedingly slow on-rate. This slow on-rate might be filtering out high frequency Ca^{2+} signals. This could also be a factor.

Blaustein: What is the link between the Ca^{2+} and the SK channel? The SK channels are not supposed to be Ca^{2+} dependent. What is going on?

Nelson: They are Ca^{2+} dependent.

Blaustein: But what are they activated by?

Nelson: They are small conductance, Ca^{2+}-sensitive K^+ channels. They are activated by Ca^{2+} and the Ca^{2+} sensor is calmodulin, which is very tightly bound in the channel. Ca^{2+} binds to calmodulin and opens the channel. The Ca^{2+} sensitivity of the SK channel is just what you'd expect for calmodulin. It is about half activated by 400 nM Ca^{2+} and completely saturated at 1 μM Ca^{2+}.

Blaustein: Do you think they are close to the Ca^{2+} channels?

Nelson: I don't know.

Brading: This is another interesting example of when we need to know just what the relationship of these experiments is to the behaviour of the bladder in real life. It is still not entirely clear what the role of the action potentials is. Certainly, when you want to empty your bladder this doesn't appear to be achieved through generation of action potentials in the smooth muscle. There is a dense innervation that activates muscarinic receptors which in some way synchronizes contraction in the bladder. What Mark is talking about is fascinating. The bladder has a very awkward position: the smooth muscle needs to be active in order that the bladder can keep a sensible shape, so if you want to urinate you can do this fast. So spontaneous activity is necessary. But what you mustn't have is intravesical pressure developing. It is an unusual situation. All of this spontaneous activity takes place without an overall pressure rise in the bladder. When you actually want to generate a pressure change in the bladder you have to synchronize the smooth muscle cells, and this doesn't appear to involve action potential mechanisms. This is fascinating because the K^+ channel blockers that Mark Nelson has been using do increase the size of the spontaneous contractions in strips of detrusor. We've looked at this in whole animal bladders and see no change in intracellular pressure in the bladder. But if we look at the surface of the bladder, the whole thing is writhing in the most extraordinary way.

Kotlikoff: Under physiological conditions, do you think that every action potential causes Ca^{2+} release which then has an effect through the BK channel? Or might it be that there is a frequency-dependent action potential?

Nelson: I think that RyR (spark) activation of BK channels is playing some role in regulating the membrane potential and repolarization.

Fry: Just to answer that, in the heart if you raise the frequency this shortens the action potential. This is a slow feedback mechanism between outward and inward current. This is not seen in bladder, where the outward and inward current are always the same duration. What you do alter is the slope of the pacemaking potential. If you give a long current-clamped pulse you see a train of action potentials and the frequency of the slope potential will alter. This is about the only thing that alters. In a way, this is what you might anticipate if there is Ca^{2+} regulation of a slowly decaying outward current. To an extent, it does beg the issue of the role outward currents serve under these circumstances. They don't underlie any normal contractile activation in at least human bladder. This is all a muscarinic response, which is independent of membrane potential. What it would do is set the level of excitability of the system. The generation of an action potential to a certain extent is not the issue. In a way, this provides a good example of the differences between studying human and animal tissues. Human has no membrane potential dependence of contraction in the bladder: it is purely a cholinergic system. There is ATP release, but it is broken down completely before it gets to the smooth muscle cell. In animals (except sheep and old world monkeys) there is a purinergic component as well.

Brading: That is a response to stimulating nerves. We don't yet know what the action potentials are actually doing in the way of producing contraction. I'm sure they are generating contraction of individual smooth muscle cells, but we don't know for certain.

Fry: I would say that they don't generate contraction, certainly in the human bladder. This illustrates the usefulness of looking at different species, because they give different answers. If someone asks why you would want a two-component system in an animal and not in a human, one explanation would be that animals often use small doses of urine to mark territories.

Brading: The appearance of spontaneous contractile activity in detrusor strips depends very much on their thickness.

Bolton: My collaborators have worked on guinea pig bladder, looking at single-cell voltage clamp and the switch-on of the Ca^{2+}-activated K^+ channels. Do your results fit well with theirs? They found that under voltage clamp with a depolarizing pulse they switch on rapidly. The time-course of the rising Ca^{2+} just under the membrane and the rise of the Ca^{2+}-activated K^+ current are almost synchronous. If you trigger an action potential under current clamp, it reaches a threshold and fires off but doesn't switch in until the repolarization phase of the action potential.

Nelson: The action potentials here are quite brief, and the BK channels are contributing to the repolarization. In fact, in order to obtain a decent estimate of the Ca^{2+} current amplitude we have to first block the BK channels.

Bolton: And this swamps the inward current.

Nelson: To a degree.

Blaustein: You have talked about getting the Ca^{2+} in, but to keep this system going we have to get the Ca^{2+} out again. Have you looked at this? What do you think will happen if you reduce Ca^{2+} exit? If you get more of the Ca^{2+} in the SR, would that increase the frequency of sparks? Could this separate BK from SK channels?

Nelson: It could. We could try to load the SR. For example, we have used the phospholamban knockout mouse where it appears that the SR Ca^{2+} load increases (Wellman et al 2001). The spark frequency increases as the load increases. At the moment we have not observed any communication from the SR to the SK channels.

Blaustein: You have quite convincing evidence: you have separated the two and they operate on different control systems.

Nelson: At the moment we have no evidence of communication from the ryanodine receptors to the SK channels, but SK channel Ca^{2+} activation does depend on Ca^{2+} influx through voltage-dependent Ca^{2+} channels.

Burdyga: Which of these K^+ channels do you think is the most dominant in repolarization?

Nelson: I don't know precisely.

Burdyga: Do you assume there is no additivity in this action?

Nelson: It's complex. But by changing SR load you could presumably not only increase the amount of Ca^{2+} released, but also decrease the latency of the first Ca^{2+} spark.

Hellstrand: Is there any idea about the sort of time range we are looking at? Could this mechanism account for bursting behaviour?

Nelson: I think it affects the bursting. There is another clock here. If we add iberiotoxin to bladder strips, the amplitudes and durations of the phasic contractions increase. We also observe a slowing of the phasic contraction frequency. This suggests that it is some sort of slow Ca^{2+}-dependent process that causes longer periods between contractions. The net effect of iberiotoxin is a large increase in force: amplitude and duration rise, the frequency goes down.

Sanders: Does it occur in smooth muscle cells?

Nelson: That's a good question.

Brading: What is the role of the action potentials in generating contraction? This is not entirely clear. We know in strips that if we add iberiotoxin or any K^+ channel blockers we do get contractions. But the frequency of contractions is nothing like the frequency of the action potentials.

Nelson: A phasic contraction presumably results from a train of action potentials.

Somlyo: Some of you may remember that there was an old-fashioned drug called TEA (tetraethyl ammonium chloride). When this was applied to smooth muscle that normally didn't generate action potentials, it caused beautiful action potentials and rhythmic contractions. I don't think this was physiological.

Eisner: I have the impression that Mark Nelson's paper has also failed the Brading–Hirst test of physiological relevance!

Brading: Not at all, it's just that we don't know what the channels are doing. I'm sure it's of physiological relevance, but I'm not sure about function.

Hirst: I'm convinced that it's clipping action potential generation and release. Alison is not sure what the action potentials are doing, but she agrees they are an essential part of the response.

Brading: That's right. What do you mean by changing excitability? That's the intriguing thing. Is it the frequency at which you produce action potentials, but does this really tell us anything about the important function of how the bladder is contracting?

Nelson: I tried to illustrate this briefly.

Fry: But ion channels don't always have to generate action potentials to be useful. You measured them as an expression of an action potential, but they might produce slow waves and still produce a function.

Brading: They certainly do produce action potentials spontaneously in humans, as well as in mice.

Nelson: There are also studies indicating that the K-ATP channel openers are useful in certain types of incontinence. This is also working on the bladder smooth muscle.

Brading: K^+-ATP channels are marvellous; they stop all of the unstable contractions in the bladder straight off. The only problem is that they do everything else as well.

Nelson: So is that where the action potentials are coming in, in excitability?

Brading: Possibly. But this is only in abnormal bladder.

McHale: In asynchronous contractions there are a lot of action potentials being fired, but not to any great effect. Casey van Breemen suggested an interesting model yesterday in which action potentials are being fired asynchronously and you don't get any contraction until there is a synchronizing signal. Perhaps this is what the nerves do: they provide a synchronizing signal.

Brading: But we don't know whether the nerves are synchronizing action potentials; they may be bypassing them altogether and generating contraction without affecting membrane potentials.

Reference

Wellman GC, Santana LF, Bonev AD, Nelson MT 2001 Role of phospholamban in the modulation of arterial Ca^{2+} sparks and Ca^{2+}-activated K^+ channels by cAMP. Am J Physiol 281:C1029–C1037

Sarcoplasmic reticulum function and contractile consequences in ureteric smooth muscles

Theodor Burdyga and Susan Wray

Department of Physiology, The University of Liverpool, Crown Street, Liverpool L69 3BX, UK

Abstract. This paper discusses the role of Ca^{2+}-induced Ca^{2+} release (CICR) and inositol-1,4,5-trisphosphate ($InsP_3$)-induced Ca^{2+} release (IICR) from the sarcoplasmic reticulum (SR) in the control of contractile activity in the ureter. The Ca^{2+} store in guinea-pig ureter has been found to be exclusively a CICR type with ryanodine receptors (RyRs) present. In the rat ureter the SR store is exclusively an IICR type with $InsP_3$ receptors ($InsP_3Rs$) present. Guinea-pig ureteric cells *in vitro* and *in situ* have been found to generate Ca^{2+} sparks — small localized, transient releases from RyRs. The sparks are enhanced by caffeine and blocked by emptying the SR. In rat cells Ca^{2+} puffs occur in response to agonists, representing the opening of $InsP_3Rs$. The puffs can be abolished by heparin or store emptying. These SR Ca^{2+}-release events affect the excitability of the ureteric cells. In guinea-pig cells, spontaneous transient outward currents (STOCs) can be recorded in response to caffeine application (an agonist for RyR), followed by a shortening of the plateau phase of the action potential. This in turn causes a decrease in the amplitude and duration of the contractions of the ureter. If the SR is inhibited then STOCs are abolished, the action potential plateau prolonged and force increased. Thus it is concluded that the SR acts to limit contraction in the guinea-pig ureter. The mechanism underlying this involves its Ca^{2+} release being directed to Ca^{2+}-activated K^+ channels on the surface membrane and causing STOCs and hyperpolarization, and controlling the duration of the action potential. In rat ureter IICR acts to potentiate force via membrane depolarization and increased L-type Ca^{2+} entry into the cells. Thus the SR can alter cell signalling and excitation–contraction coupling in the ureter, but its precise role is species dependent. The ureter, with its species-dependent expression of either IICR or CICR provides an ideal system (a natural transgenic model) for studying the SR. Eventually, we will be able to apply this knowledge to the human ureter, to increase our understanding of its functioning in health and disease.

2002 Role of the sarcoplasmic reticulum in smooth muscle. Wiley, Chichester (Novartis Foundation Symposium 246) p 208–220

The ureters

The ureters are thin, fibro-muscular tubes that transport urine from the kidneys to the bladder. The movement of urine is achieved by peristaltic contractions of the ureteric smooth muscle. The importance of this contractile activity can be seen in the consequences of contractile dysfunction: for example, ureteric obstruction, inflammation or congenital abnormality can cause kidney damage within hours and death may follow (Weiss 1986, Rose & Gillenwater 1973). Urinary tract infections are a common cause of morbidity and occasionally prove fatal. Particularly serious is pyelonephritis, which can occur from ascending infection, especially if there is reflux from the bladder back to the kidneys (vesico-ureteric reflux). Thus the health of the ureters and a sound understanding of the mechanisms controlling their functioning are of clear importance.

Mechanism of contraction

Contraction of the ureters is initiated by pacemaker activity in the renal pelvic region. This is followed by action potential propagation down the ureter and contraction (Weiss 1986, Boyarsky & Labay 1981). This contraction requires a rise in intracellular Ca^{2+} concentration ($[Ca^{2+}]_i$). The action potential, by causing depolarization, can lead to the opening of voltage-gated Ca^{2+} channels on the ureteric cell membrane, and Ca^{2+} entry (Weiss 1986, Santicioli & Maggi 1998). There is also an internal Ca^{2+} store, the sarcoplasmic reticulum (SR), which can release Ca^{2+} and that, as will be discussed later, may influence both the electrical and contractile responses of the ureter. Work in our laboratory has focused on elucidating the contribution and significance of both SR Ca^{2+} release and Ca^{2+} entry following excitation. It is now clear that both influence excitability and that to understand this, requires an appreciation of the relationship between Ca^{2+} and electrical activity in ureteric smooth muscle.

The temporal relationship between the action potential, Ca^{2+} and force in ureteric smooth muscle

The action potential of the ureter is unusually long lasting and is characterized by an initial spike followed by a plateau (Fig. 1). In the guinea-pig ureter, the action potentials further characterized by the presence of multiple oscillations (spikes) on the plateau (Fig. 1A) whereas only a single spike is observed in rat (Fig. 1B). Both the spike and plateau components contribute to the generation of the Ca^{2+} transient (Fig. 1). Depending on the duration of the plateau component of the action potential, $[Ca^{2+}]_i$ can reach a peak level and remain in a steady state. There is a close correlation between the duration of the plateau component of the action potential and the duration of the Ca^{2+} transient at its peak level. Repolarization of the action potential is followed by restoration of the Ca^{2+} transient to basal level (Fig. 1). In both species, after some delay, the Ca^{2+} signal triggers phasic contraction. The contractions lag $[Ca^{2+}]$ both during the rising and the

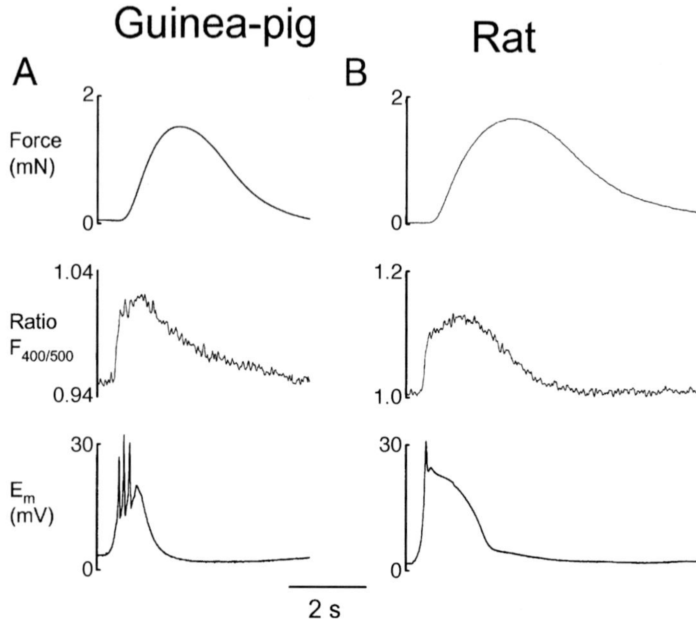

FIG. 1.   Simultaneous recording of the action potential (bottom trace), Ca^{2+} transients (middle trace) and force (top trace) in (A) guinea pig and (B) rat ureteric smooth muscle. (A, taken from Burdyga & Wray 1997; B, taken from Burdyga & Wray 1999a.)

relaxation phase (Burdyga & Wray 1999a). Modulation of the duration of the action potential was found to play a key role in modulation of the amplitude of force in ureteric muscle (Burdyga & Wray 1999a). The ionic mechanisms contributing to the action potential have been well studied in guinea-pig (Imaizumi et al 1989, Lang 1989) but not in the rat.

Agonists such as carbachol and histamine can prolong the plateau around 10-fold and thereby produce large increases in force (Fig. 4A) (Shuba 1981, Santicioli & Maggi 1998). This in turn is due to the maintenance of Ca^{2+} at its peak level. We will now discuss the SR and its contribution to the Ca^{2+} signal.

The ureteric SR

In the ureter the study of Ca^{2+} release mechanisms has proved to be extremely interesting. As shown in Fig. 1, in both rat and guinea-pig ureter electrical stimulation and depolarization leads to the generation of plateau-type action potentials and an increase in $[Ca^{2+}]_i$ and force. However, in the guinea-pig agonist application had no effect on $[Ca^{2+}]_i$ in the resting ureter but in the rat a large rise in $[Ca^{2+}]_i$ always occurred (Fig. 4). In Ca^{2+}-free solutions the rat ureter was also able to produce a transient rise of $[Ca^{2+}]_i$ upon agonist application, but the guinea-pig ureter was not. The Ca^{2+} transient in 0-Ca^{2+} solution was insensitive to nifedipine, a blocker of L-type Ca^{2+} channels (Burdyga et al 1995). These data suggest that the rat has a receptor coupled pathway from inositol-1,4,5-trisphospate ($InsP_3$) production to SR Ca^{2+} release, and that this pathway is absent in the guinea-pig. Further experiments were performed to test this hypothesis, and led to the conclusion that the guinea-pig ureter Ca^{2+} store is purely of the Ca^{2+}-induced Ca^{2+} release (CICR) type and the rat ureter store is purely $InsP_3$-induced Ca^{2+} release (IICR) type. This evidence is presented in the next section.

Species difference in SR Ca^{2+} release mechanisms — the ureter presents a natural transgenic model

Intact and chemically skinned preparations isolated from the guinea-pig and rat ureter display different mechanisms mediating the release of Ca^{2+} from the SR (Burdyga et al 1995, 1998). Direct application of carbachol or $InsP_3$ caused Ca^{2+} release from the SR and contraction of rat, but not guinea-pig ureter (Fig. 4). However, caffeine, an agonist for CICR, caused Ca^{2+} release and contraction in the guinea-pig ureter but not rat ureter. Kuemmerle et al (1994) also reported agonist-activated ryanodine-sensitive, $InsP_3$-insensitive Ca^{2+} release in intestinal smooth muscle. Ryanodine completely blocked caffeine-induced Ca^{2+} release from the SR in the guinea-pig ureter but had no effect on carbachol or $InsP_3$-induced Ca^{2+} release in rat ureter (Fig. 2) (Burdyga et al 1998). In both species cyclopiazonic acid (CPA), which was shown to selectively block SR Ca^{2+} pump in biochemical experiments (Prishchepa et al 1996), reversibly blocked Ca^{2+} and force transients induced by caffeine and carbachol in the intact and chemically skinned ureteric muscle strips of both species. These results lead us to conclude that only $InsP_3$-gated channels mediate Ca^{2+} release from the SR in rat ureteric myocytes, but in contrast, in the guinea-pig ureter, the mechanism of Ca^{2+} release is solely via CICR mediated by RyRs.

In confirmation of these conclusions Boittin et al (2000) analysed single myocytes from rat ureter for $InsP_3Rs$ and RyRs. They found by immuno-detection and binding data that there were 10–12-fold more $InsP_3Rs$ than RyRs, and that only the RyR_3 subtype was expressed, but all three subtypes of $InsP_3R$ were detected.

Thus it appears that the ureter will be a very useful tissue for examining the role and mechanisms of $InsP_3$- or RyR-dependent Ca^{2+} release and signalling. Before discussing this further, however, we will return to examining the importance of the SR Ca^{2+} store to ureteric function.

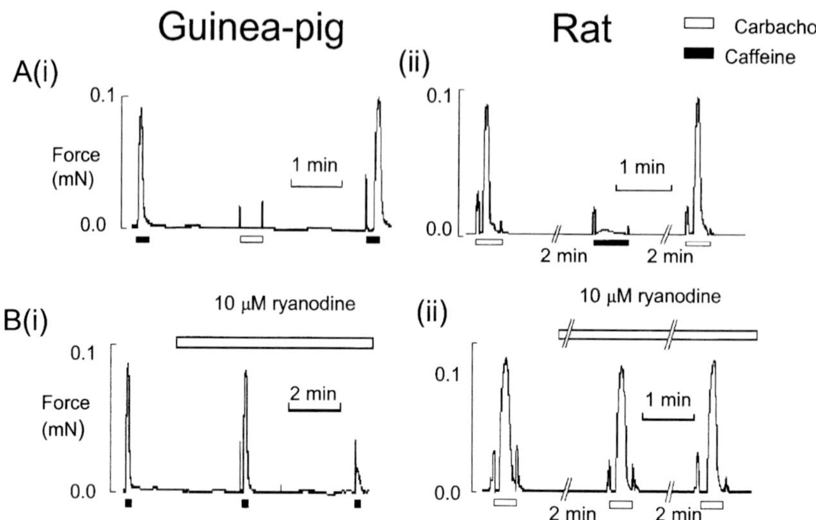

FIG. 2.   (A) The contractile responses of permeabilized guinea pig (Ai) and rat (Aii) ureteric smooth muscle to caffeine (10 mM) and carbachol (100 μM). (B) The effect of ryanodine on the response of the guinea pig ureter to caffeine (Bi) and the rat ureter to carbachol (Bii) (Taken from Burdyga et al 1998.)

SR and excitation–contraction coupling

The SR may contribute to excitation–contraction (EC) coupling in two ways; firstly, by the release of Ca^{2+} for contraction as described above, but secondly by modulating membrane excitability. As will be described elsewhere in this book, the SR is an important mediator of surface membrane ion channel activity, and hence, excitability. Spontaneous Ca^{2+} release from the SR can activate Ca^{2+}-sensitive ion channels. Both K^+ (K_{Ca}) and Cl^- (Cl_{Ca}) channels in the smooth muscle cell membrane can be activated by SR Ca^{2+}. If K_{Ca} channels are activated there will be a hyperpolarization, as K^+ ions leave the cell and spontaneous transient outward currents (STOCs) can be recorded (Carl et al 1996, Nelson & Quayle 1995). If Cl_{Ca} channels are activated then there will be a tendency for depolarization, as spontaneous inward currents result (STICs) (Large & Wang 1996). Thus the SR may alter membrane potential via K_{Ca} and Cl_{Ca}, and may thereby affect voltage-gated Ca^{2+} channel entry and force. We have recently examined the role of the SR in the guinea-pig ureter by emptying it of releasable Ca^{2+} and determining the effects.

The effects of inhibiting ureteric SR function

The SR Ca-ATPase can be inhibited by CPA (Prishchepa et al 1996). In this way the SR can be emptied of Ca^{2+} and its contribution to excitation–contraction coupling

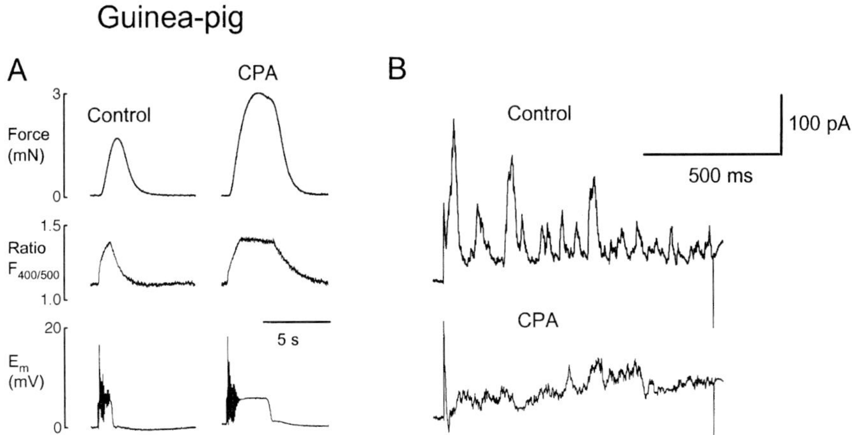

FIG. 3.   (A) Simultaneous recording of the action potential (bottom trace), Ca^{2+} transients (middle trace) and force (top trace) obtained in the presence and the absence of CPA. (B) The effect of cyclopiazonic (CPA) on spontaneous outward potassium currents (STOCs) recorded from single guinea-pig ureteric cells under voltage clamp conditions. (A, taken from Burdyga & Wray 1999a; B, from Burdyga & Wray 1999b.)

removed. We and others have found that CPA significantly enhances the amplitude and duration of phasic contractions of the ureter (Maggi et al 1995, Burdyga & Wray 1999b). This is due to an increase in the duration of the Ca^{2+} transient. Membrane recordings showed that CPA produced a small depolarization and a large increase in the duration of plateau phase of the action potential (Fig. 3). Patch-clamp studies demonstrated a marked inhibition of STOCs by CPA. Thus taken together these data show that the SR in the guinea-pig ureter plays a major role in modulating excitability. In particular, its role is to curtail the action potential, shorten the Ca^{2+} transient and hence decrease force (Burdyga & Wray 1999b). This is similar to conclusions reached for some other smooth muscles, such as uterus (Wray et al 2002, this volume) and arterial smooth muscle (Herrera & Nelson 2002, this volume). A role for K_{Ca} channels in the ureter had also been suggested from work by Imaizumi et al (1989) and Muraki et al (1994). For example, caffeine, an agonist at RyRs, had been shown to activate STOCs and decrease the duration of the plateau of the action potential (Imaizumi et al 1989), and K_{Ca} is an important target for excitatory actions of agonists in the guinea-pig ureter (Muraki et al 1994).

Much less is known about the role of the SR in excitation–contraction coupling in the rat ureter, where $InsP_3$ release predominates, but as will be discussed below its role in Ca^{2+} signalling in single cells has been studied. It is expected from this that the SR's contribution will be to potentiate force production.

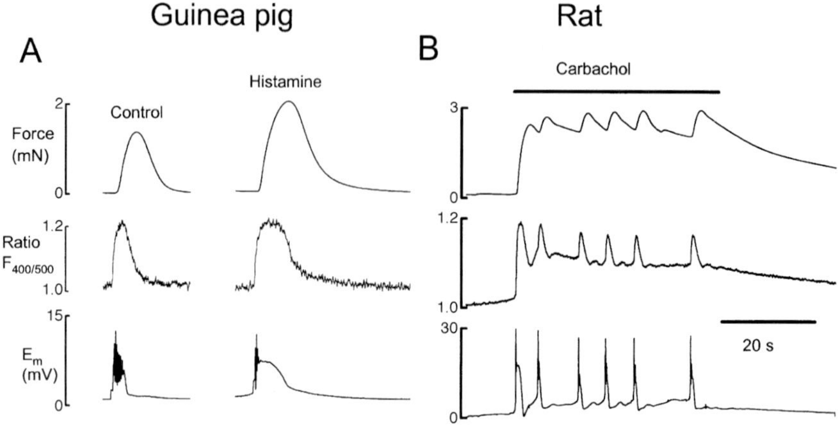

FIG. 4. (A) The effect of histamine on the evoked action potential (bottom trace), Ca^{2+} transients (middle trace) and force (top trace) in the guinea-pig ureter. (Taken from Burdyga & Wray 1999a.) (B) The effects of carbachol on the electrical activity (bottom trace), Ca^{2+} transients (middle trace) and force (top trace) in rat ureteric smooth muscle. (Modified from Burdyga & Wray 1995.)

Localized Ca^{2+} signalling events in the ureter

As mentioned already the ureter, with its separation of Ca^{2+} release mechanisms into CICR in guinea-pigs and IICR in rats, provides an opportunity for studying Ca^{2+} signalling events. Much attention has been focused on how the two release mechanisms function and interact with each other. It is generally accepted that global Ca^{2+} signals recorded in cells or tissues, arise from smaller, more localized initiating (elementary) Ca^{2+} release events (Lipp & Niggli 1996, Berridge 1997). These elementary releases have been termed Ca^{2+} sparks when they arise from the opening of a cluster of RyRs and Ca^{2+} puffs when they arise from the opening of a cluster of InsP$_3$Rs. Most smooth muscles possess both InsP$_3$Rs and RyRs on the same cells (Boittin et al 2000). As both types of receptor are sensitized by Ca^{2+} itself, it is clear that the opening of one will affect the opening of the other. Both Ca^{2+} sparks and puffs can initiate Ca^{2+} waves if the stimulus is sufficient. Otherwise the elemental events decay and the Ca^{2+} is presumably taken back into the SR by the Ca-ATPase.

In guinea-pig myocytes we have obtained preliminary data showing that sparks are present and initiated from a few sites within the cell, possibly analogues to the frequent discharge sites described by Gordienko et al (1998). Caffeine, an agonist for CICR, increased the frequency of the sparks (T. V. Burdyga & S. Wray, unpublished observations). In rat ureter, we and Boittin et al (2000) have seen

Ca^{2+} puffs in response to application of agonists to the cells, and Ca^{2+} waves. Boittin et al (2000) were also able to show these events following the photo-release of InsP$_3$ from caged precursors. They proposed that Ca^{2+} release is initiated at one site and then spreads as a wave, by 'a spatial recruitment of neighbouring Ca^{2+} release sites'.

Intact strips of ureter enable localized Ca^{2+} events in communicating cells to be studied, and by approximating more closely physiological conditions bring the mechanisms closer to those pertaining *in vivo*. We have started to make such measurements in the ureter and find sparks occurring in the guinea-pig and Ca^{2+} oscillations in individual cells from the rat (T. V. Burdyga & S. Wray, unpublished observations).

Thus much information will shortly be available looking at the spatiotemporal characteristics of receptor specific cell signalling from the SR in the ureter. This information will be compared to that found in other smooth muscles and related to the functional role of the SR in the ureter, described above.

Summary

In summary, the SR of the ureter plays an important role in EC coupling. However, any contribution of the SR directly to provide Ca^{2+} for the myofilaments is small and overwhelmed by the contribution from Ca^{2+}-activated membrane K$^+$ and Cl$^-$ channels. In particular SR Ca^{2+} will activate K$_{Ca}$ and cause a shortening of the plateau of the ureteric action potential. Such a shortening will have functional effects as the Ca^{2+} transient is abbreviated and does not maximally activate the myofilaments via myosin light chain kinase. Thus force is reduced. Therefore the SR can be viewed as providing a brake on contraction in the ureter. Recent work has focused on the patterns of Ca^{2+} signal arising from SR Ca^{2+} release. Our finding that rat ureter has only an InsP$_3$-inducible store and guinea-pig ureter has only a CICR store provides a unique model system for studying such signals. These studies should ultimately help our understanding of the human ureter, and its control in health and disease.

Acknowledgements

We are grateful to the NKRF and the Wellcome Trust for supporting this work.

References

Berridge M J 1997 Elementary and global aspects of calcium signalling. J Physiol 499:291–306

Boittin FX, Coussin F, Morel J-L, Halet G, Macrez N, Mironneau J 2000 Ca^{2+} signals mediated by Ins(1,4,5)P_3-gated channels in rat ureteric myocytes. Biochem J 349:323–332

Boyarsky S, Labay P 1981 Principles of ureteral physiology. In: Bergman H (ed) The ureter. Springer-Verlag, Berlin, p 71–104

Burdyga T, Wray S 1997 Simultaneous measurements of electrical activity, intracellular [Ca^{2+}] and force in intact smooth muscle. Pflügers Arch 435:182–184

Burdyga TV, Wray S 1999a The relationship between the action potential, intracellular calcium and force in intact phasic, guinea-pig uretic smooth muscle. J Physiol 520: 867–883

Burdyga TV, Wray S 1999b The effect of cyclopiazonic acid on excitation–contraction coupling in guinea-pig ureteric smooth muscle: role of the sarcoplasmic reticulum. J Physiol 517:855–865

Burdyga TV, Taggart MJ, Wray S 1995 Major difference between rat and guinea-pig ureter in the ability of agonists and caffeine to release Ca^{2+} and influence force. J Physiol 489: 327–335

Burdyga TV, Taggart MJ, Crichton C, Smith GI, Wray S 1998 The mechanism of Ca^{2+} release from the SR of permeabilised guinea-pig and rat ureteric smooth muscle. Biochim Biophys Acta 1402:109–114

Carl A, Lee HK, Sanders KM 1996 Regulation of ion channels in smooth muscles by calcium. Am J Physiol 271:C9–C34

Gordienko DV, Bolton TB, Cannell MB 1998 Variability in spontaneous subcellular calcium release in guinea-pig ileum smooth muscle cells. J Physiol 507:707–720

Imaizumi Y, Muraki K, Watanabe M 1989 Ionic currents in single smooth muscle cells from the ureter of the guinea-pig. J Physiol 411:131–159

Herrera GM, Nelson MT 2002 SR and membrane currents. Wiley, Chichester (Novartis Found Symp 246) p 189–207

Kuemmerle JF, Murthy KS, Makhlouf GM 1994 Agonist-activated, ryanodine-sensitive, IP$_3$-insensitive Ca^{2+} release channels in longitudinal muscle of intestine. Am J Physiol 266: C1421–C1431

Lang RJ 1989 Identification of the major membrane currents in freshly dispersed single smooth muscle cells of guinea-pig ureter. J Physiol 412:375–395

Large WA, Wang Q 1996 Characteristics and physiological role of the Ca^{2+}-activated Cl$^-$ conductance in smooth muscle. Am J Physiol 271:C435–C454

Lipp P, Niggli E 1996 Submicroscopic calcium signals as fundamental events of excitation-contraction coupling in guinea-pig cardiac myocytes. J Physiol 492:31–38

Maggi CA, Giuliani S, Santicioli P 1995 Effect of the Ca^{2+}-ATPase inhibitor, cyclopiazonic acid, on electromechanical coupling in the guinea-pig ureter. Br J Pharmacol 114: 127–137

Muraki K, Imaizumi Y, Watanabe M 1994 Effects of noradrenaline on membrane currents and action potential shape in smooth muscle cells from guinea-pig ureter. J Physiol 481:617–627

Nelson MT, Quayle JM 1995 Physiological roles and properties of potassium channels in arterial smooth muscle. Am J Physiol 268:C799–C822

Prishchepa LA, Burdyga TV, Kosterin SA 1996 Two components of sodium azide-insensitive Mg^{2+}, ATP-dependent Ca^{2+} transport in ureteral smooth muscle membrane structures (translated from Russian). Biokhimiia 61:1250–1256

Rose JG, Gillenwater JY 1973 Pathophysiology of ureteral obstruction. Am J Physiol 225:830–837

Santicioli P, Maggi CA 1998 Myogenic and neurogenic factors in the control of pyeloureteral motility and ureteral peristalsis. Pharmacol Rev 50:683–722

Shuba MF 1981 Smooth muscle of the ureter: the nature of excitation and the mechanisms of action of catecholamines and histamines. In: Bulbring E, Brading AF, Jones AW, Tomita T (eds) An assessment of current knowledge. Edward Arnold, London, p 377–384

Weiss RM 1986 Physiology and pharmacology of the renal pelvis and ureter. In: Walsh PC (ed) Campbell's Urology. Saunders, Philadelphia, PA, p 94–128

Wray S, Kupittayanant S, Shmigol T 2002 Role of the sarcoplasmic reticulum in uterine smooth
   muscle. Wiley, Chichester (Novartis Found Symp 246) p 6–25

DISCUSSION

Sanders: Is the shortening of the plateau of the action potential due to activation of the long conductance Ca^{2+}-activated K^+ (BK) channels?

Burdyga: Yes. In the guinea-pig ureter application of caffeine (1–2 mM) increases the frequency of Ca^{2+} sparks and STOCs, and under current-clamp conditions this causes a decrease of the plateau component of the action potential, thus bringing down the Ca^{2+} and force. In contrast, CPA inhibits Ca^{2+} sparks and STOCs, and increases the duration of the plateau component of the action potential. This results in an increase in the duration of the Ca^{2+} transient and amplitude and duration of the phasic contraction.

Sanders: You are suggesting that during this action potential event, the Ca^{2+} that is entering has to be loaded into the store and then released, rather than having a plateau on the action potential that would maintain membrane potential at a point where you would get Ca^{2+} entry.

Burdyga: That's my feeling. I can clearly see that when I apply caffeine it induces Ca^{2+} sparks and Ca^{2+} waves. If I stimulate the cells using high-K^+ solution to initiate action potential I can see that Ca^{2+} transient rises instantly in all parts of the cell. In most cases the amplitude of Ca^{2+} signal is smaller and the kinetics of the Ca^{2+} rise is slower in the area of the major initiating site, and CPA eliminates the difference, so it must be acting as a buffer.

Brading: This is a lesson to all of us. What seems to be happening in the ureter is that all of these mechanisms are occurring, but they have virtually nothing in the way of physiological relevance. These are structures that are not normally modulated by nerves. The action potentials are initiated up in the kidney and you want to get propagation down the ureter rapidly and synchronously to get urine moved. They are not being modulated. All of these events we can make happen in the SR under conditions which are non-physiological may show us about the sort of mechanisms that are taking place, but they probably have very little relevance to the normal physiology.

Burdyga: Let's talk about the physiology. In order to produce the peristaltic wave to drive the bolus of urine from kidney to bladder, a long-lasting action potential is necessary. This is because we need to have several centimetres of this muscle segment contracting. The nature of the action potential should be explained on the basis of the ionic mechanisms that are underlying this. Ureter is unique: it has non-inactivating L-type Ca^{2+} channels and the only repolarizing current is Ca^{2+}-activated K^+ current via the BK channels. The density of BK channels in the guinea pig ureter is low — the total K^+ current is only about 7 μA per cm^2 (in bladder it is about 200). So everything favours the plateau-type action potential. Agonists stimulate guinea-pig ureter by increasing the duration of the plateau component of the action potential. And this is achieved via inhibition of K_{Ca} current (Muraki et al 1994). So, agonists modulate force in the guinea-pig ureter via modulation of the duration of the plateau component of the action potential.

Brading: Yes, but you don't normally get agonists do you? If you add acetylcholine or histamine, these are pathologies not physiologies. There aren't nerves that are releasing transmitters as far as we know. If you try to stimulate nerves, nothing happens; transmural electrical stimulation of the ureter has no effect, and doesn't cause transmitter release.

Burdyga: What about histamine?

Brading: That might be pathology, if there is a stone or something and there are local mast cells.

Fry: This discussion is fruitless. What is being described here are attempts to use action potential manipulation to see what happens. For example, it is known that in the renal pelvis the action potential is longer than it is in the rest of the ureter. It gets shorter and shorter in the ureter, so what Dr Burdyga is doing is modulating action potential duration, which is a physiological phenomenon. This may be done by means of several alternative methods, but each involves looking at the actual event, and will produce its peristaltic wave because it will be refractory retrogradely. He is doing it by a method which can't be used *in vitro* because you can't take all the cells and put them together.

Somlyo: So which one is the better physiological method? Which one is the most effective chemical agent to contract the guinea-pig ureter?

Burdyga: Ureter has a myogenic mechanism that controls its peristaltic activity. The action potential in the main ureter is triggered by pacemaker potentials generated spontaneously by pacemaker cells from the pelvic region. Agonists are not needed to trigger this activity. However, they strongly modulate the EC coupling in the ureter via modulation of the ionic channels.

Somlyo: What about the contraction that is independent of the plateau of the action potential?

Burdyga: In the guinea-pig ureter force is exclusively controlled by action potential. Agonists just modulate it by increasing the duration of the plateau component of the action potential. Histamine for example can increase plateau component duration up to 10 times.

Somlyo: But what about if you depolarize the guinea-pig ureter? I am driving at the following. You say that it has no $InsP_3$ receptors. In that case, is histamine acting solely by electromechanical coupling? Could there be another mechanism that may have a different transmission line other than $InsP_3$, that might be interesting? Does guinea-pig ureter that is depolarized with high-K$^+$ so it no longer generates action potentials respond with contraction when stimulated by histamine or other agonists? If so, it would suggest that it has only electromechanical and no pharmacomechanical coupling mechanisms.

Burdyga: Guinea pig ureter in many ways resembles heart where the agonists modulate EC coupling on the level of ionic channels. In contrast, in rat ureter there is an action-potential-independent, nifedipine-resistant component of force evoked by agonists. So, when rat ureter is placed in high-K$^+$ nifedipine-containing solution and stimulated with carbachol it will generate both phasic (associated with Ca^{2+} release mediated by InsP$_3$) and tonic (associated with Ca^{2+} entry) contraction in response to carbachol. And in my talk I was making emphasis on species dependence of the mechanisms controlling EC coupling in ureter muscle.

Sanders: Did you say that there is no pharmacomechanical coupling? Don't you think there is a Ca^{2+} sensitization mechanism in this?

Burdyga: Not in the guinea-pig ureter, but it is present in rat ureter.

Somlyo: Skeletal muscle and cardiac muscle are probably among the few systems where there isn't an InsP$_3$-coupled system. Most non-muscle cells have this in the ER.

Eisner: Is there any difference between the way that rats and guinea-pigs urinate that might relate to this?

Paul: Rats continually spot urinate and guinea-pigs urinate in larger, less frequent bursts.

Brading: All these mechanisms are there but it doesn't necessarily mean that they are used.

Paul: Why have all these mechanisms there if they are not used? They must have a role—we just can't work out what it is. Economically, it would seem very foolish to retain all these factors if they do not have a role.

Brading: There has to be a selective advantage to generate a mechanism for not expressing them.

Raeymaekers: I think one should look at a cell type as a type of pattern of interaction of many elements: genes and proteins. A specific cell type corresponds to a specific pattern of genes that are switched on. The expression of indispensable gene products may result in the coexpression at low levels of other genes, caused by the prevailing set of transcription factors. The complexity of the system may not allow the complete suppression of the expression of all dispensable gene products. As long as their expression does no harm, there is no disadvantage for the cell having them there, or it may be less costly than changing their regulatory mechanisms.

Paul: I am maintaining they are doing harm. Maintenance of those proteins is causing the organism to waste a lot of metabolism. There should be an evolutionary pressure to lose them if they aren't used. There's an energetic cost in maintaining vestigial systems.

Raeymaekers: What is selected is a whole pattern of interactions, which includes some residual expression.

Brading: Everywhere you look there are non-functional factors. Why do the dorsal root ganglia have hundreds of receptors on their cell membrane? You can't assume that just because they are present they will be used.

Reference

Muraki K, Imaizumi Y, Watanabe M 1994 Effects of noradrenaline on membrane currents and action potential shape in smooth muscle cells from guinea pig ureter. J Physiol 481:617–627

General discussion II

The physiological significance of smooth muscle Ca^{2+} stores

Hirst: I want to present the argument that the Ca^{2+} stores in gastric antral smooth muscle cells have no physiological significance as a source of Ca^{2+} for excitation, but are present and they can be activated by a range of non-physiological mechanisms, so providing a source of Ca^{2+}. I would like to suggest that this might be true with much of the intestine. Gastric antrum is rhythmically active and generates slow waves, in the absence of stimulation slow waves occur regularly. Each has a long-lasting membrane potential change that is associated with an increase in Ca^{2+} and a contraction. If L-type Ca^{2+} channels are blocked in the tissue, the slow wave continues totally unabated. The associated increase in Ca^{2+} contraction is reduced as is the contraction. So Ca^{2+} is entering smooth muscle cells and triggering a contraction. Tomita showed a long time ago that if ryanodine receptors (RyRs) are blocked in this tissue, by soaking the preparation for several hours in ryanodine, the slow waves don't alter, there is a slight increase in baseline tone and the contractions are little changed. This would suggest that RyRs are not important in the generation of slow waves or the associated contractions. In mutant mice that lack inositol-1,4,5-trisphosphate (InsP$_3$) type 1 receptors, slow waves are totally absent. It could be argued from this that the InsP$_3$-dependent stores are needed in the smooth muscle cells to give slow waves. This turns out to be completely wrong. For slow waves, we need InsP$_3$ type 1 receptors to be present in some cells in the intestine. The question is, is the important InsP$_3$-dependent Ca^{2+} store mechanism present in smooth muscle cells, or is it present in a different type of cell? It turns out that as well as smooth muscle cells in the intestine, there are interstitial cells of Cajal (ICCs). These can be identified using an antibody to Kit. Interstitial cells form a network in the myenteric region and are also distributed through the circular muscle bundle. Kenton Sanders and his colleagues have shown that in a mutant mouse, which lacks ICCs in the myenteric region, that pacemaking is absent. Thus the InsP$_3$ receptors that are involved in the generation of slow waves are not those in smooth muscle cells: it must be InsP$_3$ receptors in the ICCs. I have taken a different approach. I have recorded from these interstitial cells and the smooth muscle cells that they communicate with. Interstitial cells generate pacemaker activity which flows passively into the circular muscle and triggers a slow wave. The pacemaker activity produces initial depolarization, and then the circular muscle bundle augments this. Perhaps we are using InsP$_3$ receptors in smooth muscle cells to augment the depolarization reaching the circular layer. If, again, we look at the mutant mouse used by Kenton Sanders, he used the intestine where myenteric pacemaker ICCs are absent, and the intramuscular ICCs are present. In the stomach of the same mutant mice, the situation is reversed. Myenteric pacemaker ICCs are present and intramuscular ICCs are absent. In the stomach we detect the initial depolarization coming through from the pacemaker ICCs but the second component is absent. We conclude that the whole of the slow wave is generated not by smooth muscle cells but by interstitial cells. These cells depend critically on the handling of InsP$_3$-dependent Ca^{2+} that the smooth muscle cells have no excuse for using.

What are we left with? The innervation. Perhaps this drives and uses Ca^{2+} stores within the muscle layer. Shaun Ward and Kenton Sanders have recorded the response to stimulating the nerves in the stomach of a wild-type mouse. This produces an excitatory response and an inhibitory response. Inhibition can be blocked by taking out the nitric oxide pathway, which leaves an excitatory junction potential. This could be using internal stores in smooth muscle cells. However, in the W/W^V mutant mouse where there are no intramuscular ICCs, nerve stimulation is without effect. The smooth muscles of the gastric antrum has stores present but physiologically we can see no access to the stores; the pacemaker mechanisms don't need the stores in the smooth muscle; transmission doesn't need the stores in the smooth muscle. I am making these points because I think there is good evidence that there are stores around, but I don't think we are asking the right physiological questions about how these are used. I have tried to convince you that we don't use stores for normal rhythmic activity and we don't need it for normal neuronal control mechanisms.

Bolton: Did you try this in response to acetylcholine on the WW prime mutant?

Hirst: That's not my work; it's Kenton Sanders and Sean Ward's. Yes, they did try this, and the depolarization is much reduced.

Bolton: How do you know that getting rid of the ICCs doesn't prevent the nerves coming into the muscle? There could be some sort of trophic effect.

Hirst: If you eserinise the preparation and stimulate the nerves, then a response is restored.

Bolton: They may not be making contacts. They may not be functional.

Sanders: They release acetylcholine to the same extent. We measured the output of acetylcholine from the nerves, and this was the same in wild-type and mutant animals. It appears that the close association between the nerve terminals and the interstitial cells is very important. As acetylcholine is released, it is broken down by the esterase. If the esterase is inhibited, a response in the smooth muscle can be seen, but this is not physiological.

Bolton: Perhaps when you get rid of the ICCs, which have Kit, a tyrosine kinase involved in developmental cell–cell association, you are doing something to the nerves that makes them non-functional.

Sanders: But the nerves can still be stimulated and they still release acetylcholine, so they are not non-functional.

Bolton: How do you measure the release of acetylcholine?

Sanders: Tritiated choline overflow.

Hirst: We can demonstrate that the nerves are releasing acetylcholine. If we block the esterases we get a slow response in the muscle. The acetylcholine can get there; it just doesn't normally get there.

Bolton: Why doesn't the acetylcholine depolarize the smooth muscle when it is released from the nerves?

Hirst: Because it doesn't get there.

Bolton: But you said that you can detect it in the bath.

Brading: That's when cholinesterase is added.

Sanders: The esterase destroys acetylcholine as an effective agonist, so you can still see the consequences of acetylcholine release by measuring tritiated choline, but it is not an effective agonist.

Somlyo: A technical question. Is this a single line of knockouts? If so, what is the genetic background? Are they compared with congenic mice?

Sanders: That's a good question. When we did these experiments we used siblings as the control. There are heterozygotes and wild-types from the same family.

Somlyo: It takes about a year, at very best, doing the speed congeneic line. We have been burned with knockouts in the past.

Sanders: The WW^{V}s are natural mutants. They have been around for a long time.

Paul: How does the smooth muscle contract? I'm missing something here.

Sanders: The slow wave conducts into the smooth muscle cell, which activates voltage-dependent Ca^{2+} entry.

Paul: And that's not modulated by any of the stores or long-conductance Ca^{2+}-dependent K^+ (BK) channels?

Hirst: As far as we can tell it isn't.

Somlyo: Is there any Ca^{2+}-induced Ca^{2+} release (CICR)?

Sanders: Not in gastric cells.

Somlyo: Not in the knockouts? Remember, one of the confusing things is the redundancy in pathways. Under normal conditions there is no CICR, but perhaps in this knockout it may occur.

Hirst: For any species that is known, if you do what you can to knock out ryanodine stores, this doesn't affect rhythmicity and contractility.

Kotlikoff: It would be nice to do the other experiment where $InsP_3$ is selectively removed in the muscle, but allowing it to be maintained in the ICCs.

Hirst: That would be nice. It is possible that there is Ca^{2+} entry triggering release from an $InsP_3$ store not linked to ryanodine. As far as I can gather that is unlikely to happen.

McCarron: What is the nifedipine-resistant increase in Ca^{2+} and contraction?

Hirst: That reflects Ca^{2+} cycling in the interstitial cells.

McCarron: But you still had contraction on the smooth muscle cells.

Hirst: This is where Kenton Sanders and I disagree. I would argue that ICCs can generate $InsP_3$ and pass it to smooth muscle cells. Some people think this isn't possible, but if it is, then we are starting to use the intracellular Ca^{2+} stores in the smooth muscle.

Fry: Are you saying that the gastric cells have no internal release mechanisms?

Hirst: No, I said there is no evidence from simple physiological experiments that these are used.

Fry: I don't think you can write them off. Fine, you have one model system where you say they don't play a significant role, but what about in the other cells where they do? Can you extrapolate your results to the entire subset of smooth muscle cells?

Hirst: No. Professor Iino has evidence in the vasculature, where noradrenaline is a transmitter, that the stores are activated. Mark Nelson has evidence that stretch depolarization involved the ryanodine stores. It is not clear that with most of the pharmacological experiments you are doing you are in any way mimicking a physiological response. I have seen people putting noradrenaline on resistance vessels, but as far as I know this never happens in the body. Noradrenaline release in sympathetic nerves doesn't have access to the noradrenaline receptors in the arterioles. The amounts of catecholamine needed are far higher than you ever get physiologically. We are talking about concentrations of $10\,\mu M$.

Somlyo: It is possible in smooth muscle preparations to do high frequency stimulation of nerves, and get a contraction that is abolished by TTX. This suggests that the transmitter does get to the smooth muscles. This is a physiological experiment.

Hirst: That is an example. There is no doubt that some transmitters get to arterioles, but it is equally not clear to me that in a mesenteric arteriole noradrenaline accesses the α receptors. ATP might access purine receptors and trigger voltage-dependent Ca^{2+} entry, but we have been told that CICR doesn't necessarily occur there. It is not clear that this pathway is used. It is, however, used to modulate depolarization.

Sanders: We did a frequency–response curve in the knockout mice. We were still unable to elicit responses.

Somlyo: I wasn't talking about your knockout; I was responding to the rather strong statement that noradrenaline in vascular smooth muscle cells is not a natural transmitter.

Hirst: I didn't make that statement. I said it didn't act as a post-junctional transmitter in a fine resistance artery. In the vein I would say that noradrenaline activates an α receptor and goes through an $InsP_3$-dependent pathway. If you look physiologically there are clear examples where internal stores are involved in neurotransmission. What I am saying is that the places where internal stores are present seem to me to exceed the physiological requirement for them.

Somlyo: Is there anyone here who has done old-fashioned hind limb vascular resistance experiments.

Hirst: Yes, I have. I have stimulated the sympathetic changes with physiological frequencies of stimulation, loaded the animals with benextramine, an irreversible α blocker, and I can see no changes in the neuronal responses although the responses to circulating catecholamines are abolished.

Somlyo: Do you get a response in the absence of α blocker?

Hirst: The neuronal response isn't changed by α blockade. We do get a neuronal response, which is purinergic. This suggested that any noradrenaline released reaches the smooth muscle in such a low concentration that it is without effect.

Brading: You don't have to be using the mechanisms that are there. There are many cases where normal physiology doesn't seem to be using these mechanisms, and in other cases it is.

Somlyo: Have you shown that the purinergic receptor does not activate these pathways that release Ca^{2+}?

Hirst: If you take a piece of arteriole, stimulate the purinergic nerves to evoke an excitatory junction potential, measure the diameter of the arteriole, and apply a voltage clamp, you get no contraction. If you let the membrane potential run into levels where voltage-dependent Ca^{2+} channels are activated, there is a contraction.

Eisner: You are saying that in these gut smooth muscles the sarcoplasmic reticulum (SR) is not relevant to normal physiological contraction. Earlier, Sue Wray showed that in the uterus SR release wasn't involved in normal contraction (Wray et al 2002, this volume). This is a meeting about the SR. In which smooth muscle is there unequivocal evidence that the SR plays a significant role in contraction?

Brading: What about bladder?

Nelson: Certainly bladder.

Hirst: Veins. Pulmonary arteries.

Lompré: Ashida et al (1988) showed a long time ago that the SR was important in large arteries, and much less in smaller vessels.

Blaustein: A question: is the aorta an artery? It is a big tube that doesn't have to do much a lot of the time. It might have less SR than smaller arteries.

Somlyo: No, it has quite a large volume of SR.

Lompré: There is more SR in the aorta and large conduit vessels than in muscular arteries.

Somlyo: To my knowledge there have been very few experiments done under what David Hirst would accept as physiological conditions.

Wray: Can we take the discussion one step back? We have heard about the horrors of cultured cells, and the difficulty of doing anything in intact tissue. Ted Burdyga, you have comparisons of single cells and intact bundles. How robust a model are single cells for the intact muscle preparations?

Burdyga: When you isolate single cells there is the risk of some remodelling. Caffeine responses in single cells exceed those produced by high K^+, and the situation is reversed when you work with multicellular preparations. I think there are some changes, but they are not too drastic if the cells are freshly isolated and used straight away. They retain all the basic functions. But if you keep them for several hours something might change.

Wray: But by and large you saw the same processes in the intact bundle?

Burdyga: Yes, I did.

Brading: The problem here is that if you just look at bits of ureter, on the whole they don't generate spontaneous action potentials unless they are very damaged. The mechanisms may be there in isolated cells and bundles of cells, but whether or not they are used is a completely different matter.

Burdyga: There's one issue we'd both agree on: when we blocked Na^+ pump activity there was a rise of Na^+ and the Na^+/Ca^{2+} exchanger went to the 'Ca^{2+} entry' mode. In the guinea-pig ureter there is a complete abolition of the action potential under these conditions. The SR is overloaded, and instead of positive inotropic action it produces inhibition of phasic contraction. Excitability almost goes to zero. This doesn't happen in rats. Also, at high Ca^{2+} in the guinea-pig there is a decreased excitability of the cells, but in rats it is fine. In the guinea-pig zero Na^+ abolishes the plateau component of the action potential, but in rat it potentiates it. I don't know how physiological it is, but this is the way it is.

Brading: They are both doing exactly the same thing: they are both conducting action potentials normally, and the fact that there is a difference in the basic mechanisms may be relevant to pathology rather than physiology. It may become quite important for clinical situations, but in this case we really only need to know about humans.

McHale: Isn't it true, nonetheless, that redundancy is what physiology is all about? For example, we have found that using two different species of animal (cattle and sheep) the same end result can be achieved with two completely different neurotransmission systems. It is also helpful to work on isolated tissues and on isolated cells, because of the much greater control afforded by reductionist techniques.

Brading: I don't think anyone is disputing this. But we want to eventually get to the stage where we can manipulate the system to produce treatments for human disease.

References

Ashida T, Schaeffer J, Goldman WF, Wade JB, Blaustein MP 1988 Role of the sarcoplasmic reticulum in arterial contraction: comparison of ryanodine's effect in a conduit and muscular artery. Cic Res 62:854–863

Wray S, Kupittayanant S, Shmigol A 2002 Role of the sarcoplasmic reticulum in uterine smooth muscle. In: Role of the sarcoplasmic reticulum in smooth muscle. Wiley, Chichester (Novartis Found Symp 246) p 6–25

The sarcoplasmic reticulum and smooth muscle function: evidence from transgenic mice

R. J. Paul*, G. E. Shull† and E. G. Kranias‡

Departments of *Molecular and Cellular Physiology, †Molecular Genetics, Biochemistry and Microbiology, ‡Pharmacology and Cell Biophysics, University of Cincinnati College of Medicine, 231 Albert Sabin Way, Cincinnati, OH 45267-0576, USA

Abstract. Smooth muscle Ca^{2+} handling is of major importance to understanding its function. A new approach utilizes molecular biology to develop transgenic mouse models in which the protein constituents of the various Ca^{2+} regulatory subsystems have been altered. Gene-targeted or gene-ablated (knockout) mice have been reported for the sarcoplasmic reticulum (SR) Ca^{2+} pump isoforms SERCA2, SERCA2a and SERCA3, the plasma membrane Ca^{2+} pump isoforms, PMCA1, PMCA2 and PMCA4, and the SR-associated protein, phospholamban (PLB), an inhibitor of SERCA2. A mouse line carrying a transgene for the smooth muscle specific expression of PLB has been reported. Evidence from studies using these mice combined with the classical pharmacological approaches has provided new insight into the relative role of the SR. We review this field with particular emphasis on PLB, since its modulation of SR function and smooth muscle contractility has the largest database. PLB via modulation of SERCA can play a major role in regulation of both phasic and tonic smooth muscle contractility. The use of transgenic mice has yielded surprises such as PLB modulation of endothelial cell Ca^{2+} homeostasis, and the demonstration that PLB is the major site for A-kinase-mediated relaxation of mouse bladder. The use of these gene-altered models has provided evidence clearly implicating a major role for the SR in modulating smooth muscle Ca^{2+} and contractility, with the caveat that this modulation is tissue specific.

2002 Role of the sarcoplasmic reticulum in smooth muscle. Wiley, Chichester (Novartis Foundation Symposium 246) p 228–243

Ca^{2+} is the major second messenger in the activation of smooth muscle contraction. Thus smooth muscle Ca^{2+} handling is of major importance to understanding its function. Ca^{2+} homeostasis is a balance of Ca^{2+} influx and extrusion. Influx is generally through channels, such as L- or T-type Ca^{2+} channels or the so-called capacitive entry pathway, through stretch activated channels, leak pathways and reversed mode Na^{+}/Ca^{2+} exchanger, which may also play roles under certain conditions. Mechanisms for Ca^{2+} extrusion are all energy-dependent processes for there is a very large electrochemical gradient favouring Ca^{2+} influx. These include the plasma membrane Ca^{2+} pump or Ca-ATPase (PMCA) and Na^+/Ca^{2+} exchange coupled with the Na/K-ATPase. In addition, the intracellular Ca^{2+} stores, the endoplasmic reticulum (ER) or sarcoplasmic reticulum (SR) are in a position to modulate the intracellular Ca^{2+} concentration ($[Ca^{2+}]_i$). Uptake of Ca^{2+} into these stores is powered by an associated Ca-ATPase called the sarco(endo)plasmic Ca-ATPase or SERCA. The SR has also been postulated to play a role in Ca^{2+} extrusion from the cell, by vectorial release into a subsarcolemmal space, and subsequent extrusion via Na^+/Ca^{2+} exchange. Mitochondria have also been suggested to play a role in regulation of $[Ca^{2+}]_i$, but this remains controversial. Due to space limitations, citations cannot be exhaustive and we apologize in advance for any omissions (for excellent reviews see Jaggar et al 2000, Kiriazis & Kranias 2000, Paul 1998, Raeymaekers & Wuytack 1996, Shull 2000, van Breemen et al 1995, Wray et al 2001).

There is considerable interest as to the role that each of these Ca^{2+} handling pathways play in cellular Ca^{2+} homeostasis. This is not only of fundamental importance at the basic science level, but also in terms of development of therapeutics. The relative contributions of these Ca^{2+} handling subsystems are likely to be specific to the type of smooth muscle, thus potentially permitting the therapeutic targeting of different organs. Several approaches have been used to understand the relative contributions of these pathways. The most common approach is to disrupt each system pharmacologically then assess changes in Ca^{2+} and smooth muscle contractility. In the earlier literature, the latter was often used as an index of $[Ca^{2+}]_i$ and must be viewed cautiously, as it is now apparent that changes in Ca^{2+} sensitivity of the activation of the contractile proteins is a major factor in determination of contractility. Perhaps the most successful example of this approach is the use of Ca^{2+} antagonists, such as nifedipine or diltiazem, which block the L-type Ca^{2+} channel and have been used to determine the role of Ca^{2+} influx under various conditions. Much attention has also focused on the SR, since the most specific of our pharmacological tools target either: (1) SERCA, for example, with cyclopiazonic acid (CPA), thapsigargin (TG), or UBQ; (2) the SR ryanodine Ca^{2+} release channel, with caffeine or ryanodine; or (3) the SR inositol-1,4,5-trisphosphate ($InsP_3$) Ca^{2+} release channel, with 2-APB (2-aminoethoxydiphenyl borate) or xestospongin. Specificity of these drugs is a central shortcoming of this approach, and though it is reasonably well defined for CPA, TG and ryanodine, the action of the latter drugs is not as well documented. The lack of well characterized drugs targeting PMCA or the Na^+/Ca^{2+} exchange is also a limiting factor.

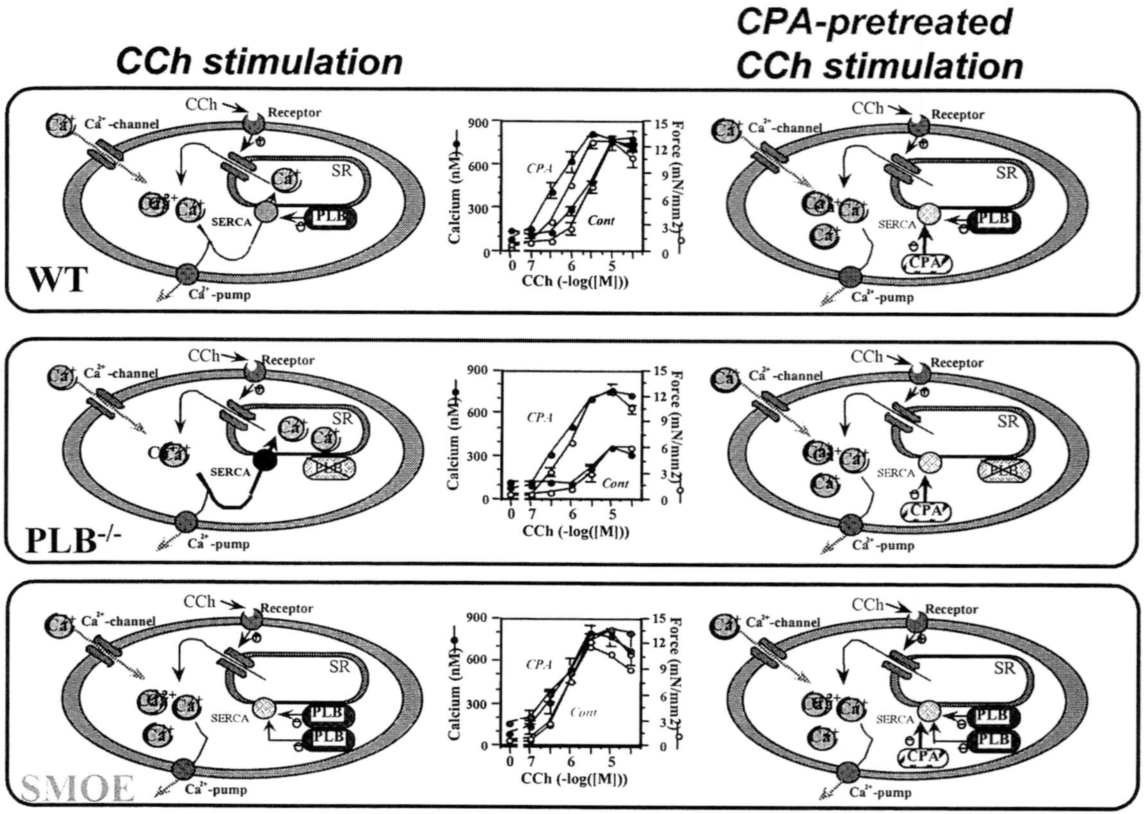

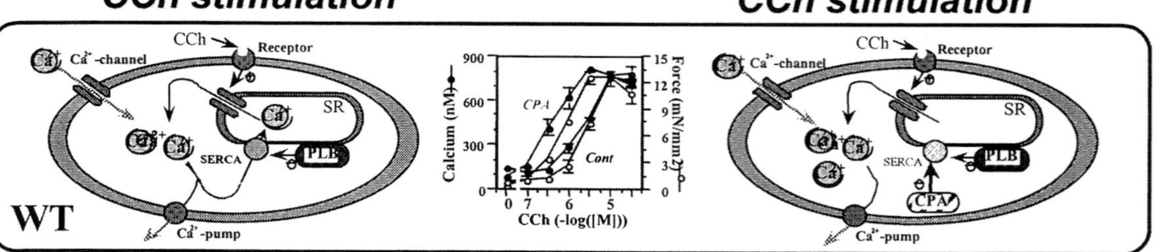

CCh stimulation
CPA-pretreated CCh stimulation
CCh
Receptor
Ca²⁺-channel
Ca²⁺-pump
SERCA
PLB
SR
CPA
Cont
Calcium (nM)
Force (mN/mm2)
CCh (-log([M]))
WT
PLB⁻/⁻
SMOE

Transgenic mice models

A new approach based on the use of molecular biology allows for the development of transgenic mouse models in which the protein constituents of the various Ca^{2+} regulatory subsystems have been altered. Gene-targeted or gene-ablated (knockout) mice have been reported for the SR Ca^{2+} pump isoforms SERCA2 (Periasamy et al 1999), SERCA2a (Ver Heyen et al 2001) and SERCA3 (Liu et al 1997), the plasma membrane Ca^{2+} pump isoforms, PMCA1, PMCA2 and PMCA4, and the SR associated protein, phospholamban (PLB) (Luo et al 1994), an inhibitor of SERCA2. In addition, a mouse line carrying a transgene for the smooth muscle specific expression of PLB has been developed (Nobe et al 2001). Evidence from studies using these mice combined with the classical pharmacological approaches has provided new insight into the relative role of the SR. Moreover these studies provide a unique line of attack to the understanding of the specific functions of Ca^{2+} pump isoforms. Our goal is to review the reported data from these recent studies in light of the symposium topic of the role of the SR in smooth muscle.

Phospholamban

PLB is a 52 amino acid protein associated with the SR. This was the first SR related gene-altered mouse model and has been intensively studied (Luo et al 1994). It is an ideal model to test the protein–function relation in that the function of PLB is modulatory and apparently not critical in the sense of not being lethal. The heterozygous null animals show no gross anatomical or functional alterations. There is considerable evidence on the biochemical level which shows that PLB is

---

FIG. 1. (*Opposite*)   Schematic model of phospholamban (PLB) modulation of SERCA and consequent smooth muscle contractility based on data from wild-type (WT) and gene-altered mice, phospholamban knockout (PLB$^{-/-}$), and smooth muscle-specific overexpression (SMOE). Graphic data modified from Nobe et al (2001). Ordinate: left, intracellular calcium concentration; right axis, isometric force. Abcissa: carbachol (CCh) concentration. (Upper panel) wild-type: $[Ca^{2+}]_i$ (filled circles) and force (open circles) are both shifted leftward by the SERCA inhibitor, cyclopiazonic acid (CPA). Inhibition of a major Ca^{2+}-uptake system is proposed to lead to higher $[Ca^{2+}]_i$ for any given level of stimulation and Ca^{2+} input. (Middle panel) PLB$^{-/-}$: in the absence of PLB inhibition of SERCA, the CCh–force relation lies to the right of the wild-type. Thus an increased SR Ca^{2+}-uptake yields a lower $[Ca^{2+}]_i$ for any given level of Ca^{2+} input compared to the wild-type. Importantly, inhibition of SERCA yields CCh–force relations identical to the wild-type (upper panel), indicating that the differences in contractility originated at the SR level. (Bottom panel) PLB overexpression: increased PLB inhibition of SERCA leads to a leftward shift in the CCh–force relation relative to the wild-type, similar to that elicited by CPA. Further addition of CPA is ineffective as SERCA is already significantly inhibited by the increased expression of PLB. Schematic courtesy of Dr Koji Nobe.

an inhibitor of cardiac SR and that this inhibition can be relieved by its phosphorylation. There was substantial information in the cardiac literature indicating a correlation between PLB phosphorylation and functional changes associated with β-adrenergic stimulation. The development of the PLB knockout mouse provided the first evidence, which showed at the *in vivo* level that this correlation could be considered causal.

While several studies reported that PLB was present in smooth muscle, very little was known about its role in Ca^{2+} handling. Some evidence suggested that in addition to A-kinase pathway phosphorylation, activation of the G-kinase pathway was associated with PLB phosphorylation. The latter was of particular interest to vascular smooth muscle, for which endothelium-dependent relaxation via nitric oxide (NO) made the mechanism of G-kinase-mediated relaxation of considerable physiological significance (Karczewski et al 1998).

The PLB knockout mouse also demonstrated a major role for PLB modulation of SR function and contractility in smooth muscle. In the tonic aorta (Lalli et al 1997) and bladder (Nobe et al 2001), isometric force vs. agonist relations from the knockout mouse were rightward of those in the wild-type. This decrease in sensitivity was mirrored by that of $[Ca^{2+}]_i$, as measured with Fura-2 fluorescence methodology (Lalli et al 1999). Theoretically, one would anticipate that in the absence of PLB, SERCA would operate without any inhibition and consequently a greater rate of Ca^{2+} uptake and potentially greater Ca^{2+} loading of the SR. It would appear that in the case of these tonic smooth muscles, increased Ca^{2+} uptake predominated, as $[Ca^{2+}]_i$ in the PLB knockout smooth muscles was lower than that of the wild-type for any given level of agonist. Our hypothesis is that for any given level of stimulus and concomitant Ca^{2+} input, Ca^{2+} uptake in the PLB knockout smooth muscles is greater than in the wild-type leading to a lower $[Ca^{2+}]_i$ and force. Figure 1 shows our model based on this hypothesis to explain the observations made with the PLB transgenic mice models.

An alternate hypothesis is that a greater SR Ca^{2+} load will lead to greater production of Ca^{2+} 'sparks', which in smooth muscle is postulated to lead to hyperpolarization of the membrane potential due to activation of Ca^{2+}-dependent K^+ channels, and consequently a rightward shift in the agonist–force relationship (Jaggar et al 2000). We also observed the decrease in sensitivity of the PLB knockout aorta when high KCl levels were used. Since K^+ channels would not be expected to hyperpolarize in high K^+ solutions, this suggests that the former hypothesis likely holds for these conditions. These hypotheses are not mutually exclusive. Further experimentation is needed to distinguish whether either or some combination underlie the observed lower sensitivity of force to stimulation in the PLB knockout smooth muscles.

The underlying assumption for interpretation of studies on gene-altered models is that only the protein of interest was modified. It is important to verify this, as compensatory mechanisms may in fact modify the interpretation of the data. Here classic pharmacology can be used as an important tool to study the extent of compensatory pathways. If the only changes in the PLB knockout were due to altered SR function, then we hypothesized that functional removal of the SR should lead to identical force–agonist relations in the knockout and wild-type. Treatment with CPA to block SERCA led to a leftward shift in the force–agonist relations in the wild-type. This might be anticipated as a major Ca^{2+} uptake system was functionally removed. However, the relations in the PLB knockout smooth muscles were shifted even further to the left, so that they were indistinguishable from those of the wild-type. This was of particular importance in the bladder, as not only the sensitivity, but also the maximum force was significantly depressed in the PLB knockout. Importantly, maximum force and $[Ca^{2+}]_i$ was returned to wild-type levels after CPA treatment. It should be noted that for receptor-mediated stimulation in the bladder, force versus carbachol post-CPA was restored to wild-type levels only in the generation of the initial force–carbachol concentration relation; a second concentration–force relation could not be generated, presumably due to depletion of SR Ca^{2+} in the continued presence of CPA. Another interesting point raised in these studies was that the depressed sensitivity of the aorta to KCl was maintained for long durations in the continued presence of KCl. One might have anticipated that the greater SR Ca^{2+} uptake in the PLB-KO could be nullified as the SR filled to capacity and SERCA itself became inhibited by the increase in the free Ca^{2+} content of the SR. The fact that this was not observed suggests that the SR indeed may not be saturated due to its vectorial unloading of Ca^{2+}. If so, then the SR may have to be considered not only a system for modulating $[Ca^{2+}]_i$ via Ca^{2+} uptake, but also as a major Ca^{2+} extrusion mechanism in these smooth muscles.

In the phasic portal vein, the PLB knockout mouse was used to show that PLB modulation of the SR also plays an important role in determination of its characteristic spontaneous mechanical activity. In the PLB knockout portal vein, the duration of the periods of activity and quiescence were both prolonged by approximately twofold compared to that in the wild-type (Sutliff et al 1997). Though speculative at this stage, these findings could be attributed to a combination of both greater SR Ca^{2+} loading (increased period of activity) and SR Ca^{2+} uptake (prolonged quiescent period). Again, treatment with CPA to remove SR function led to near equalization of the activities in PLB knockout and wild-type portal vein. Clearly the SR can modulate the spontaneous activity of phasic smooth muscles.

In view of the literature suggesting that both A- and G-kinase-mediated phosphorylation of PLB were correlated with relaxation of vascular smooth muscle, it was of considerable interest to assess the roles of PLB and the SR in cyclic nucleotide-mediated relaxation. It was thus somewhat of a surprise that activation of either pathway in aorta led to a relaxation in the PLB knockout that was indistinguishable from the wild-type. There were also only moderate differences in the portal vein from PLB knockout mice. The basis of this is not known, but clearly other mechanisms must dominate. Potentially the control of the PLB:SERCA ratio may be a method for long term regulation of SERCA. It has been reported that the PLB:SERCA ratio varies over a wide range in smooth muscles (Eggermont et al 1990), and may reflect this type of modulation.

Most interesting in this light are our initial experiments on mouse bladder showing the absolute dependence of β-adrenergic relaxation in the bladder to the presence of PLB. Isoproterenol or forskolin activation of the A-kinase pathway led to complete relaxation of bladder from wild-type mice, with little or no response in the PLB knockout bladder (K. Nobe & R. J. Paul, unpublished observations). On the other hand, relaxation via G-kinase pathway activation was identical in the PLB knockout and wild-type bladder. PLB modulation of SR appears to be the dominant pathway for A-kinase mediated relaxation in mouse bladder in contrast to its lesser role in this pathway for the vascular tissues studied.

Another interesting and surprising finding was that PLB was present in vascular endothelium and was a modulator of endothelium-dependent relaxation of the aorta (Sutliff et al 1999). This is one of the unanticipated and novel findings that are often associated with the use of gene-altered models. In the literature, PLB is reported to be present in cardiac, slow-twitch skeletal and smooth muscles. Though never explicitly stated, the prevailing opinion was that these represented the sole loci for PLB. We reported that the vascular endothelium also was a site for PLB. We found that the endothelium-dependent relaxation to acetylcholine in the aorta from PLB knockout mice was less sensitive than that of the wild-type. This led us to investigate whether PLB was present in the endothelium, which we confirmed first at the message and then protein levels. Since the production of NO by endothelial NO synthase is dependent on Ca^{2+}, modulation of endothelial $[Ca^{2+}]_i$ would affect NO production. Using similar arguments to those proposed for smooth muscle, endothelium in the PLB knockout with an uninhibited SERCA would likely have lower $[Ca^{2+}]_i$ and generally lower stimulated NO production. This also appeared to be true for basal NO production as the response to A-kinase pathway activation with forskolin was also decreased in the intact PLB knockout aorta. While the exact mechanism needs to be verified it is clear that the loss of PLB alters endothelial cell mediated relaxation of vascular smooth muscle.

Gene-targeted (knockout) mouse models provide one avenue for assessing the *in vivo* functional significance of particular proteins. A complementary approach is the expression of the protein of interest using transgenic methodology. There are a number of reasons for developing these models. Using tissue specific promoters, one can target smooth muscle for the expression of the transgene. We utilized the α actin promoter developed by Dr A. Strauch (Cogan et al 1995) to target PLB to smooth muscle tissues. This promoter is very robust, producing significant 4–10-fold increases in PLB expression in aorta and bladder (Nobe et al 2001). The transgenic mouse lines (PLB-SMOE) produced have varied levels of PLB expression thus permitting rigorous tests of the effects of various PLB:SERCA ratios to further delineate the functional significance of PLB. In addition the PLB-SMOE mice can be crossed with the PLB knockouts, yielding mice in which the smooth muscle PLB is similar to the wild-type ('rescued') while the other PLB-containing tissues remain PLB deficient. This leads to models in which tissue-specific effects of PLB can be tested. Finally, structure–function relations through expression of mutant PLBs can be studied (Zhai et al 2000).

The contractile function observed for smooth muscles of PLB-SMOE mice is complementary to those of the PLB knockouts (Fig. 1). The bladder shows a high level of transgene expression and is of sufficient size to permit relatively easy biochemical measurements, particularly Western blot analysis. Interestingly, the expression of the PLB transgene led to a down-regulation of SERCA. One might have anticipated that to maintain Ca^{2+} homeostasis, an increase in the inhibitor would lead to an up-regulation of SERCA. On the other hand, expression of ligand can lead to down-regulation of receptor, as the bound pair is often targeted for proteolysis. The latter may occur with increased PLB:SERCA ratios and consequent higher fraction of SERCA bound to PLB.

In functional terms, the sensitivity of isometric force of the PLB-SMOE smooth muscles in response to activation was increased, the opposite of the desensitization observed in the PLB-knockout. The $[Ca^{2+}]_i$ responses were also found to parallel force, suggesting that modulation of $[Ca^{2+}]_i$ was the underlying factor. Removal of SR function with CPA did not significantly shift the stimulus–response relations. Our interpretation is that with high PLB:SERCA ratios, SERCA is already significantly inhibited and further inhibition is less effective. These experiments suggested that the expressed PLB was properly incorporated into the SR and inhibited SERCA, although this assumption needed to be validated. If the expressed PLB was a functional inhibitor of SERCA, then phosphorylation by activation of the A-kinase pathway should remove this inhibition. Treatment with forskolin decreased the Ca^{2+} and force responses to levels similar to those observed in the forskolin-treated wild-type bladder. This indicated that the expressed PLB was not only inhibiting SERCA but displayed the hallmark relief of inhibition by A-kinase mediated phosphorylation. Expression of mutant PLB will be the next level in assessing the phosphorylation sites and the relative impact of the SR in the associated relaxation pathways.

Other gene-targeted mice models

SERCA isoforms arise from three genes (SERCA1, 2, and 3) and alternate splice products from SERCA2 (a and b). SERCA1 is expressed in fast skeletal muscle, SERCA2a is the sole isoform in the heart and slow skeletal muscle, and SERCA2b is ubiquitously expressed. Smooth muscle is reported to contain SERCA2a and 2b, but the evidence for the exact distribution is limited (Paul 1998). SERCA3 has a very restricted tissue distribution, and for our purposes is reported to be found in vascular endothelium and tracheal epithelium. The physiological bases for the existence of SERCA isoforms and their tissue-specific distribution is not known with certainty. The ubiquitous distribution of SERCA2b has led to speculation that it has a 'housekeeping' function, whereas the limited and tissue-specific distribution of the other isoforms suggests a more defined function. Knockout mice for SERCA2, SERCA2a and SERCA3 have been developed and have provided initial functional evidence for their role in smooth muscle contractility.

The SERCA2 homozygous null mouse (−/−) is lethal, but the heterozygous mouse has been useful in assessing the cardiovascular effects of a reduction of SERCA2 protein. Significant alterations in cardiac function were observed (Periasamy et al 1999). However, neither mouse aorta nor portal vein contractility were affected (Weber 1999). We verified that the SERCA protein was indeed reduced so the negative results may indicate that there is a substantial reserve of Ca^{2+} pump activity.

The SERCA3 knockout is viable and neither aortic nor tracheal smooth muscle exhibited any phenotype. These negative data validated the studies indicating that SERCA3 is not found in smooth muscle. On the other hand both endothelium-dependent relaxation of the aorta (Liu et al 1997) and epithelium-dependent relaxation of the trachea (Kao et al 1999) were altered in the SERCA3 knockout. Thus while the smooth muscle itself was not affected, the function of the integrated smooth muscle-containing tissue was dependent on SERCA3 function.

Finally, in attempting to assess SR function in light of cellular Ca^{2+} homeostasis, it is useful to consider transgenic mouse models of the other major Ca^{2+}-handling systems. PMCA isoform knockouts have been developed and our initial evidence for a major functional role is suggestive (G. E. Shull, G. J. Pyne, G. Okunade & R. J. Paul, unpublished observations). PMCAs arise from four genes with several potential isoforms attributable to alternative splice sites. So far, PMCA1, PMCA2 (Kozel et al 1998) and PMCA4 knockouts have been developed. PMCA1 shows a ubiquitous tissue distribution and not surprisingly the −/− mutation is lethal. The PMCA1$^{+/-}$ mouse is viable and our initial data indicate that smooth muscle tissues show a higher sensitivity to stimulation than the wild-type. This suggests that PMCA1 is a major player in smooth muscle Ca^{2+} homeostasis. This is in contrast to the negative results with the SERCA2$^{+/-}$ mouse. Interestingly, the PMCA4$^{-/-}$ mouse is viable. This isoform is reported to be present in smooth muscle, however our preliminary data indicated no obvious phenotypic differences in aorta or bladder.

Modulation of the SR Ca^{2+}-ATPase by PLB perhaps provides the clearest evidence of the significance of the SR to smooth muscle Ca^{2+} homeostasis and contractility. It also offers some cautions since the effects of modulation of SR function were clearly dependent on the smooth muscle tissue studied. In summary, genetically altered mouse models provide a new approach for assessing the physiological significance of the SR to smooth muscle function *in vivo*.

Acknowledgements

Supported by HL54829 (RJP), HL61974 (GES), & P40RR12358 (EGK).

References

Cogan JG, Sun S, Stoflet ES, Schmidt LJ, Getz MJ, Strauch AR 1995 Plasticity of vascular smooth muscle alpha-actin gene transcription. Characterization of multiple, single-, and double-strand specific DNA-binding proteins in myoblasts and fibroblasts. J Biol Chem 270:11310–11321

Eggermont JA, Wuytack F, Verbist J, Casteels R 1990 Expression of endoplasmic-reticulum Ca^{2+}-pump isoforms and of phospholamban in pig smooth-muscle tissues. Biochem J 271:649–653

Jaggar JH, Porter VA, Lederer WJ, Nelson MT 2000 Calcium sparks in smooth muscle. Am J Physiol 278:C235–C256

Kao J, Fortner CN, Liu LH, Shull GE, Paul RJ 1999 Ablation of the SERCA3 gene alters epithelium-dependent relaxation in mouse tracheal smooth muscle. Am J Physiol 277:L264–L270

Karczewski P, Hendrischke T, Wolf WP, Morano I, Bartel S, Schrader J 1998 Phosphorylation of phospholamban correlates with relaxation of coronary artery induced by nitric oxide, adenosine, and prostacyclin in the pig. J Cell Biochem 70:49–59

Kiriazis H, Kranias EG 2000 Genetically engineered models with alterations in cardiac membrane calcium-handling proteins. Annu Rev Physiol 62:321–351

Kozel PJ, Friedman RA, Erway LC et al 1998 Balance and hearing deficits in mice with a null mutation in the gene encoding plasma membrane Ca^{2+}-ATPase isoform 2. J Biol Chem 273:18693–18696

Lalli J, Harrer JM, Luo W, Kranias EG, Paul RJ 1997 Targeted ablation of the phospholamban gene is associated with a marked decrease in sensitivity in aortic smooth muscle. Circ Res 80:506–513

Lalli MJ, Shimizu S, Sutliff RL, Kranias EG, Paul RJ 1999 [Ca^{2+}]$_i$ homeostasis and cyclic nucleotide relaxation in aorta of phospholamban-deficient mice. Am J Physiol 277:H963–H970

Liu LH, Paul RJ, Sutliff RL et al 1997 Defective endothelium-dependent relaxation of vascular smooth muscle and endothelial cell Ca^{2+} signaling in mice lacking sarco(endo)plasmic reticulum Ca^{2+}-ATPase isoform 3. J Biol Chem 272:30538–30545

Luo W, Grupp IL, Harrer J et al 1994 Targeted ablation of the phospholamban gene is associated with markedly enhanced myocardial contractility and loss of beta-agonist stimulation. Circ Res 75:401–409

Nobe K, Sutliff RL, Kranias EG, Paul RJ 2001 Phospholamban regulation of bladder contractility: evidence from gene-altered mouse models. J Physiol 535:867–878

Paul RJ 1998 The role of phospholamban and SERCA3 in regulation of smooth muscle-endothelial cell signalling mechanisms: evidence from gene-ablated mice. Acta Physiol Scand 164:589–597

Periasamy M, Reed TD, Liu LH et al 1999 Impaired cardiac performance in heterozygous mice with a null mutation in the sarco(endo)plasmic reticulum Ca^{2+}-ATPase isoform 2 (SERCA2) gene. J Biol Chem 274:2556–2562

Raeymaekers L, Wuytack F 1996 Calcium Pumps. In: Barany M (ed) Biochemistry of smooth muscle contraction. Academic Press, New York, p 241–253

Shull GE 2000 Gene knockout studies of Ca^{2+}-transporting ATPases. Eur J Biochem 267:5284–5290

Sutliff RL, Hoying JB, Kadambi VJ, Kranias EG, Paul RJ 1999 Phospholamban is present in endothelial cells and modulates endothelium-dependent relaxation. Evidence from phospholamban gene-ablated mice. Circ Res 84:360–364

Sutliff RL, Kranias EG, Paul RJ 1997 Phospholamban gene ablation is associated with alterations in portal vein contractility. FASEB J 11:A25

van Breemen C, Chen Q, Laher I 1995 Superficial buffer barrier function of smooth muscle sarcoplasmic reticulum. Trends Pharmacol Sci 16:98–105

Ver Heyen M, Heymans S, Antoons G et al 2001 Replacement of the muscle-specific sarcoplasmic reticulum Ca^{2+}-ATPase isoform SERCA2a by the nonmuscle SERCA2b homologue causes mild concentric hypertrophy and impairs contraction–relaxation of the heart. Circ Res 89:838–846

Weber CS, Sutliff RL, Liu LH, Periasamy M, Shull GE, Paul RJ 1999 Vascular smooth muscle function in SERCA2 gene-ablated mice. Biophys J 76:A286

Wray S, Kupittayanant S, Shmygol A, Smith RD, Burdyga T 2001 The physiological basis of uterine contractility: a short review. Exp Physiol 86:239–246

Zhai J, Schmidt AG, Hoit BD, Kimura Y, MacLennan DH, Kranias EG 2000 Cardiac-specific overexpression of a superinhibitory pentameric phospholamban mutant enhances inhibition of cardiac function in vivo. J Biol Chem 275:10538–10544

DISCUSSION

Eisner: I would have thought that in the steady state the Ca^{2+} in the cytoplasm is determined by what happens at the surface membrane, but you have apparently got a PLB knockout decreasing the Ca^{2+} in the cytoplasm.

Paul: This is particularly the case in vascular tissue where this suppression can be maintained for hours. We have to get back to the calculation of how much the SR can hold. When I did the KCl contractions, my expectation was that I would saturate the SR in a reasonable period of time: that it couldn't take up any more Ca^{2+} because it would be self-inhibited. This was going to be my control. It was both disappointing and exciting to find out that the shift was maintained and disappeared with CPA. I am going back to the sort of vectorial unloading hypothesis, but it would be very consistent with Casey van Breemen's data if that were the case in this mouse tissue.

Blaustein: I was going to suggest the same thing: there's no reason to think that the SR is holding less Ca^{2+}, because the CPA experiment shows that when you unload this you see basically the same amount of Ca^{2+} here. The Ca^{2+} must have gone somewhere. Is it transported out simply because as more is put into the SR more gets extruded?

Eisner: So why is the second response lost after CPA?

Paul: It isn't lost in the aorta. The bladder, at least in our hands, will retain 80% of its contractility after nifedipine treatment, whereas the aorta would be down to 25% under similar conditions. My assumption is that this has a lot to do with recycling.

Eisner: Are you saying that most of the Ca^{2+} for contraction in the bladder doesn't come via the L-type channel?

Paul: Yes.

Eisner: Where is it coming from then? A store?

Paul: That would be my guess. If you leave it in the CPA for long enough you can get rid of the contraction. If you leave the Ca^{2+} free long enough you won't get a contraction either.

Brading: It doesn't take that long, either.

Fry: It would take about 20 min or so. If you put CPA on for that time the contraction will disappear.

Paul: I'm assuming that in the absence of a functional SR, excitation–contraction (EC) coupling is lost. The plasma membrane comes to play an important role in Ca^{2+} extrusion in the bladder.

Eisner: In that case, shouldn't the PLB overexpression give you a much smaller expression? There is presumably less Ca^{2+} in the SR in this case, and since 80% of the Ca^{2+} is coming from the SR, isn't this inconsistent with your data?

Paul: We are talking about decreased SR Ca^{2+} release and an impaired SR Ca^{2+} uptake and extrusion. I do not have the figures, but I hypothesize that the inhibited uptake is more important to the final steady-state intracellular Ca^{2+} concentration. There is 20% that is inhibitable by nifedipine. We have to keep in mind that these are not necessarily linear systems. We want to add together this contribution and the next one, but it depends on the Ca^{2+} affinities and the Ca^{2+} concentration. We may have different systems in at different points.

Blaustein: It seems to me to be a question of kinetics versus the steady-state condition. In the steady-state you will probably have the same amount of Ca^{2+} there, if you wait long enough, even though uptake is slower.

Eisner: But you wouldn't expect the PLB knockout to have less Ca^{2+} in the SR.

Blaustein: Does it have less Ca^{2+} in the SR? I think not. The CPA experiment would suggest that it has the same amount of Ca^{2+}.

Paul: It is just sucking it up faster. It is interesting. The reason it is inhibitory is that the Ca^{2+} uptake is a lot more powerful, and dominates in the steady-state. In just about every tissue it increases the force. How can it increase the force if there is no Ca^{2+} in the SR? The idea is that it is just sucking up and extruding Ca^{2+} to the extracellular space via the Na^+/Ca^{2+} exchanger.

Somlyo: The amount of loading really depends not only on the affinity of the Ca-ATPase/PLB system or cytosolic Ca^{2+}, but also on the internal Ca^{2+}. A long time ago Annemarie Weber showed that the uptake into the SR depends on the luminal Ca^{2+} concentration.

Paul: It will inhibit the SERCA pump. Again, we don't know at what level, nor do we know what the binding capacity is. Even to model it at this stage might be difficult.

Somlyo: I can make it worse! It will depend on the phosphate concentration in the cytosol and the availability of ATP for phosphate for uptake.

van Breemen: You can see this effect of potentiation by CPA really strongly if you first deplete the SR and then add Ca^{2+} back. Quite often you will see no contraction because all the Ca^{2+} that comes in is being sucked into the SR. If this experiment is repeated in the presence of CPA, then there is a huge contraction. In this particular case the Ca^{2+} doesn't get sucked back into the SR but goes to the myofilaments.

Paul: That is certainly the hypothesis that we are proposing.

Hellstrand: Are the effects of PLB overexpression identical to the effects of SERCA inhibition or those seen in the SERCA knockouts?

Paul: Gary Shull's SERCA knockout, which takes out both SERCA2A and 2B is lethal. The heterozygote nulls don't show much of a smooth muscle phenotype. One of the things that differs between inhibition of SERCA by overexpression of PLB and CPA is that the PLB inhibition can be overridden at high enough Ca^{2+} concentrations. With CPA the V_{max} of the pump is being competitively inhibited, so this gets rid of it totally. In the overexpressor the pump is still functional once you get to high enough intracellular Ca^{2+}.

Nelson: I would like to return to what David Eisner mentioned about the plasma membrane determining the steady-state free Ca^{2+}, and what Rick Paul said about sparks and long-conductance Ca^{2+}-dependent K^+ (BK) channels. We have looked at cerebral arteries from PLB knockout mice. The spark frequency and the associated transient BK current frequency are elevated by about a factor of three. SR load goes up, the membrane potential hyperpolarizes and the artery relaxes. It would be useful to measure membrane potential under all the conditions as well as determine the voltage dependence of tone, to make sure that your manipulations are not simply changing the membrane potential.

Paul: It is not clear that the bladder smooth muscle is dependent on voltage.

Nelson: Carbachol-induced constrictions are nicely relaxed by membrane hyperpolarization. I would say that measuring the membrane potential under these conditions is important. If it is more hyperpolarized this has to be taken into account.

van Breemen: It probably wouldn't explain the high K^+ effect.

Nelson: At high K^+ there is a convergence, but at lower K^+ the K^+ equilibrium potential and the membrane potential are different. A K^+ channel opener can still work at low (e.g. 20 mM) K^+.

Paul: I show a voltage difference of 4 mV, from 63 to 59, and I just don't have a good enough qualitative feel for these kinds of measurements to be able to know whether this could account for all the difference. If it does account for all the difference, then this is a super-sensitive system.

Nelson: 4 mV could certainly have a significant effect.

Young: Referring back to the uterus, we need to remember that the uterus is a secretory tissue and that prostaglandins are produced whenever the subplasmalemmal Ca^{2+} is raised. One of the effects of the CPA may be to raise the subplasmalemmal Ca^{2+} and increase the secretion of prostaglandins. By this mechanism the duration and force of the contractions will be increased, possibly without affecting the frequency.

Sanders: Do non-cumulative dose–response studies have any effect on your Ca^{2+} force curves? In other words, do you bias your data by doing cumulative dose–response curves towards Ca^{2+} sensitization? When there is a short exposure to the compound do you get a shift?

Paul: That's an interesting question. I have never done it in a non-cumulative fashion. My assumption was that it was always Ca^{2+} sensitized. The Ca^{2+} force curve seemed to be steeper than with the KCl. I was assuming that any level of receptor-mediated interaction would lead to sensitization.

Brading: We never use cumulative dose–response curves. If these are used, they are different to curves based on non-cumulative addition.

Paul: They tend to be greater. If you give a dose of 10 μM by itself and then work up to 10 μM, it is always much greater in the cumulative dose–response curve.

Fry: Your EC_{50} was about 5.5 for carbachol. In the hands of people who do it sequentially it is about the same value. This is probably a slightly less sensitive system than yours.

Paul: That is interesting, because the ED_{50}s may be the same, but the maximums may be different in the cumulative.

Fry: If there was a desensitization then you would expect the EC_{50}s to shift to higher concentrations, because you have flattened the response of the higher concentration. The fact that it doesn't change the EC_{50} suggests that there isn't much desensitization, fortunately.

Paul: There is at really high levels.

Burdyga: I don't think using agonists for looking at the steady-state force–Ca^{2+} relationship is the most appropriate model. The release of Ca^{2+} is there. Also, agonists can sensitize contractile machinery. I think it is better to use high K^+ responses and to see how the force–Ca^{2+} relationship is affected during the high K^+ depolarization. What is your feeling about this?

Paul: We have done both. I showed the muscarinic, hoping that someone would think there was some physiological relevance there. I thought that the KCl was going to be a control, and would saturate the SR, but it didn't. To be honest, I have not done this in detail in the bladder.

Burdyga: If spark activity is high this should affect the steady-state by changing the membrane potential. The contribution of BK channels to the resting potential would be greater. And with reference to the knockout experiments: is there any way that these animals could escape, and then in 200 years time people carry out experiments on them asking what was going on 200 years earlier in Cincinnati?

Paul: The intruiging thing with the knockout is that people thought this would always be under β adrenergic control. But these mice seem to live longer and they are better on treadmills.

Kotlikoff: Am I correct in thinking that there are three different genetic backgrounds here?

Paul: That's an excellent point. The background of these mice is extremely important. FVB/n is the overexpressor background mouse, as opposed to the SVJ129/CF1 background of the knockout. With three or four of these models people have suggested making them congenic, breeding the knockout against knockout. If we do this we see tremendous drift. The only safe thing is to work on siblings.

Somlyo: I refuse to work on any knockout that does not have a congenic control. We have been burned in the past, and I don't want to do it again.

Kotlikoff: These are all very polygenic responses. Have you looked carefully at something like K^+ depolarization in all three of your backgrounds?

Paul: We have taken a different tack in that we bred the overexpressor onto the knockout background. We crossed it back to make an FVB/n knockout. There don't seem to be many major differences there. There is a shift in the absolute $ED_{50}s$ but there is still a suppression in the knockout.

Somlyo: The trouble is that to make a true congenic by conventional means takes two to three years. You can do a speed congenic in about eight months. But for those of us who like to look at force development it is a rather long time to wait. I have a mechanistic question. Has anyone looked at the leak from the SR? There was an old debate as to whether PLB did something to the uptake mechanism: is it direct or is it via creating a leak?

Paul: At one stage it was postulated that the pentameric PLB formed a leak channel. There is a fair amount of evidence against this. Perhaps most striking is work by the Kranias laboratory in which the transgene for a PLB mutant which remains monomeric and does not form the pentamers was expressed in the heart.

The monomeric PLB mutant was as potent (actually even more so) than the native PLB, which indicated that a PLB pentamer forming a leak channel was unlikely. There is a reported backflux through the Ca-ATPase which is lower in the PLB knockout. In the heart you can make a really good SR vesicle preparation for direct evaluation of SR Ca^{2+} uptake, but one of the limitations here is that this is difficult for smooth muscle tissues, though the bladder is useful for some quantitative biochemistry. So the question of a PLB leak is still open for smooth muscle but, on the basis of cardiac data, it is unlikely.

The sarcoplasmic reticulum in disease and smooth muscle dysfunction: therapeutic potential

A. F. Brading

University Department of Pharmacology, Mansfield Road, Oxford OX1 3QT, UK

Abstract. The functions of the sarcoplasmic reticulum (SR) in diseased smooth muscle can be investigated by measuring Ca^{2+} transients in response to agonist application, and through cell homogenization, isolation of microsomes and measurements of Ca-ATPase activity (SERCA). Such measurements have indicated that contractile dysfunction may be associated with degradation of SERCA in some systems, such as hypertrophied bladder smooth muscle. However, the postulated roles of the SR in smooth muscle function vary from one tissue to another and SR may mediate relaxation as well as contraction. Function seems to depend on the precise location of the SR with respect to the plasma membrane and its Ca^{2+}-activated ion channels, the Ca^{2+} transporters, the cavaeoli, the mitochondria, and the contractile machinery. In diseases characterized by smooth muscle dysfunction, the size of the smooth muscle cells is frequently altered, as occurs in the hypertrophy seen in gut and bladder obstruction and hypertension. This will inevitably lead to alterations in the morphology and the function of the SR. Any therapeutic potential awaits considerable advances in our understanding of the systems in individual smooth muscles and the development of selective drugs.

2002 Role of the sarcoplasmic reticulum in smooth muscle. Wiley, Chichester (Novartis Foundation Symposium 246) p 244–257

This symposium highlights the enormously important and varied role of the sarcoplasmic reticulum (SR) in smooth muscles, and there can be little doubt that anything disrupting the functions of these Ca^{2+} stores will lead to changes in the properties and contractile behaviour of the muscles. There is also good evidence that in dysfunctional smooth muscles, changes can be seen in the behaviour of the Ca^{2+} stores. What is, however, open to question is whether alterations in SR function in smooth muscles are in fact responsible for disease, or whether drugs targeting the molecular machinery involved in SR function have potential therapeutic value in treating smooth muscle dysfunction.

In this chapter, I will consider the various pieces of molecular machinery that are involved in SR function in turn, give clinical examples of diseases caused by mutations in these molecules, and discuss what is known about their functions in smooth muscle, particularly anything with relevance to disease.

The main mechanisms that are important are: (1) SERCA pumps; (2) phospholamban; (3) Ca^{2+} binding proteins; (4) inositol-1,4,5-trisphosphate (InsP$_3$) receptors and mechanisms involved in InsP$_3$ production; (5) ryanodine receptors and cADP ribose production; and (6) the cytoskeleton.

SERCA pumps

Ca^{2+} is pumped into the SR stores by specific ATPases, the SERCA pumps. Three *SERCA* genes have been identified and several alternatively spliced gene products. In muscles SERCA1 is predominantly expressed in fast-twitch striated muscles and SERCA 2 isoforms are dominant in slow twitch, heart and smooth muscles, with SERCA2a in the former two, and SERCA2b predominant in smooth muscle and non-muscle cells. SERCA3 has a widespread distribution.

The considerable importance of these pumps to cells in general is exemplified in humans by some naturally occurring syndromes in which the underlying problem is a mutation in these Ca^{2+} transport ATPases. Hailey–Hailey and Darier's syndrome are rare diseases of the skin. These are sometimes difficult to distinguish from each other and are characterized by the eruption of small greasy papules that coalesce to give yellowish–brown scaling sheets. In Darier's disease there is also a characteristic damage to the nail bed resulting in white lines and notching of the nail. It is thought that an underlying problem is in some loss of adhesion between keratinocytes, possibly due to lack of expression of certain adhesion molecules. The link between this and SERCA pumps is the suggestion that endoplasmic reticulum (ER) Ca^{2+} may be important in fundamental cell processes such as protein trafficking. In Darier's disease the mutation is to SERCA2a (Sakuntabhai 1999).

Mutations in the SERCA1 pump of skeletal muscle lead to Brody's disease, in which there is delayed muscle relaxation, particularly after exercise, leading to cramps. It is thought to be due to reduced efficiency of the SERCA1 pumps reducing the rate at which Ca^{2+} can be removed from the cytoplasm (Odermatt et al 2000).

Most work on the SR and diseased smooth muscle has concerned vascular smooth muscle in hypertensive animals, and bladders from animal models of outflow obstruction. The tools used to study SR function are mainly indirect, and include recording tension or intracellular $[Ca^{2+}]$ with fluorescent probes, measuring Ca^{2+} fluxes with ^{45}Ca, and investigating the effects of drugs known to block SERCA or activate store release. More directly, some measurement of the activity of SERCA in microsomal preparations has been undertaken (e.g. Zderic et al 1996).

Work on vascular smooth muscle has not led to any real consensus as to the precise differences in SR function between normal and hypertensive animals, although there is evidence, summarized in a review by Raeymaekers & Wuytack (1993) for diminished SR Ca^{2+} transport in hypertensive animals. More recent work has also provided evidence for increased Ca^{2+} influx from the extracellular space in vascular smooth muscles from various rat models of hypertension (Nomura et al 1997, Arii et al 1999).

Much work has been carried out on SR and Ca^{2+} stores in the bladder by Levin and co-workers, and the effects of outflow obstruction observed. Technically, the ability of ryanodine to inhibit contractions in response to field stimulation of muscle strips has been examined, and the binding of ryanodine to microsomal preparations (Levin et al 1994). The effects of a combination of thapsigargin and ryanodine on the ability of the whole isolated rabbit bladder to respond to field stimulation and the application of bethanechol have also been studied (Levin et al 1997). In other studies this group have examined the sensitivity of the response to agonists to changes in extracellular Ca^{2+} and how this is affected by outflow obstruction and inhibition of Ca^{2+} store function with thapsigargin and ryanodine (Rohrmann et al 1996). They have also looked at ATPase activity of the SERCA pumps measured in normal and obstructed bladders along with the responsiveness of the tissues to bethanechol (Zderic et al 1996). The results of these studies suggest that there is a loss of SR function with obstruction, which parallels the reduced ability of the tissues to contract in response to agonists. The authors also show a loss of immunoreactivity to a monoclonal antibody raised against the slow form of SERCA. The interpretation of these experiments is complicated by the fact that there is hypertrophy of the bladder which is progressive after outflow obstruction, but can show marked variation in different animals, and also a 'decompensation' occurs some time after obstruction in which contractile function is markedly impaired. In the hypertrophied and 'compensated' bladders, which still respond reasonably well to applied agonists, it is probable (and has been shown in other obstructed animal models) that the smooth muscle cells themselves hypertrophy, and Levin and colleagues have suggested that the reduction in contractile activity may be because the SR does not grow sufficiently to make up for the increased cell volume.

Phospholamban

Phospholamban is a protein that inhibits the SERCA pumps by decreasing their affinity for Ca^{2+}. It can be phosphorylated by protein kinase A, for instance in response to β-adrenoceptor activation, resulting in inhibition of its effects and enhanced SERCA activity. The effects of phospholamban on contraction depend on the relative importance of Ca^{2+} uptake or release in the smooth muscle in question. Lalli et al (1997, 1999) investigated the role of phospholamban in aortic smooth muscle using ablation of the phospholamban gene in mice. They found a reduced contractile response to agonists and a reduced level of $[Ca^{2+}]_i$. Nobe et al (2001) have investigated the effects on bladder smooth muscle of both over-expressing and knocking out the phospholamban gene in mice. The results show that in both bladder and aorta the role of the SR in uptake of Ca^{2+} is predominant, thus if a particular stimulation is applied, leading to a certain amount of Ca^{2+} release, the level of $[Ca^{2+}]_i$ achieved and thus the contraction is dependent on the rate of re-uptake into the store. In the urinary bladder responding to muscarinic agonist stimulation, the increased SERCA activity in the knockout mice resulted in reduced $[Ca^{2+}]_i$ and contraction in response to carbachol, and this effect could be reversed in the short term by blocking the SERCA pump with cyclopiazonic acid (CPA). Over-expressing phospholamban, and thus inhibiting SERCA resulted in an enhanced contraction to carbachol, and further blocking the pump with CPA was ineffective. Nobe et al (2001) suggest that therapeutic targeting of phospholamban might be a new approach to ameliorating bladder dysfunction.

Ca^{2+} binding proteins

Smooth muscle SR contains Ca^{2+} binding proteins such as calsequestrin and calreticulin. The amount of Ca^{2+} that can be held by the stores, and the kinetics of Ca^{2+} release will be affected by these proteins. Calreticulin was shown by Villa et al (1993) to be uniformly distributed throughout the SR components of different smooth muscles, whereas the distribution of calsequestrin varied not only between smooth muscles in the amounts present (Raeymaekers et al 1993), but also within a single smooth muscle cell (Volpe et al 1994). The effects of over-expression or reduction of the levels of these binding proteins can be investigated in cultured cell lines, showing as would be expected effects on Ca^{2+} storage and agonist-induced Ca^{2+} transients, but modulation of gene expression in transgenic mice can have profound effects, for instance knocking out the calreticulin gene in mice is embryonically lethal, whereas over expression of calsequestrin in mice allows survival into adulthood, but with a phenotype of severe cardiac hypertrophy (Jones et al 1998). It has been suggested that the $[Ca^{2+}]$ in the SR lumen may have very important effects on protein trafficking during early embryonic development (Sorrentino & Rizzuto 2001).

InsP$_3$ receptors

InsP$_3$ receptors are tetrameric Ca^{2+} channels in the SR membrane, and play an important role in smooth muscles in receptor–effector coupling, particularly that classified as pharmacomechanical coupling. Receptor stimulation leads to activation of phospholipase C and the production of InsP$_3$, which can bind to and open the InsP$_3$ receptors. Three main classes of InsP$_3$ receptor have been identified, types 1–3, and in many cell types all three receptors may be expressed. In smooth muscles, Morgan et al (1996) have demonstrated the presence of the mRNA for all three classes in human myometrium, but in some vascular smooth muscle the mRNA for only type 1 and type 3 InsP$_3$ receptors has been detected (Tasker et al 1999). These latter authors have also demonstrated that there is a switch in subtype expression during postnatal development of neonatal vascular smooth muscle in the rat. The developing smooth muscle has relatively high expression of type 3 receptors with a low level of type 1 whereas the fully developed smooth muscle has a low expression of type 3 and a relatively higher level of type 1 InsP$_3$ receptors. Since the affinities of the three receptor subtypes for InsP$_3$ have been shown to differ markedly in their sensitivities to Ca^{2+}, it is likely that alterations of the relative expression of different subtypes may have considerable effects on Ca^{2+} signalling. At present, however, I am unaware of any reports of changes in subtype expression associated with particular dysfunctions of smooth muscle.

In those tissues controlled by transmitters utilizing the InsP$_3$ pathway, the signal has to be terminated rapidly in order to achieve good control of the level of cytoplasmic Ca^{2+} through this pathway, and this is achieved by dephosphorylation of InsP$_3$ and its re-incorporation into the membrane phospholipids. There is a human disease—the oculocerebrorenal syndrome of Lowe (OCRL)—which is a rare X-linked recessively inherited disease characterized by congenital cataract, mental retardation, renal tubular dysfunction and progressive renal insufficiency. The gene responsible for OCRL encodes an abnormal inositol polyphosphate-5-phosphatase (Attree et al 1992), and thus interferes with the metabolism of InsP$_3$ to inositol-4,5-bisphosphate, preventing the normal termination of the InsP$_3$ signal, with extensive adverse consequences.

Ryanodine (RY) receptors

There are three genes coding for these large complex tetrameric SR Ca^{2+} channels. RY$_1$ channels are present in striated muscle and involved in excitation–contraction coupling between the T tubules and the SR. The RY$_2$ channels are present and necessary for function in heart muscle, in which their activation is through a Ca^{2+}-activated Ca^{2+} release mechanism, a local rise in free Ca^{2+} entering the cell through voltage sensitive Ca^{2+} channels activating the channels. RY$_3$ channels are co-expressed with other RY channels in most cells. All three channels have been demonstrated in smooth muscles, but there can be marked differences in their expression between smooth muscles. Most smooth muscles express both InsP$_3$ receptors and RY receptors, although this is not always the case: for instance in the rabbit intestine, the circular smooth muscle contains predominantly $InsP_3$ receptors and the longitudinal smooth muscle RY receptors (Murthy et al 1991, Kuemmerle et al 1994).

Mutations in the RY_1 receptor lead to identified diseases in man and animals. The main one is malignant hyperthermia, and the pig equivalent porcine stress syndrome, in which individuals are susceptible to halothane anaesthesia, developing the sometimes fatal symptoms of a severe rise in body temperature due to hypermetabolism, with muscle rigidity, metabolic and respiratory acidosis, and tachycardia. Also seen is central core disease which is manifest in infancy as a hypotonia and a delayed motor development. The mutations present seem to alter the sensitivity of the channels to Ca^{2+}, and to other sensitizing agents such as caffeine and halothane. Whether or not the individual is normal in the absence of halothane, or has central core disease seems to depend on the degree to which the changed responsiveness of the channels leads to reduced levels of stored Ca^{2+} and increased cytoplasmic Ca^{2+}, and the adaptation of the cells to these changes.

In smooth muscle the RY receptors can be activated by a rise in $[Ca^{2+}]$ subsequent to activation of voltage-sensitive calcium channels on the plasma membrane. However, there is also a second messenger system present in cells that can elicit release of Ca^{2+} through the RY receptors, and that is cADP ribose (Galione 1993, Galione & Sethi 1996). This molecule was first identified in sea urchin eggs by Lee et al (1989) and is synthesized from NAD^+ by ADP ribosyl cyclase. Recently it has become apparent that cADP ribose may play an important role in smooth muscles. Kuemmerle et al (1998) suggest the involvement of cADP ribose in the contractile responses of rabbit longitudinal smooth muscle in which transmitters elicit contraction in a phosphoinositide-independent pathway. In vascular smooth muscles ADP ribosyl cyclase has been shown to be up-regulated by various hormones of the steroid super family such as retinoids, calcitriol and T(3) (de Toledo et al 2000), and the contractile response of small resistance arteries to endothelin 1 is due to receptor-mediated production of cADP ribose (Giulumian et al 2000). Recently, it has been suggested that the primary trigger for hypoxic pulmonary vasoconstriction in the rat lung is cADP ribose (Dipp & Evans 2001). It is proposed that the cellular redox state may be coupled via an increase in β NADH levels to enhanced cADP ribose synthesis, activation of RY receptors and SR Ca^{2+} release (Wilson et al 2001).

Cytoskeleton

I have included cytoskeletal elements for completion. By 'cytoskeletal', I mean any molecular mechanisms that play a role in holding the various elements of the SR in their correct relationship with the plasma membrane and its various channels, pumps and enzymes, mitochondria or the nuclear envelope. Already there are numerous studies of the effects of drugs that disrupt microtubules or filaments on smooth muscle function, such as colchicine (prevents polymerization of microtubules) and the cytochalasins (prevent polmerization of actin filaments). Cytoskeletal proteins have been implicated in the control of kinase activation in vascular smooth muscle (Abedi & Zachary 1998, Govindarajan et al 2000). Ion channel activity can be affected by disruption of the cytoskeleton (P2$_X$ receptor non-selective cation channels [Parker 1998]; L-type Ca^{2+} channels [Nakamura et al 2000]) as can activation of inducible NO synthase in aortic smooth muscle (Marczin et al 1996) and receptor-mediated release of Ca^{2+} from intracellular stores (Samain et al 1999). Actin filaments also appear to link the plasma membrane to the Ca^{2+} stores in smooth muscles cells dispersed from rabbit colon (Young et al 1997).

Observations and conclusions

Although studies on the molecular machinery involved in SR/ER functioning in cells is a fast-developing area of research, relatively few of these studies have involved smooth muscles. Various cell lines have been used for studying the elimination or over-expression of particular molecules and gene knockout mice have been created. In most experimental animals with genetically altered ER/SR function, interest centres on the more obviously important organs such as the brain, heart and skeletal muscle, with little if anything recorded on the effects on smooth muscle function.

One serious problem for studying SR function in smooth muscles with gene knockout animals is the remarkable versatility and variability of smooth muscles. There are several alternative routes through which Ca^{2+} uptake and release can be accomplished, and for most pieces of molecular machinery there are several genes, and probably also different splice variants for each gene product. Which genes and which splice variants are expressed can vary between different smooth muscles in a single individual, between the same smooth muscles in different species, and also may change in a single smooth muscle in response to altered use or changed environment. Often there is a degree of redundancy so that knocking out a gene whose product plays an important role in normal tissue, may have surprisingly little effect in the knockout animal, the function being taken over by other available pathways. On the other hand, the function of the SR during the embryonic development of an animal may be quite different from its role in the adult, and mutations of genes whose products may only play a minor role in the adult may prove unexpectedly lethal, or have very widespread consequences. The study of naturally occurring mutations has not been much help in elucidating smooth muscle SR dysfunction. As far as I am aware, in January 2002 no particular genetic dysfunction of the Ca^{2+} release machinery has been implicated as a cause of any smooth muscle disorder, although several other human syndromes are caused by such mutations.

The studies cited above on the SR function in diseased smooth muscles not surprisingly indicate altered function. However, smooth muscles are very responsive to changes in their conditions and environment, and in particular to changes in the activity of the efferent nerves (e.g. Westfall 1981). In several instances of smooth muscle dysfunction partial denervation of the tissue has also been observed (for instance in the unstable bladder, Brading & Turner 1994). Relevant to this are the interesting observations of Lehotsky et al (1993) in striated muscles. These authors showed that the early response to the loss of motor innervation in rabbit striated muscle, is an altered Ca^{2+} homeostasis. The study indicated that denervation influenced expression of some sarcoplasmic Ca^{2+}-modulating proteins including an increase in the level of calsequestrin and of the putative ryanodine receptor paralleled with a slight decrease in the total amount of Ca^{2+} pump protein.

In the hypertrophic smooth muscles seen in bladder outflow obstruction and blood vessels in hypertensive subjects, alterations in the smooth muscle phenotype are seen, leading to changes in the morphology of the cell, and in their secretion of extracellular matrix. The increasing evidence that the precise functions of the SR in normal smooth muscles depend critically upon its location with respect to other cell components, means that there are difficulties in assessing whether any smooth muscle dysfunction is caused by a change in the SR function, or whether SR dysfunction results from alterations in the spatial arrangement or changes in the nature of the SR and the relative amounts of SR membrane and cell volume. In other words, if there are signals that result in the growth of individual cells, then this growth alone could result in alterations in the function of the SR simply through a spatial disruption of the normal Ca^{2+} signalling pathways. If growth factors also alter the function of the SR and increase protein production and trafficking, again this may result in alterations in the contractile role of the SR. It is thus impossible to assign any functional impairment seen in diseased smooth muscles simply to a reduced amount or activity of the SR.

For the rational design of drug treatment for a particular functional disorder of the SR, the routes and molecules used in the particular smooth muscle would have to be established, and what needs to be blocked or enhanced to correct the dysfunction determined. Then the main problem will be how to selectively treat the muscle in question without generating dysfunction in other cells. The difficulties in this route make it likely that therapeutic advance will come less through rational design and more through serendipitous observations. Another approach might be to develop new ways of targeting particular accessible organs such as the lungs and bladder by the method of drug delivery.

References

Abedi H, Zachary I 1998 Cytochalasin D stimulation of tyrosine phosphorylation and phosphotyrosine-associated kinase activity in vascular smooth muscle cells. Biochem Biophys Res Commun 246:646–650

Arii T, Ohyanagi M, Shibuya J, Iwasaki T 1999 Increased function of the voltage-dependent calcium channels, without increase of Ca^{2+} release from the sarcoplasmic reticulum in the arterioles of spontaneous hypertensive rats. Am J Hypertens 12:1236–1242

Attree O, Olivos IM, Okabe I et al 1992 The Lowe's oculocerebrorenal syndrome gene encodes a protein highly homologous to inositol polyphosphate-5-phosphatase. Nature 358:239–242

Brading AF, Turner WH 1994 The unstable bladder: towards a common mechanism. Br J Urol 73:3–8

Dipp M, Evans AM 2001 Cyclic ADP-ribose is the primary trigger for hypoxic pulmonary vasoconstriction in the rat lung in situ. Circ Res 89:77–83

de Toledo FG, Cheng J, Liang M, Chini EN, Dousa TP 2000 ADP-Ribosyl cyclase in rat vascular smooth muscle cells: properties and regulation. Circ Res 86:1153–1159

Galione A 1993 Cyclic ADP-ribose: a new way to control calcium. Science 259:325–326

Galione A, Sethi J 1996 Cyclic ADP-ribose and calcium signaling. In: Bárány M (ed) Biochemistry of smooth muscle contraction. Academic Press, San Diego p 295–305.

Giulumian AD, Meszaros LG, Fuchs LC 2000 Endothelin-1-induced contraction of mesenteric small arteries is mediated by ryanodine receptor Ca^{2+} channels and cyclic ADP-ribose. J Cardiovasc Pharmacol 36:758–763

Govindarajan G, Eble DM, Lucchesi PA, Samarel AM 2000 Focal adhesion kinase is involved in angiotensin II-mediated protein synthesis in cultured vascular smooth muscle cells. Circ Res 87:710–16

Jones LR, Suzuki YJ, Wang W et al 1998 Regulation of Ca^{2+} signaling in transgenic mouse cardiac myocytes overexpressing calsequestrin. J Clin Invest 101:1385–1393

Kuemmerle JF, Murthy KS, Makhlouf GM 1994 Agonist-activated, ryanodine-sensitive IP3-insensitive Ca^{2+} release channels in longitudinal muscle of intestine. Am J Physiol 266: C1432–C1439

Kuemmerle JF, Murthy KS, Makhlouf GM 1998 Longitudinal smooth muscle of the mammalian intestine. A model for Ca^{2+} signaling by cADPR. Cell Biochem Biophys 28: 31–44

Lalli J, Harrer JM, Luo W, Kranias EG, Paul RJ 1997 Targeted ablation of the phospholamban gene is associated with a marked decrease in sensitivity in aortic smooth muscle. Circ Res 80:506–513

Lalli MJ, Shimizu S, Sutliff RL, Kranias EG, Paul RJ 1999 $[Ca^{2+}]_i$ homeostasis and cyclic nucleotide relaxation in aorta of phospholamban-deficient mice. Am J Physiol 277:H963–H970

Lee HC, Walseth TF, Bratt GT, Hayes RN, Clapper DL 1989 Structural determination of a cyclic metabolite of NAD^+ with intracellular Ca^{2+}-mobilizing activity. J Biol Chem 264:1608–1615

Lehotsky J, Bezakova G, Kaplan P, Raeymaekers L 1993 Distribution of Ca^{2+}-modulating proteins in sarcoplasmic reticulum membranes after denervation. Gen Physiol Biophys 12:339–348

Levin RM, Levin SS, Zderic SA, Saito M, Yoon JY, Wein AJ 1994 Effect of partial outlet obstruction of the rabbit urinary bladder on ryanodine binding to microsomal membranes. Gen Pharmacol 25:421–425

Levin RM, Yu HJ, Kim KB, Longhurst PA, Wein AJ, Damaser MS 1997 Etiology of bladder dysfunction secondary to partial outlet obstruction. Calcium dysregulation in bladder power generation and the ability to perform work. Scand J Urol Nephrol Suppl 184:43–50

Marczin N, Jilling T, Papapetropoulos A, Go C, Catravas JD 1996 Cytoskeleton-dependent activation of the inducible nitric oxide synthase in cultured aortic smooth muscle cells. Br J Pharmacol 118:1085–1094

Morgan JM, De Smedt H, Gillespie JI 1996 Identification of three isoforms of the InsP3 receptor in human myometrial smooth muscle. Pflüger's Arch 431:697–705

Murthy KS, Grider JR, Makhlouf GM 1991 InsP$_3$-dependent Ca^{2+} mobilization in circular but not longitudinal muscle cells of intestine. Am J Physiol 261:G937–G944

Nakamura M, Sunagawa M, Kosugi T, Sperelakis N 2000 Actin filament disruption inhibits L-type Ca^{2+} channel current in cultured vascular smooth muscle cells. Am J Physiol 279:C480–C487

Nobe K, Sutliff RL, Kranias EG, Paul RJ 2001 Phospholamban regulation of bladder contractility: evidence from gene-altered mouse models. J Physiol 535:867–878

Nomura Y, Asano M, Ito K, Uyama Y, Imaizumi Y, Watanabe M 1997 Potent vasoconstrictor actions of cyclopiazonic acid and thapsigargin on femoral arteries from spontaneously hypertensive rats. Br J Pharmacol 120:65–73

Odermatt A, Barton K, Khanna VK et al 2000 The mutation of Pro789 to Leu reduces the activity of the fast-twitch skeletal muscle sarco(endo)plasmic reticulum Ca^{2+} ATPase (SERCA1) and is associated with Brody disease. Hum Genet 106:482–491

Parker KE 1998 Modulation of ATP-gated non-selective cation channel (P2X1 receptor) activation and desensitization by the actin cytoskeleton. J Physiol 510:19–25

Raeymaekers L, Wuytack F 1993 Ca^{2+} pumps in smooth muscle cells. J Muscle Res Cell Motil 14:141–157

Raeymaekers L, Verbist J, Wuytack F, Plessers L, Casteels R 1993 Expression of Ca^{2+} binding proteins of the sarcoplasmic reticulum of striated muscle in the endoplasmic reticulum of pig smooth muscles. Cell Calcium 14:581–589

Rohrmann D, Zderic SA, Wein AJ, Levin RM 1996 Effect of thapsigargin on the contractile response of the normal and obstructed rabbit urinary bladder. Pharmacology 52:119–124

Sakuntabhai A, Burge S, Monk S, Hovnanian A 1999 Spectrum of novel ATP2A2 mutations in patients with Darier's disease. Hum Mol Genet 8:1611–1619

Samain E, Bouillier H, Perret C, Safar M, Dagher G 1999 ANG II-induced Ca^{2+} increase in smooth muscle cells from SHR is regulated by actin and microtubule networks. Am J Physiol 277:H834–H841

Sorrentino V, Rizzuto R 2001 Molecular genetics of Ca^{2+} stores and intracellular Ca^{2+} signalling. Trends Pharmacol Sci 22:459–464

Tasker PN, Michelangeli F, Nixon GF 1999 Expression and distribution of the type 1 and type 3 inositol 1,4, 5-trisphosphate receptor in developing vascular smooth muscle. Circ Res 84:536–542

Villa A, Podini P, Panzeri MC, Soling HD, Volpe P, Meldolesi J 1993 The endoplasmic–sarcoplasmic reticulum of smooth muscle: immunocytochemistry of vas deferens fibers reveals specialized subcompartments differently equipped for the control of Ca^{2+} homeo-stasis. J Cell Biol 121:1041–1051

Volpe P, Martini A, Furlan S, Meldolesi J 1994 Calsequestrin is a component of smooth muscles: the skeletal- and cardiac-muscle isoforms are both present, although in highly variable amounts and ratios. Biochem J 301:465–469

Westfall DP 1981 Supersensitivity of smooth muscle. In: Bülbring E, Brading AF, Jones AW, Tomita T (eds) Smooth muscle: an assessment of current knowledge. Arnold, London p 285–309.

Wilson HL, Dipp M, Thomas JM, Lad C, Galione A, Evans AM 2001 ADP-ribosyl cyclase and cyclic ADP-ribose hydrolase act as a redox sensor. A primary role for cyclic ADP-ribose in hypoxic pulmonary vasoconstriction. J Biol Chem 276:11180–11188

Young SH, Ennes HS, Mayer EA 1997 Mechanotransduction in colonic smooth muscle cells. J
    Membr Biol 160:141–150
Zderic SA, Rohrmann D, Gong C et al 1996 The decompensated detrusor II: evidence for loss
    of sarcoplasmic reticulum function after bladder outlet obstruction in the rabbit. J Urol
    156:587–592

DISCUSSION

Hellstrand: Given that there is a correlation between trophic disease states and a
reduced SR function, can you distinguish between long-term interventions that
would affect SR function and the function of the ER? For instance, if we want to
do something about the Ca^{2+} handling of the SR, can we be sure that we are not
affecting protein synthesis by the ER?

Brading: No, I don't think we can. This is one of the problems we face. There is
good evidence that when the smooth muscles are responding to changes in their
environment, they nearly all will change their phenotypes and alter the secretion of
matrix. And there will be increased expression of ER. It is extremely difficult to
disentangle. You can say that the evidence suggests that there is reduced SR
function in this hypertrophic tissue, but we can't tell whether this is because the
tissue has changed its phenotype or not.

Hellstrand: That's an interesting phenomenon in itself. It is something we would
like a handle on.

Nixon: Isn't that the point? If you get more ER and you change the Ca^{2+} channel
expression, there will be changes in Ca^{2+} oscillations and release. This will then
affect gene expression. This is what drives part of the change.

Brading: Yes. Frequently, it is change in various growth factors that triggers a
whole cascade of changes. However careful the experiments are, I don't think
assessment of SR function can be interpreted meaningfully. To say that SR
function is altered, is not saying anything useful.

Hellstrand: But if you have an idea that this is an important control station, how
can we take the standpoint that the experiments don't say anything? They may be
difficult to disentangle, though.

Brading: They say that things have changed, but this is about all.

Paul: But we can measure what has changed. In fact, one of the more modern
approaches is to use gene chips or proteomics, to find out exactly which genes or
proteins are being changed and what they do. I'm not as pessimistic as you are.

Brading: Where does this leave us?

Sanders: Take the obstructive bladder model. If you can do this in a mouse, you
can do gene array work looking at the time course of the gene changes that occur.
Then you could look at some end-stage situation in the human and find out
whether the same genes that change in the end-stage of the mouse also apply in

human. This is a rapid way to address what is going on in the human and get time-course information of how that might develop.

Brading: Where does that leave you?

Sanders: You may get some signal that can be turned off to avoid the damage that is caused by obstruction.

Brading: The problem is that many of the diseases we are dealing with have developed over a long time. Perhaps we could work out what is leading to these. I have strong feelings that what we are dealing with in bladder obstruction is anoxic damage, in the long run.

Paul: The other major approach is to take diseased human tissue, for example in ischaemic heart disease, and then use a gene chip or proteomic approach to identify some of the genes or proteins that have changed. In half the cases we don't know what these genes or proteins do, so we use animal knockouts to find out their function to give some suggestion of what they might be doing in humans.

Fry: The important thing with gene chips is that you need a hypothesis. If you get 36 000 dots then you don't know what they do. The advantage of having a pathological model is that there is a hypothesis to aim at.

Sanders: I disagree. This is a hunting expedition, and you have to assume there are unknown genes or proteins waiting to be found.

Somlyo: It's a fishing expedition in very murky waters.

Paul: But with these new techniques our nets are a suddenly a lot bigger. It really is a paradigm change. I have talked to some of the drug companies using this approach. Modern nets are a lot more effective.

Brading: I'm still not convinced it is taking us anywhere.

Paul: We'll see.

Brading: The only drugs at the moment that are of any use in treating one of the biggest bladder problems are antimuscarinics. And no one has a clue as to why they are working. People have been trying for years to develop other drugs.

McHale: You made a remark about targeting drugs via the urine. How effective is this? We have tried in the past to give drugs to lymphatics intralumenally, and we found that blood-borne drugs are a lot more effective. Have you a different experience?

Brading: Clinically, one of the populations that use the antimuscarinics (which are the only drugs that are useful for these unstable bladders) is people with spinal injuries who do self-catheterization to empty their bladders. These people take antimuscarinics because if they don't, they get spontaneous contractions of the bladder and leak urine or get urgency. There are a range of side-effect profiles which make these drugs unpleasant. They found that intravasical application eliminated a lot of the side effects, even though the plasma levels look just as high as with orally delivered drugs.

Bolton: The bladder is designed so that substances aren't absorbed from it. When drugs are given through a catheter these are lipophilic substances that will cross epithelia readily. The kidney is designed to create water-soluble substances to be excreted. They tend to be ionized and will not readily be absorbed from the bladder.

Brading: Drugs can be designed to be absorbed through the bladder. There are K^+ channel openers that are selectively secreted by the kidney. These are weak acids and weak bases, and their lipid solubility can be manipulated by changing urine pH. The problem at the moment is that altering the structures of these drugs so that they are handled by the organic molecule transporter does reduce their efficacy.

Wray: In your review of the literature, were you saying that the best examples of the importance of the SR are during ontogeny? What detailed mechanism were you able to pick up from that literature?

Brading: The Ca^{2+} levels in the ER seem to be very important for protein trafficking. Presumably this is why so many of the spontaneous mutations are lethal. The role of the ER is so different from the adult SR that any mutations that affect this have far wider effects than on smooth muscle.

Young: I'd like to jump on the side of being positive. The only mild criticism I would have is that I think it is probably not productive to look at the end product of human genetic diseases in order to go backwards and find out how and when they can be fixed. The paradigm that we have used clinically has been the opposite: we have tried things and seen what happens. Traditionally, that has been thousands of years of history. But without the underlying scaffolding of how things work and the normal physiology, we are doomed never to understand what anything does. I can think of many examples of chronic diseases where we have had major impacts on health. This is not only chronic hypertension (which I would emphasize is a syndrome of many diseases), but I'd also like to turn the page to acute diseases. We have mainly been having this discussion couched in the sense of chronic diseases, and smooth muscle relaxation in the respiratory tree is not either acute or chronic but has elements of both. We have had major advances in understanding the state of anaesthesia, specifically the regulation of blood pressure. The paradigm of looking at chronic diseases from people with genetic defects may not be the optimistic way to take it at all. Perhaps we should instead take it from an underlying knowledge point. I appreciate we are not at the point of being able to design the drug from first principles. But we need to understand what we do when we do therapeutic manoeuvres.

Brading: I agree, but unfortunately this logical approach hasn't worked.

Paul: What examples of any medicine that has been developed fail to satisfy your criteria?

Young: I have a specific example about a clinical area I know about, which is pre-term labour. I have treated patients with pre-term labour, and here we know that we get a tachyphylaxis when we use β-adrenergic agonists. We know why, because we have the basic studies to support this, and we know how to convert to another medication. We also know which medications are slightly more effective in each clinical condition. But we are trying to refine this. Without this information we would have been stuck. It is very positive that we have this information.

Brading: I agree, but we need it on humans rather than animals, because the two often differ. But so far this approach hasn't yet led to the development of good treatments. Hopefully, one day it will.

Sanders: I disagree with you about the value of chronic human disease as a place to start. I think a perfectly reasonable approach that is being done is to take a certain number of candidate genes that are really important in smooth muscle function: perhaps 200. Then we could go to a population that suffer from hypertension, looking for polymorphisms. It would take a huge amount of data crunching, but if you find out that 90% of people with hypertension have polymorphisms in certain genes, you are on your way to discovering the cause. It's a good place to start.

Young: It still doesn't get you away from the plasticity of smooth muscle, which is the underlying problem.

Sanders: I agree, but perhaps some of those polymorphisms are what set this all off. Then you can go to the animal model, make those polymorphisms and see whether they cause hypertension.

The sarcoplasmic reticulum:
then and now

Andrew P. Somlyo and Avril V. Somlyo

Departments of Molecular Physiology and Biological Physics, Medicine and Pathology, University of Virginia School of Medicine, PO Box 800736, Jordan Hall, Charlottesville, VA 22908-0736, USA

Abstract. Structural and functional studies indicate the important role of the sarcoplasmic reticulum (SR) in excitation–contraction coupling in smooth and striated muscles, as well as a similar Ca^{2+} signalling function of the endoplasmic reticulum (ER) in non-muscle cells. Electron probe analysis directly established the SR/ER of smooth muscle as a sink and source of Ca^{2+}, while immunoelectron and immunofluorescence microscopy showed both inositol-1,4,5-trisphosphate ($InsP_3$) and ryanodine receptors localized to its membranes. Structural relationships, some yet to have fully determined functions, occur between the mitochondria and the SR, and the junctional SR and plasma membrane. Ca^{2+} is released by stimuli that generate $InsP_3$ indicating the primary role of $InsP_3$ receptors in Ca^{2+}-release in smooth, although not in striated, muscle. Pathological mitochondrial Ca^{2+} uptake occurs at high $[Ca^{2+}]_i$ similarly in both muscle and non-muscle cells. Based on newer evidence, earlier experimental results obtained with fluorescent Ca^{2+} indicators and related to phasic and tonic components of contraction can now be reinterpreted. Electron energy loss spectroscopy for high-resolution Ca^{2+} imaging and flash photolysis of caged agonists for exploration of the rapid kinetics of Ca^{2+} release from the SR are currently being explored.

2002 Role of the sarcoplasmic reticulum in smooth muscle. Wiley, Chichester (Novartis Foundation Symposium 246) p 258–271

The recognition of the critical role of Ca^{2+} in intracellular signalling (review in Ebashi 1991) led, like the pursuit of Peer Gynt's identity, to a series of questions with answers, that, like layers of an onion, revealed multiple, interconnected mechanisms. This brief review will deal with some of the answers, such as the sources and sinks of activator Ca^{2+} and the channels and transporters through which Ca^{2+} moves, with an emphasis on results obtained with methods having relatively high spatial and temporal resolution. The views presented are somewhat personal, and interested readers will find more comprehensive bibliographies and a history of the subject going further back in time in other reviews (Ebashi 1991, Bolton et al 1999, Karaki et al 1997, Somlyo et al 1999, Somlyo & Somlyo 1994, 2000). Because smooth muscle, our main focus, is also an excellent experimental paradigm for signal transduction in general, we shall attempt to place the answers obtained through its studies in the broader context of the biology of skeletal, cardiac and non-muscle cells.

The sarcoplasmic reticulum (SR) was first identified as the major mobilizable intracellular store of Ca^{2+} in skeletal muscles through the work of S. Ebashi, W. Hasselbach and A. Weber (review in Ebashi 1991). Identification of the SR and its role in smooth muscle met some early difficulties, partly due to the destructive effects of osmium fixation. Eventually the SR of smooth muscle was also identified, quantitated and its spatial distribution, peripheral and central, determined (Somlyo et al 1971, Devine et al 1971). Strontium (Sr), used as an electron opaque analogue of Ca^{2+}, permitted direct, electron microscopic visualization of divalent cation transport into the SR (Somlyo & Somlyo 1971).

The narrow gap traversed by bridging structures between the junctional SR and the plasma membrane (Fig. 1) (Somlyo et al 1971, Devine et al 1971) gave rise to the dual concepts that these are specialized sites of excitation–contraction coupling, analogous to the triadic junctions of skeletal muscle, while also forming a surface barrier to Ca^{2+} influx (Janssen et al 1999, Lee et al 2002, this volume, Blaustein et al 2002, this volume). The junctional SR is also the preferred site from which Ca^{2+} sparks originate (Gordienko et al 2001, review in Laporte & Laher 1997), and activate K^+ channels, leading to hyperpolarizing 'negative feedback' (review in Jaggar et al 2000, Nelson 2002, this volume), recalling early demonstrations of α-adrenergic hyperpolarization of intestinal smooth muscle (Jenkinson & Morton 1967).

The luminal continuity between peripheral/central SR and the perinuclear space, is well visualized through selective infiltration with osmium ferricyanide in stereoviews of electron micrographs (Lesh et al 1998, Tasker et al 1999) or in confocal images of SR containing Fluo-3 (Fig. 2). It indicates that Ca^{2+} can diffuse throughout this system, albeit at a rate slowed by intraluminal Ca^{2+}-binding proteins (calsequestrin, calreticulin) and the tortuousity factors imposed by its geometry. In some instances, the SR consists of large fenestrated sheets reminiscent of SR in the A-band of striated muscles (Fig. 3). That Ca^{2+} is released by agonists from both the central and the peripheral SR was directly demonstrated by electron probe X-ray microanalysis (Bond et al 1984, Kowarski et al 1985).

Several of the proteins that mediate Ca^{2+} flow in and out of SR have been identified. Oxalate-facilitated Ca^{2+} uptake into the SR and *in vitro* biochemical studies of purified SR identified it as an ATP-driven Ca^{2+} pump (SERCA pump; reviewed in Himpens et al 1995) that is inhibited by thapsigargin and cyclopiazonic acid and regulated, at least in some smooth muscles, by phosphorylation of phospholamban by cyclic nucleotide-activated protein kinase(s) (Karczewski et al 1998).

Molecular mechanisms of Ca^{2+} release operate during both electromechanical and pharmacomechanical coupling (review in Somlyo & Somlyo 1994, Somlyo

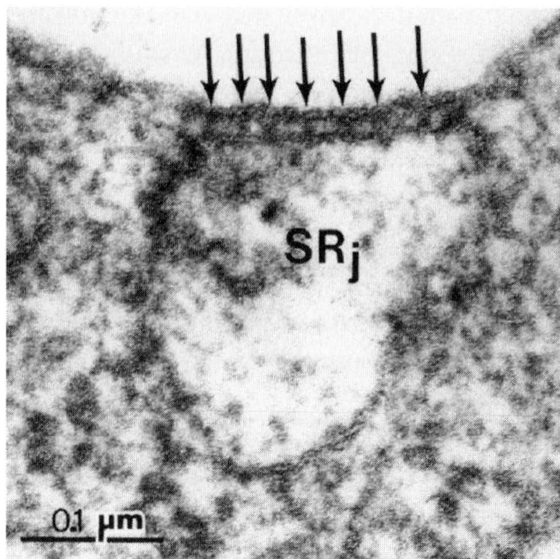

FIG. 1.   Surface coupling in a portion of a smooth muscle cell from the chicken amnion. An element of junctional sarcoplasmic reticulum (SR_j) separated by an 18 nm junctional gap between the plasma membrane and the SR is traversed by periodic bridging structures.

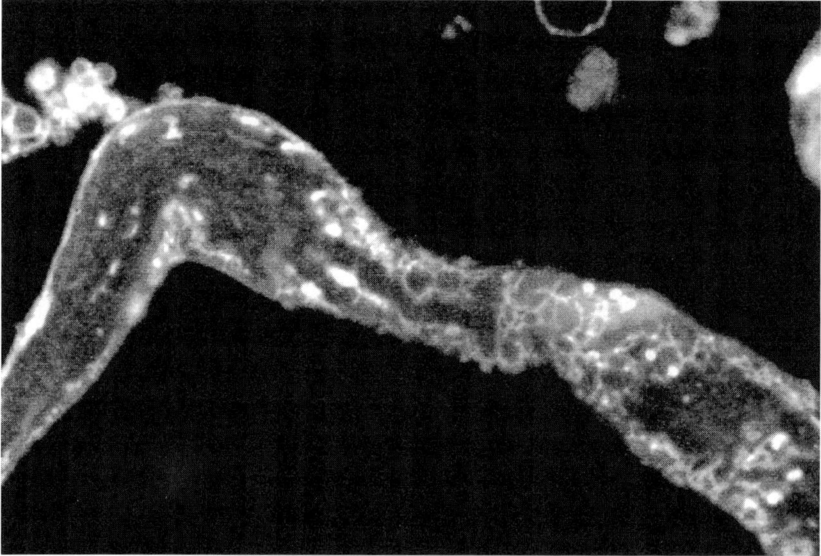

FIG. 2.   Confocal image of an isolated smooth muscle cell from guinea-pig ileum which has been permeabilized with staphylococcal α-toxin and incubated with 150 μM Fluo-3 acid which stains the SR. The SR is predominantly localized to the periphery of this type of smooth muscle as seen on the left hand side where the image plane is through the centre of the cell, whereas an extensive network is seen where the image plane is adjacent to the plasma membrane as seen in the right hand portion of the cell.

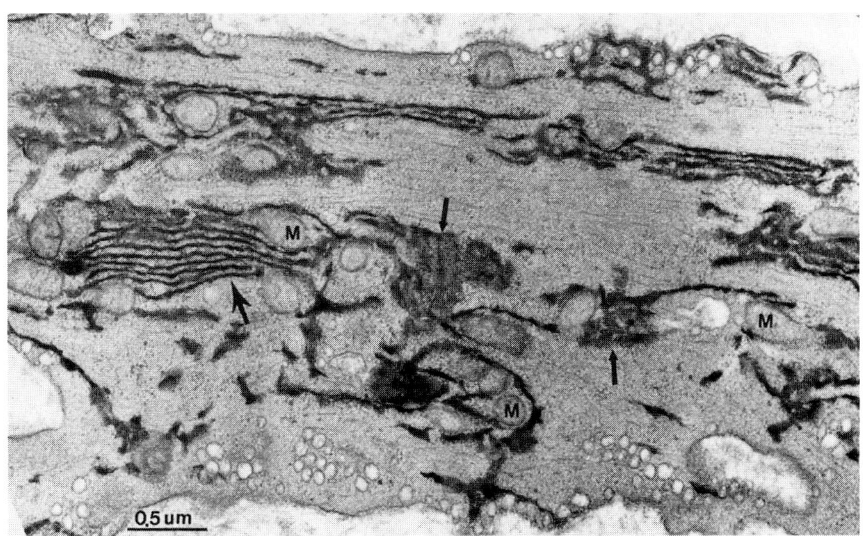

FIG. 3.   Extensive SR network stained with osmium ferricyanide in a smooth muscle cell of the rabbit pulmonary artery. Note the variable morphology of the network with tubules, fenestrated sheets seen enface (small arrows) or stacks of fenestrated sheets (large arrow) as well as surface couplings where the SR apposes the plasma membrane. M, mitochondria. (From Nixon et al 1994.)

et al 1999). The recognition that inositol-1,4,5-trisphosphate (InsP$_3$), a product of phosphatydilinositol bisphosphate (PIP$_2$) hydrolysis induced by stimulation of G protein-coupled receptors released Ca^{2+} from the endoplasmic reticulum (ER) of non-muscle cells (review in Berridge 1997), was followed by the demonstration that InsP$_3$ could release sufficient Ca^{2+} from the SR to activate contraction of smooth muscle (Somlyo et al 1985, 1992, Walker et al 1987) and isolation of InsP$_3$ receptors from smooth muscle (reviewed in Iino 2000). These and the finding that InsP$_3$ receptors are localized throughout the peripheral and central SR (Nixon et al 1994, Tasker et al 1999) indicated that the same InsP$_3$-mediated Ca-signalling mechanisms operate in smooth muscle as in non-muscle cells, although not to a significant extent in vertebrate striated muscles.

Ryanodine receptors are also localized to both the peripheral and central SR (Fig. 4) in smooth muscle (Lesh et al 1993), and extensive evidence indicates that caffeine releases Ca^{2+} through ryanodine receptors in smooth muscles, just as in striated muscle (reviewed in Iino 2000). Ca^{2+} influx can also induce Ca^{2+} release from the SR in smooth muscle (Ganitkevich & Isenberg 1995, Kamishima & McCarron 1997), suggesting that Ca^{2+}-induced Ca^{2+} release (CICR) is at least one of the mechanisms, first shown in cardiac muscle (Fabiato 1983), of electromechanical coupling in smooth muscle.

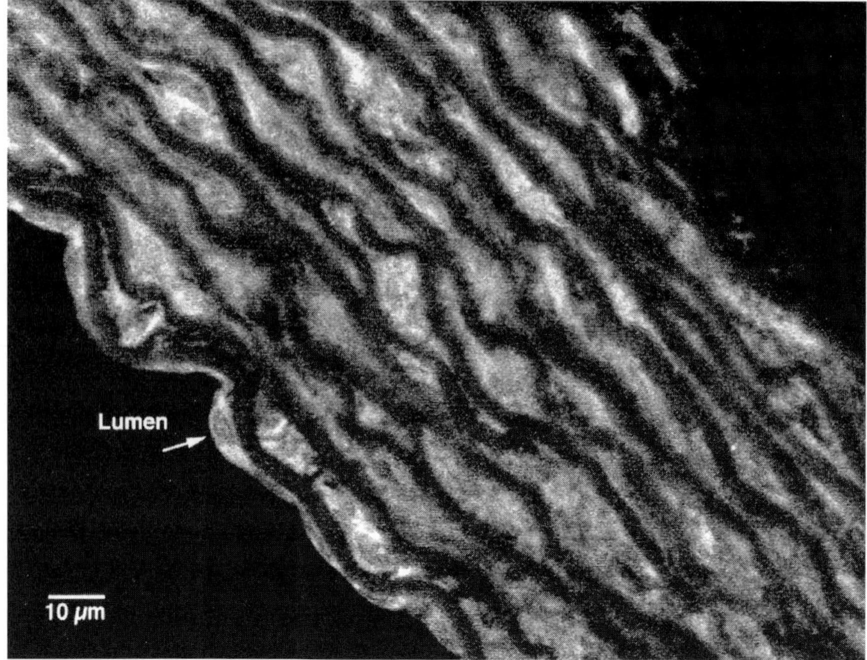

FIG. 4.   Confocal photomicrograph of guinea pig aorta labelled with anti-RyR8-2 (binds to ryanodine receptors) and then with a TRITC-conjugated secondary antibody. There is cytoplasmic labelling of both endothelial (arrow) and aortic smooth muscle cells, and the label correlates with the distribution of the SR in both cell types. A mesh-like staining pattern is suggestive of the SR network. (From Lesh et al 1998.)

Mitochondria have long been recognized as low affinity, high-capacity sites of Ca^{2+} uptake that protect cells from the detrimental effects of Ca^{2+} overload. Massive mitochondrial Ca^{2+} uptake (largely through the uniporter) occurs in practically all cells, even in frog striated muscle stimulated with caffeine, when mitochondria are exposed to high $[Ca^{2+}]_i$ (Yoshioka & Somlyo 1984). Whether mitochondria accumulate significant amounts of Ca^{2+} under physiological conditions in smooth muscles and other cells, remains a debatable matter. The use of fluorescent and luminescent indicators led to the suggestion that mitochondria may take up Ca^{2+} even under physiological conditions, when their low affinity is overcome by localized high $[Ca^{2+}]_i$, as a result of their proximity to the cell membrane or to the SR/ER. Unfortunately, rigorous quantitation of mitochondrial Ca^{2+} related to measurements of total (bound and unbound) mitochondrial Ca in the same tissue is rarely available. There appears to be general agreement that in resting cells free mitochondrial Ca^{2+} $[Ca^{2+}]_{mt}$ is $\sim$100 nM whereas total mitochondrial Ca is $\sim$0.4–1.0 mmoles/kg mitochondrial

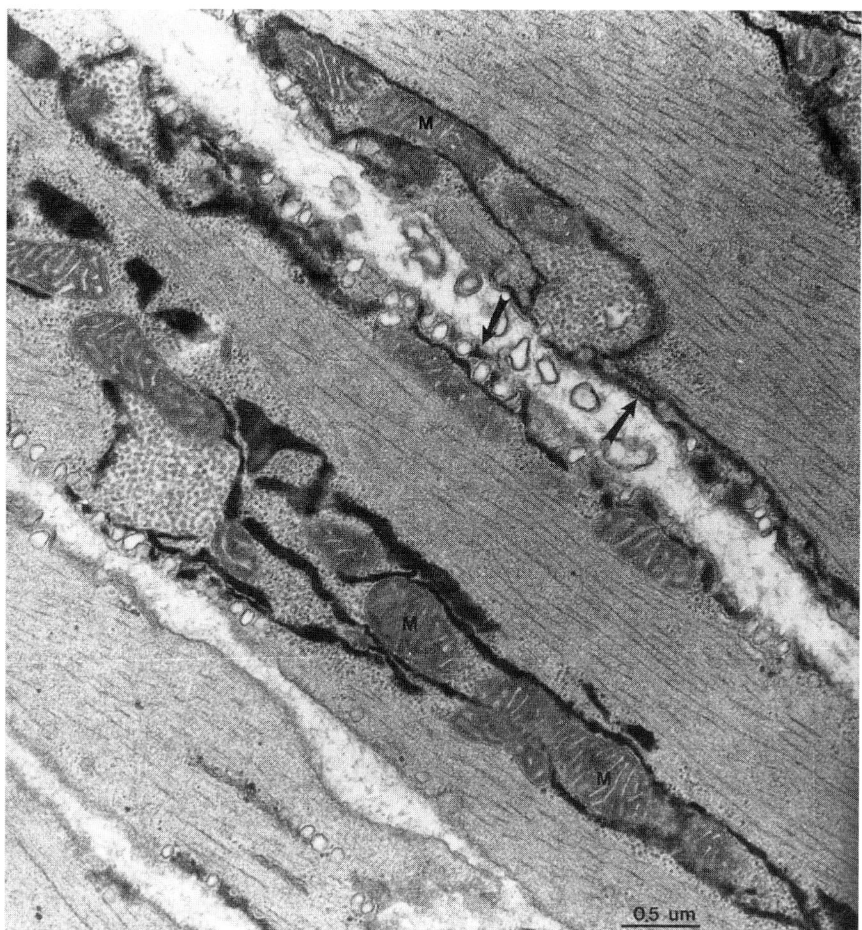

FIG. 5. View of the SR network in a longitudinally oriented rabbit portal vein smooth muscle cell stained with osmium ferricyanide. Note the close apposition of the SR to the plasma membrane (small arrows) and caveolae, as well as its relationship to mitochondria (M); in some instances the SR completely surrounds mitochondria.

dry weight (Horikawa et al 1998, and references therein), yielding an estimate of total/free mitochondrial Ca^{2+} of at least 4000 to 1. Accepting published $[Ca^{2+}]_{mt}$ values as high as $10\,\mu M$ as reliable, one would arrive to an estimate of total mitochondrial Ca^{2+} of at least $40\,mmoles/kg$ mitochondrial dry weight: an unlikely result, considering the partial mitochondrial volume and total Ca^{2+} content of most cells, and never found in normal cells. Whether the discrepancy between some $[Ca^{2+}]_{mt}$ estimates and total $[Ca]_{mt}$ measurements can be reconciled by the kinetics of mitochondrial Ca^{2+} buffers remains to be determined.

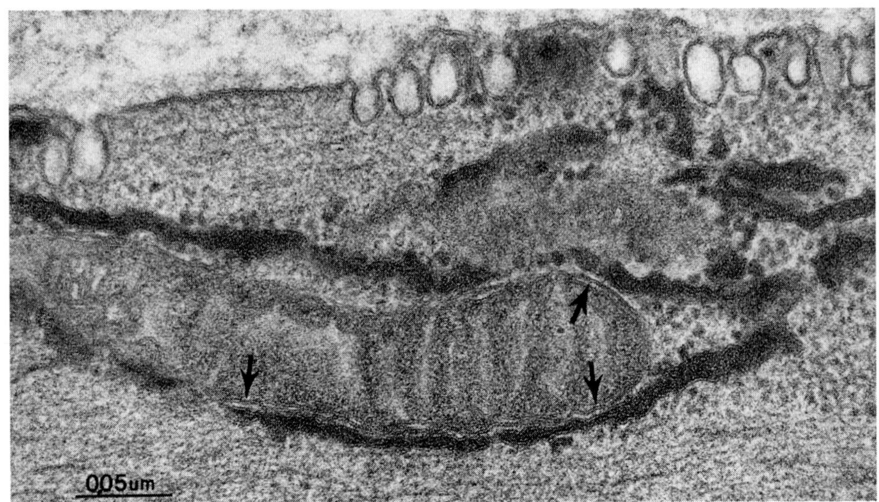

FIG. 6.   High-magnification view of the relationship of osmium ferricyanide-stained SR and a mitochondrion in a guinea-pig portal vein smooth muscle cell. Where resolved, a gap of approximately 20 nm (arrows) occurs between the outer leaflets of the mitochondrial and SR membranes.

Furthermore, in more recent studies high $[Ca^{2+}]_{mt}$ signals are seen in only a few mitochondria within a given cell, and reports claiming very high $[Ca^{2+}]_{mt}$ under physiological conditions are based on cells isolated by enzymatic dispersion. This, coupled with the uncertainty of calibration of luminescent and fluorescent Ca^{2+} indicators within the mitochondrial matrix, (for nuclei see Perez-Terzic et al 1997) should raise serious questions about the correct values of $[Ca^{2+}]_{mt}$. It is unfortunate that, with rare exceptions, very few available studies compare free with total mitochondrial Ca in the same cell type observed under the same condition.

The SR, like probably all components of cell signalling mechanisms, is coupled to other pathways through cellular localization, shared receptors of signalling molecules (e.g. $InsP_3$, ryanodine and $[Ca^{2+}]$ itself), and transporters. Thus, the localization of the junctional SR near the surface membrane (Fig. 2) may allow it to function, respectively, as a 'protective barrier', a source of Ca^{2+} sparks that open K^+ channels to hyperpolarize and relax smooth muscle or, when emptied, the activator of 'store-operated' Ca^{2+} channels. While such channels are often opened following depletion of the SR/ER by activation of $InsP_3$ receptors, they can also open upon inhibition of Ca^{2+} uptake by cyclopiazonic acid and, even in striated muscles, through activation of ryanodine receptors (Kurebayashi & Ogawa 2001).

The proximity of the SR/ER to mitochondria (Figs 5 & 6) (e.g. Nixon et al 1994) led to recent suggestions that release from the SR/ER could be a source of high local $[Ca^{2+}]_i$ giving rise to Ca^{2+} uptake by the low-affinity mitochondrial uniporter. However, similar arguments, based on structure alone, could also be made for the high-affinity SR protecting the mitochondria from Ca^{2+} influx. In fact, in liver cells analysed *in situ*, the release of Ca^{2+} from the ER by vasopressin is not accompanied by an increase in total mitochondrial Ca (Bond et al 1987).

The kinetics of Ca^{2+} release from the SR continue to be investigated by detecting the force and $[Ca^{2+}]$; transients initiated by laser flash photolysis of caged phenylephrine. Comparison of this time course with that of the force transient evoked by photolysis of caged GTP-loaded RhoRhoGDI complex resulting in Ca^{2+}-sensitizing inhibition of smooth muscle myosin phosphatase (reviewed in Somlyo & Somlyo 2000) identifies the kinetics of the two signalling pathways and their respective contributions to the phasic and tonic phases of contraction. This approach also determines the time course of upstream reactions that precede $InsP_3$-induced Ca^{2+} release and agonist-induced activation of Rho kinase.

The early force transient triggered by increases in $[Ca^{2+}]_i$, whether initiated by photo release of α-adrenergic agonists (Somlyo et al 1988, Muralidharan et al 1993, Walker et al 1993), $InsP_3$ (Somlyo et al 1992) or by Ca^{2+} itself released from caged Ca^{2+} (Zimmermann et al 1995) or through photolytic activation of Ca^{2+} influx (Malmqvist & Arner 1999), is rate limited, primarily, by myosin light chain phosphorylation reactions preceding and associated with myosin light phosphorylation (Horiuti et al 1989, Zimmermann et al 1995). The upstream reactions that precede agonist-induced Ca^{2+} release are probably rate limited by the kinetics of phospholipase C activation and $InsP_3$ production, and consume several hundred milliseconds (about 400–500 ms at 21–22 C) of the 1 s delay that precedes force development induced by photolysis of a caged α-adrenergic agonist. The lag phase of activation of Ca^{2+} sensitization by the Rho/Rho kinase pathway is of the order of several seconds, and the tonic component of contractile response evoked by agonists probably represents a combination of the relatively slow time course of RhoA-mediated Ca^{2+} sensitization with the residual, slight elevation of in $[Ca^{2+}]_i$ (Fujihara et al 1997). Thus, contrary to earlier belief, sustained Ca^{2+} influx is not the sole mechanism of prolonged force maintenance, at least in phasic smooth muscles.

We regret that due to limitations imposed by the publication, numerous valuable contributions to the literature could not be cited.

Acknowledgements

Supported by NIH PO1 HL48807 and PO1 HL19242.

References

Berridge 1997 Elementary and global aspects of calcium signalling. J Physiol 499:291–306

Blaustein MP, Golovina VA, Juhaszova M 2002 Organization of SR Ca^{2+} stores in vascular smooth muscle. In: Role of the sarcoplasmic reticulum. Wiley, Chichester, (Novartis Found Symp 246) p 125–141

Bolton TB, Prestwich SA, Zholos AV, Gordienko DV 1999 Excitation–contraction coupling in gastrointestinal and other smooth muscles. Annu Rev Physiol 61:85–115

Bond M, Kitazawa T, Somlyo AP, Somlyo AV 1984 Release and recycling of calcium by the sarcoplasmic reticulum in guinea-pig portal vein smooth muscle. J Physiol 355:677–695

Bond M, Vadasz G, Somlyo AV, Somlyo AP 1987 Subcellular calcium and magnesium mobilization in rat liver stimulated in vivo with vasopressin and glucagon. J Biol Chem 262:15630–15636

Devine CE, Somlyo AV, Somlyo AP 1971 Sarcoplasmic reticulum and excitation–contraction coupling in mammalian smooth muscles. J Cell Biol 52:690–718

Ebashi S 1991 Excitation–contraction coupling and the mechanism of muscle contraction. Annu Rev Physiol 53:1–16

Fabiato A 1983 Calcium-induced release from the cardiac sarcoplasmic reticulum. Am J Physiol 245:C1–C14

Fujihara H, Walker LA, Gong MC et al 1997 Inhibition of RhoA translocation and calcium sensitization by in vivo ADP-ribosylation with the chimeric toxin DC3B. Mol Biol Cell 8:2437–2447

Ganitkevich VY, Isenberg G 1995 Efficacy of peak Ca^{2+} currents (I_{Ca}) as trigger of sarcoplasmic reticulum Ca^{2+} release in myocytes from the guinea-pig coronary artery. J Physiol 484: 287–306

Gordienko DV, Greenwood IA, Bolton TB 2001 Direct visualization of sarcoplasmic reticulum regions discharging Ca^{2+} sparks in vascular myocytes. Cell Calcium 29:13–28

Herrera GM, Nelson MT 2002 SR and membrane currents. In: Role of the sarcoplasmic reticulum in smooth muscle. Wiley, Chichester (Novartis Found Symp 246) p 189–207

Himpens B, Missiaen L, Casteels R 1995 Ca^{2+} homeostasis in vascular smooth muscle. J Vasc Res 32:207–219

Horikawa Y, Goel A, Somlyo AP, Somlyo AV 1998 Mitochondrial calcium in relaxed and tetanized myocardium. Biophys J 74:1579–1590

Horiuti K, Somlyo AV, Goldman YE, Somlyo AP 1989 Kinetics of contraction initiated by flash photolysis of caged adenosine triphosphate in tonic and phasic smooth muscles. J Gen Physiol 94:769–781

Iino M 2000 Regulation of IP_3 receptor Ca^{2+} release channels. In: Endo M, Kurachi Y, Mishina M (eds) Pharmacology of ionic channel function: activators and inhibitors. Springer-Verlag, Berlin Heidelberg, p 605–624

Jaggar JH, Porter VA, Lederer WJ, Nelson MT 2000 Calcium sparks in smooth muscle. Am J Physiol 278:C235–C256

Janssen LJ, Betti PA, Netherton SJ, Walters DK 1999 Superficial buffer barrier and preferentially directed release of Ca^{2+} in canine airway smooth muscle. Am J Physiol 276:L744–L753

Jenkinson DH, Morton IK 1967 The effect of noradrenaline on the permeability of depolarized intestinal smooth muscle to inorganic ions. J Physiol 188:373–386

Kamishima T, McCarron JG 1997 Regulation of the cytosolic Ca^{2+} concentration by Ca^{2+} stores in single smooth muscle cells from rat cerebral arteries. J Physiol 501.3:497–508

Karaki H, Ozaki H, Hori M et al 1997 Calcium movements, distribution, and functions in smooth muscle. Pharmacol Rev 49:157–230

Karczewski P, Hendrischke T, Wolf WP, Morano I, Bartel S, Schrader J 1998 Phosphorylation of phospholamban correlates with relaxation of coronary artery induced by nitric oxide, adenosine, and prostacyclin in the pig. J Cell Biochem 70:49–59

Kowarski D, Shuman H, Somlyo AP, Somlyo AV 1985 Calcium release by norepinephrine from central sarcoplasmic reticulum in rabbit main pulmonary artery smooth muscle. J Physiol 366:153–175

Kurebayashi N, Ogawa Y 2001 Depletion of Ca^{2+} in the sarcoplasmic reticulum stimulates Ca^{2+} entry into mouse skeletal fibres. J Physiol 533:185–199

Laporte R, Laher I 1997 Sarcoplasmic reticulum–sarcolemma interactions and vascular smooth muscle tone. J Vasc Res 34:325–343

Lee CH, Poburko D, Kuo KH, Seow C, van Breemen C 2002 Relationship between the sarcoplasmic reticulum and the plasma membrane. In: Role of the sarcoplasmic reticulum in smooth muscle. Wiley, Chichester (Novartis Found Symp 246) p 26–47

Lesh R, Marks A, Somlyo AV, Fleischer S, Somlyo AP 1993 Anti-ryanodine receptor antibody binding sites in vascular and endocardial endothelium. Circ Res 72:481–488

Lesh RE, Nixon GF, Fleischer S, Airey JA, Somlyo AP, Somlyo AV 1998 Localization of ryanodine receptors in smooth muscle. Circ Res 82:175–185

Malmqvist U, Arner A 1999 Kinetics of contraction in depolarized smooth muscle from guinea-pig taeniae coli after photodestruction of nifedipine. J Physiol 519:213–221

Muralidharan S, Maher GM, Boyle WA, Nerbonne JM 1993 'Caged' phenylephrine: development and application to probe the mechanism of α-receptor-mediated vasoconstriction. Proc Natl Acad Sci 90:5199–5203

Nixon G, Mignery GA, Somlyo AV 1994 Immunogold localization of inositol 1,4,5-trisphosphate receptors and characterization of ultrastructural features of the sarcoplasmic reticulum in phasic and tonic smooth muscle. J Mus Res Cell Motil 15:682–700

Perez-Terzic C, Stehno-Bittel L, Clapham DE 1997 Nucleoplasmic and cytoplasmic differences in the fluorescence properties of the calcium indicator Fluo-3. Cell Calcium 21:275–282

Somlyo AP, Somlyo AV 1994 Signal transduction and regulation in smooth muscle. Nature 372:231–236

Somlyo AP, Somlyo AV 2000 Signal transduction by G-proteins, Rho-kinase and protein phosphatase to smooth muscle and non-muscle myosin II. J Physiol 522:177–185

Somlyo AP, Devine CE, Somlyo AV, North SR 1971 Sarcoplasmic reticulum and the temperature-dependent contraction of smooth muscle in calcium-free solutions. J Cell Biol 51:722–741

Somlyo AP, Walker JW, Goldman YE et al 1988 Inositol trisphosphate, calcium and muscle contraction. Philos Trans R Soc Lond B Biol Sci 320:399–414

Somlyo AP, Wu X, Walker LA, Somlyo AV 1999 Pharmacomechanical coupling: the role of calcium, G-proteins, kinases and phosphatases. Rev Physiol Biochem Pharmacol 134:201–234

Somlyo AV, Somlyo AP 1971 Strontium accumulation by sarcoplasmic reticulum and mitochondria in vascular smooth muscle. Science 174:955–958

Somlyo AV, Bond M, Somlyo AP, Scarpa A 1985 Inositol-trisphosphate induced calcium release and contraction in vascular smooth muscle. Proc Natl Acad Sci 82:5231–5235

Somlyo AV, Horiuti K, Trentham DR, Kitazawa T, Somlyo AP 1992 Kinetics of Ca^{2+} release and contraction induced by photolysis of caged D-myo-inositol 1,4,5-trisphosphate in smooth muscle: the effects of heparin, procaine, and adenine nucleotides. J Biol Chem 267:22316–22322

Tasker PN, Michelangeli F, Nixon GF 1999 Expression and distribution of the type I and type 3 inositol 1,4,5-trisphosphate receptor in developing vascular smooth muscle. Circ Res 84:536–542

Walker JW, Somlyo AV, Goldman YE, Somlyo AP, Trentham DR 1987 Kinetics of smooth and skeletal muscle activation by laser pulse photolysis of caged inositol 1,4,5-trisphosphate. Nature 327:249–251

Walker JW, Martin H, Schmitt FR, Barsotti R J 1993 Rapid release of an α-adrenergic receptor ligand from photolabile analogues. Biochemistry 32:1338–1345

Yoshioka T, Somlyo AP 1984 Calcium and magnesium contents and volume of the terminal cisternae in caffeine-treated skeletal muscle. J Cell Biol 99:558–568

Zimmermann B, Somlyo AV, Ellis-Davies CR, Kaplan JH, Somlyo AP 1995 Kinetics of prephosphorylation reactions and myosin light chain phosphorylation in smooth muscle. J Biol Chem 270:23966–23974

DISCUSSION

Hirst: I'm in total agreement that there is some part of the neuronal response that is associated with sensitization. I examined force production and Ca^{2+} measurement following nerve stimulation in the annococcygeus muscle, and I could start to see sensitization after about 15 s of 2 Hz stimulation. If this is a reasonable physiological stimulus, you certainly start to see this in the same time-course as you are seeing there.

Nelson: Did you see any ryanodine receptors in the nuclear envelope? There are obviously $InsP_3$ receptors present.

Nixon: We didn't see any on the nuclear envelope close to the nucleus SR. But this doesn't mean that they are not there.

Burdyga: What about $InsP_3$ receptors?

Nixon: We saw the same with immunogold labelling, but looking with confocal microscopy we saw some $InsP_3$ receptors around the nuclear envelope: they look to be closely associated with it. When we do osmium ferrocyanide staining of the SR, the nuclear envelope does stain the same as the SR, so you would predict that there probably would be some there.

Somlyo: Regarding the nuclear envelope and Ca^{2+}, there is an old picture in a book showing strontium in the nuclear envelope within the perinuclear space (Somlyo & Somlyo 1975). The trouble is, when we load with calcium oxalate, we can get an 'egg shell' of calcium oxalate around the nucleus, but when we do electron probe analysis under normal physiological conditions, there doesn't seem to be detectable Ca^{2+} in the nuclear space. However, this is probably because there is no calsequestrin or other Ca^{2+} binding protein present.

Bolton: I want to ask your opinion on a quantitative point. How far are the caveolae and their associated SR from the plasma membrane? My point is, if it is a small distance, let's say some $0.2\,\mu m$, which is just below the resolution of the confocal microscope so that the caveolae, their SR and the plasma membrane cannot be separately resolved, then the plasma membrane ought to stain strongly with $DiOC_6$ or labelled ryanodine. Is this the case?

Somlyo: DiI_{18} stains the whole plasma membrane and we can't tell the difference between the caveolae and the non-caveolar membrane. We also have a pretty good idea that when we see dense bodies this pretty much excludes SR and caveolae. At this resolution we would not want to distinguish caveolae. The best lateral resolution a confocal will give is about $0.2\,\mu m$, and by the time you get close to this resolution, unless you have a fantastic fluorophore there is so much noise that it becomes difficult.

van Breemen: When we look at the inferior vena cava, we see that the SR is next to the cell membrane and the caveolae actually poke through holes in the SR.

Somlyo: That's true; we also see this. It depends on the plane of the section.

Walsh: Given the distribution and quantity of the SR that one can see in the micrographs, I am having a hard time understanding issues such as propagating Ca^{2+} waves. Does the structure support such a mechanism?

Somlyo: We can see that a very large fraction of the SR is continuous. I remember that the first oscillations that were reported, in muscle, were published in a paper in *Nature* by Professor Endo (Endo et al 1970). If we see the same mechanisms here, then the presence of the SR is at least sufficient: I am not saying that it is necessary. If this is the case, then in some smooth muscles we see sufficient interconnected SR to indicate that it could play a role. In the case of skeletal muscle the overloaded SR is most likely to exhibit oscillatory phenomena.

Walsh: With regards to the issue of Ca^{2+} sensitization, could you share your thoughts on events downstream of Rho? Let's say Rho kinase is activated. What is the relative importance of myosin phosphatase versus CPI17 as a downstream target of this pathway?

Somlyo: I am not against CPI17, but I think the evidence so far is strongest for phosphatase phosphorylation. CPI17 clearly has the pharmacological effects that one predicts when it is phosphorylated. CPI17 is more highly expressed in some tonic smooth muscles but is barely present in phasic smooth muscles. We are now trying to use GDI as a specific Rho inhibitor to sort out whether CPI17 plays a role.

Kotlikoff: I have a question about $InsP_3$ receptors and ryanodine receptors. There is good evidence for wave propagation with both systems in smooth muscle. Can you say anything in experiments where you have looked with antibodies about the relationship between $InsP_3$ receptors and ryanodine receptors? And does this provide any insight into the interaction between these Ca^{2+}-sensitive intracellular Ca^{2+} release channels?

Somlyo: I don't know enough to give a fair answer. All I can say is that we see $InsP_3$ and ryanodine receptors on the SR, and they were localized wherever there was SR.

Nixon: The problem is that we lack sufficient resolution. Immunogold labelling should give the required resolution, but technically we have been unable to find evidence for them being closely localized.

Paul: I wanted to come back to the issue of mitochondria and Ca^{2+}. I happen to have friends on both sides of that issue. We have been promoting the idea, which seems to work in a variety of tissues, that subsarcolemmal space is highly populated by glycolytic enzymes. This is true in skeletal muscles also. One of the first things seen with ouabain is that it kills the lactate production without touching oxygen consumption. If you do a similar experiment, stimulating the muscle with KCl (pouring Ca^{2+} in) and then adding cyclopiazonic acid (CPA), this kills lactate production, reducing it by about 50%. It doesn't touch oxygen consumption. There is a resolution issue here in that we get four times as much lactate as oxygen in terms of the ATP. We are fairly careful: the first evidence that we see when we remove with the Na^+/K^+ ATPase or the SERCA pump is that lactate goes away.

Somlyo: What does this mean in terms of mitochondrial Ca^{2+} uptake?

Paul: John McCarron, I thought you had some evidence from the inhibition of oxidative metabolism that this messes up the uncoupling?

McCarron: Yes, uncouplers increase the depolarization of evoked Ca^{2+} transients and slow the rate of decay. This was with CCCP (carbonyl cyanide *m*-chlorophenylhydrazone).

Somlyo: I would refer people to an interesting paper by Ganitkevitch (1999) on cell Ca^{2+}.

McCarron: The experiments we did involved whole cell patch-clamp studies. We were aware of the pH changes that may occur with CCCP so we buffered protons with 30 mM HEPES rather than the usual 10 mM.

Somlyo: Did you measure the ATP concentration?

McCarron: We clamped the ATP via the patch pipette filling solution.

Somlyo: That's the usual statement people make, but I have yet to see a report where they actually measured ATP.

Paul: We had some pretty interesting data that without glycolytic substrates, even with ATP in a patch pipette it doesn't get where it is supposed to go (Lorenz & Paul 1997). We can see really big changes with the glycolytic cocktails in the presence of 5 mM ATP.

McCarron: Is the idea that the ATP synthesis will be running in reverse, and that there will be local ATP depletion?

Somlyo: Yes.

McCarron: We had 3 mM Mg-ATP in the pipette, and in addition to using the uncoupler, oligomycin was present which should prevent ATPase activity.

Somlyo: I would like to point out one other thing. People are using CPA, CCCP and so on to show the importance of mitochondrial uptake. But many years ago we did a study (Yoshioka & Somlyo 1984) looking at the effect of caffeine on frog skeletal muscle. When we did a caffeine contraction we got massive mitochondrial Ca^{2+} uptake. We did not consider this to be physiological, and thankfully up to now no one has suggested that frog mitochondria have a major role in excitation–contraction (EC) coupling, but we concluded that when the SR is knocked out and cytosolic Ca^{2+} is increased, there is a redundant, non-physiological protective pathway by which the mitochondria take up Ca^{2+}. The question is, what happens under physiological conditions?

Eisner: Why don't the mitochondria take up Ca^{2+} during a tetanus?

Somlyo: With a big enough tetanus they will. When we did a tetanus in rat cardiac muscle we added ryanodine and raised extracellular Ca^{2+} to 10 mM, tetanized the muscle: sure enough, there was mitochondrial Ca^{2+} uptake.

Blaustein: We published a paper a couple of years ago on aortic smooth muscle stimulated with serotonin. Using moderate doses for short periods, which cause modest contraction, we saw no effect on Ca^{2+} in individual mitochondria. However, a big dose for a longer period causes the muscle to go into contracture, and there is a significant rise in Ca^{2+} in mitochondria.

Burdyga: I'm not playing for the mitochondria team, but I'd like to mention some results on secretory cells from Ole Petersen's lab. The secretory cell is polarized, and the Ca^{2+} wave begins in the basolateral membrane and then propagates to the secretory part of it. They found that there is a mitochondrial belt that damps this Ca^{2+} wave. They see elevation of mitochondrial Ca^{2+} as well.

Somlyo: Did they see the SR–mitochondrial relationship? This is very common. In secretory cells in particular there is a great deal of ER. I don't think he had the resolution to distinguish this. People who claim that mitochondria take up Ca^{2+}, under physiological conditions, have an open invitation to come to our lab and make the measurements calibrating mitochondrial matrix Ca^{2+}, and we will make the measurement of total Ca^{2+}.

References

Endo M, Tanaka M, Ogawa Y 1970 Calcium induced release of calcium from the sarcoplasmic reticulum of skinned skeletal muscle fibres. Nature 228:34–36

Ganitkevich VYa 1999 Clearance of large Ca^{2+} loads in a single smooth muscle cell: examination of the role of mitochondrial Ca^{2+} uptake and intracellular pH. Cell Calcium 25:29–42

Lorenz JN, Paul RJ 1997 Dependence of Ca^{2+} channel currents on endogenous and exogenous sources of ATP in portal vein smooth muscle. Am J Physiol 272:H987–H994

Somlyo AP, Somlyo AV 1975 Ultrastructure of Smooth Muscle. In: Daniel EE, Paton DM (eds) Methods in pharmacology. Plenum Press, New York, p 3–45

Yoshioka T, Somlyo AP 1984 Calcium and magnesium contents and volume of the terminal cisternae in caffeine-treated skeletal muscle. J Cell Biol 99:558–568

Final general discussion

Paul: I have a general question that I'd like to discuss. What is the point of Ca^{2+} waves and oscillations? I want to put this question in the context of metabolism and muscle energetics. There seems to be a lot of fuel cycling going on. The pathway components are phosphorylating, dephosphorylating and passing Ca^{2+} back and forth. Why wouldn't it be easier just to do a graded contraction?

Iino: I can only speculate, but perhaps if the cell maintained high intracellular Ca^{2+} concentrations for a long time this would be toxic.

Paul: A high Ca^{2+} concentration isn't needed with the Rho kinase pathway.

Iino: The Rho kinase system is rather slow. There might also be a need for a more rapid system. Take blood pressure as an example, which is continually changing. The sympathetic nerve tone is changing in phase with the respiration. Some sort of rapid system is needed.

Brading: Indeed, we wouldn't be able to stand up if there wasn't a rapid system: we would faint unless our smooth muscles responded quickly.

Paul: I would contend that most of this is done by the vessel wall.

van Breemen: Even as we are talking and looking around, the bloodflow in our brains is shifting constantly. This is due to rapid motions in the smooth muscle of our cerebral resistance vessels.

Blaustein: There are two systems superimposed on each other. There are acute changes of that kind, occurring rapidly. Superimposed on that are the constant second-to-second changes that are going on. We have to accommodate both. With the skeletal muscle we don't; it's a different phenomenon.

Paul: But we recruit different motor units and they are going on and off.

Blaustein: This is using different cells that are adjacent to one another that are out of phase in order to maintain that tone.

Paul: It still seems like an expensive way of doing things.

Nelson: There are three separate roles for Ca^{2+} waves. First, delivering Ca^{2+} for contraction. Second, modulating Ca^{2+}-dependent ion channels that control the membrane potentials. Depending on the tissue this can involve BK channels, SK channels or Ca^{2+}-activated Cl^- channels. Third, controlling Ca^{2+}-dependent transcription factors. There is quite clear evidence that the frequency and the amplitude components of the Ca^{2+} signals can determine which Ca^{2+}-dependent transcription factors are activated. This can encode both short and long-term information to control smooth muscle function.

Paul: That's probably where I would be coming down. The speed of a smooth muscle is not set by the Ca^{2+} release. I would maintain that the speed of the muscle is set by its intrinsic unloaded shortening velocity, as modulated by its elasticity and whatever force it is facing. Generally, the Ca^{2+} transient is largely over before the smooth muscle contracts. I don't believe that the speed of the smooth muscle is regulated by means of Ca^{2+} waves.

van Breemen: It is possible that activating Ca^{2+} waves deposit Ca^{2+} nearer to the intracellular Ca^{2+} receptor sites and thereby enhance the efficiency of activation.

Young: In the myometrium, the Ca^{2+} waves are from the deep cytoplasmic Ca^{2+} to the plasma membrane. The intracellular Ca^{2+} waves should be considered separately from either phasic or tonic smooth muscle types. The myometrium is phasic smooth muscle. And intercellular Ca^{2+} waves are an entirely different phenomenon altogether.

Brading: There is a problem about this. You only established this in non-excitable tissues. In fact, in excitable tissues what is more important for the fast modulation of ion channels is Ca^{2+} coming in through the plasma membrane, rather than waves. A lot of people who are looking at waves are not measuring membrane potential and ignore spikes altogether.

Burdyga: Where localized high Ca^{2+} spiking occurs, we never see this translated into the mechanical response. These spikes can be as high as $10\,\mu M$ locally. We have to consider that without a wave we are unable to switch on the contractile machinery.

Brading: Do you really mean this? Without a big rise in cytoplasmic Ca^{2+} you may not activate the contractile machinery, but does this necessarily require a wave?

Nelson: Ca^{2+} entry doesn't. In pressurized arteries very few cells produce waves.

Paul: It depends on the size of the cell. There are large smooth muscle cells and small ones.

Burdyga: There is always a delay between the Ca^{2+} rise and initiation of the mechanical activity. Tom Bolton reported a 1 s delay between the rise of Ca^{2+} and cell shortening. In intact preparations the delay is a maximum of 300 ms. This is something to think about when we are talking about localized and global events and translating them into functional responses.

van Breemen: Also, if there is a fast on and a slow off, then with an average lower Ca^{2+} concentration you can get more activation with waves.

Hellstrand: If you are worried about energetics, perhaps it is more efficient to get more force out for less average Ca^{2+}, even at the expense of having to cycle Ca^{2+}. Ca^{2+} is a low abundance ion, but if you overload mitochondria with Ca^{2+} by having a constant KCl contraction, this would not be very physiological and not very efficient.

Paul: I was thinking that when there is a high level of myosin phosphorylation, with maximum activation and then turning it off, that there would be a lot of cycling between phosphorylation/dephosphorylation, as opposed to having a low level of phosphorylation for the same 20% level of activity.

Kotlikoff: We seem to be talking about waves as though there was just one type of wave. It has been clearly demonstrated that waves can vary substantially in terms of their speed. Certainly, InsP$_3$ waves are much slower (a third of the rate of transmission). In terms of the amplitude of the wave itself, one would expect that one mediated by ryanodine receptors would be different from one mediated by InsP$_3$ receptors. They may have different functions.

Somlyo: In doing kinetics, one should also consider the temperature. The Q$_{10}$ for myosin phosphate is about 5, and it makes quite a difference as to how physiological experiments are.

McHale: I would suggest that waves are more important in pacemaker cells than in smooth muscle cells. Our model for the way the urethra works is that the smooth muscle cells don't have waves but the pacemaker cells do. The oscillations of the pacemaker cells can then drive the smooth muscle cells.

Nelson: Definitions are important. We have tried to characterize these Ca^{2+} signals. We define a wave as something that propagates throughout the cell, usually starting at one end of the cell and then petering out about half way through, with a velocity decrease. We also see cells that oscillate, undergoing global fluctuations in the Ca^{2+} level throughout the whole cell. A wave should be something that is propagating and regenerative.

Brading: Has anyone here really got evidence that waves underlie contractions in any smooth muscle that they have studied?

van Breemen: Yes, if you block the waves you don't get a contraction.

Brading: That isn't quite the same.

Nelson: We can induce waves, and add a Ca^{2+} channel blocker. The average Ca^{2+} goes down, the arteries relax and the waves continue on.

Burdyga: If I apply caffeine, it produces a regenerative wave that has a limited speed of propagation and culminates in cell shortening. Then I use cyclopiazonic acid (CPA) or ryanodine and disable this wave of propagation, and I apply high K$^+$. Instead of seeing slow contractions, I see it going much faster. This means that we don't need wave propagation during influx of Ca^{2+} through L-type channels.

Brading: Perhaps what I meant is to ask whether anyone has seen waves underlying contraction in physiological conditions. I would love to see someone switching on a muscle to contract and showing that this contraction was triggered by a Ca^{2+} wave.

Iino: The depolarization cannot cause contraction directly. There is no direct coupling between depolarization and myosin light chain phosphorylation.

Brading: But do you need a propagated wave using the SR to generate the Ca^{2+} release?

Iino: Without an increase in Ca^{2+}, how can you phosphorylate myosin?

Brading: Do you need a wave, though?

Hellstrand: But if wave activity is a normal way for the cell to increase Ca^{2+} concentration, and you know that increased Ca^{2+} concentration causes contraction, we can discuss whether waves are directly engaged in the contractile activation, but why is this a problem?

Kotlikoff: I guess the other point is that in some of these cells it is almost impossible to raise Ca^{2+} without a wave. Other than by saturating things non-physiologically, it is hard to imagine that some of these systems don't propagate.

Eisner: Gil Wier, you spend your life measuring waves. Can you comment?

Wier: There is no doubt that agonist stimulation of intact pressurized artery, in a system buffered with CO_2 and otherwise physiological, elicits these aysnchronous Ca^{2+} waves. Then at higher agonist concentrations one can get vasomotion, accompanied by spatially uniform oscillations of Ca^{2+} that we think have a different mechanism (they are membrane potential dependent, for example). But what would be direct evidence relating these asynchronous Ca^{2+} waves to contraction? We don't see the individual cells contract at the time that the waves are seen. There's no easy answer. What we can dream about is detecting a heterogeneity of myosin light chain phosporylation that correlated with the heterogeneity of Ca^{2+}. Technically this is extremely difficult, if not impossible.

Somlyo: Waves may be sufficient, but they are certainly not necessary. We can elevate Ca^{2+} without making waves.

Wier: It seems reasonable to think that these waves may serve to spike up the phosphorylation of myosin light chain. In combination with the Ca^{2+} sensitizing mechanisms, this force can be maintained.

Brading: They may be doing something quite different though.

van Breemen: There is a simple experiment I have done that answers your question. You have the waves and you have a large contraction. Then you add CPA and the waves stop, but the Ca^{2+} concentration, if anything, goes up a little bit. The contraction then comes down. In this case, the presence of the waves enhanced force development.

Brading: What did you do to make your wave in the first place? You applied agonists, so that's not physiology. If you stimulate the sympathetic nerves, what happens is that every now and again one varicosity will release a vesicle of transmitter, which will act on one cell. Phenylephrine application activates all of the cells, not just one. This is quite different. In real life with a 1–2 Hz signal the probability of releasing a vesicle of transmitter onto a cell is very low. On average, in a population of 1000 cells, just 10 of them at any one time will have transmitter released onto them. This is not just noradrenaline but also ATP. If the functional receptors are present, ATP will open channels and cause local depolarization. This will spread to the neighbouring cells, but noradrenaline concentration will not stay up long enough to diffuse to neighbouring cells. It may activate something in the cell immediately underneath, but it won't affect the others. If the frequency rises from 1–2 Hz, more cells are recruited and the probability of release rises. At 5 Hz, probably quite a large number of cells will be receiving transmitter. The way physiology actually happens is that you are not bathing things in transmitters and synchronously activating these processes. Physiology is asynchronously activating the cells.

Wier: The fact remains that Dr Iino's original study was done with nerve stimulation at about 5 Hz. The results that people get with bath-applied agonists are essentially the same.

Nelson: Were those waves, or oscillations? Was the whole cell lighting up, or could you see a propagating wave?

Iino: They were waves.

Nelson: Gil Wier, in your paper with Mordy Blaustein, you applied Ca^{2+} channel blockers, the arteries relaxed and the waves continued, didn't they?

Wier: Yes.

Hirst: So you can have waves without contractions.

Brading: What you can't tell is what the individual cells are doing when a wave occurs.

Iino: If the depolarization is the main cause of contraction in those rat tail artery cells, we should be able to see the continuous increase in the Ca^{2+} throughout the cell. The space constant is such that the entire smooth muscle cell will depolarize, and Ca^{2+} is coming from all over the place. However, what we see is a localized Ca^{2+} increase that is propagating through the cells. This cannot be brought about by depolarization.

Blaustein: It is not very different from cardiac muscle and some of the early experiments that Gil Wier and others did there. A wave of Ca^{2+} starts in one place and spreads through the rest of the cell. To try to dump Ca^{2+} uniformly throughout the cell is very unusual. Skeletal muscle cells may be unusual in this regard, because of the structure of the T-tubules and SR.

One of the things that I think has come out of this meeting is that when we talk about things, every one of us has been talking about some specialization or other. But the cells really are integrated in that one has to think about Ca^{2+} entry, Ca^{2+} release from the SR, Ca^{2+} uptake into the SR and Ca^{2+} extrusion from the cell. If you interfere in any one of those places, a large number of things can go wrong. We have been looking at myogenic tone. There was a dogma that all you had to do was to take away or block the L-type Ca^{2+} channels, and this would abolish myogenic tone, and myogenic tone was therefore due to L-type Ca^{2+} channels and Ca^{2+} entry. But a whole variety of other phenomena are seen: interfering with SR function, the frequency of sparks, or the rate at which Ca^{2+} is extruded, these all alter myogenic tone. The SR plays a central role in this because it is an integrator of some of these functions.

Index of contributors

Non-participating co-authors are indicated by asterisks. Entries in bold indicate papers; other entries refer to discussion contributions.

Subject index